Lecture Notes of the Institute for Computer Sciences, Social-Informatics and Telecommunications Engineering 23

Athanasios V. Vasilakos   Roberto Beraldi
Roy Friedman   Marco Mamei (Eds.)

Autonomic Computing and Communications Systems

Third International ICST Conference, Autonomics 2009
Limassol, Cyprus, September 9-11, 2009
Revised Selected Papers

 Springer

Volume Editors

Athanasios V. Vasilakos
University of Western Macedonia
Department of Telecommunications Engineering
50100 Kozani, Greece
E-mail: vasilako@ath.forthnet.gr

Roberto Beraldi
Universita'di Roma "La Sapienza"
00100 Rome, Italy
E-mail: beraldi@dis.uniroma1.it

Roy Friedman
Computer Science Department
Technion, Haifa 32000, Israel
E-mail: roy@cs.technion.ac.il

Marco Mamei
University of Modena and Reggio Emilia
Dipartimento di Scienze e Metodi dell'Ingegneria
42100 Reggio Emilia, Italy
E-mail: mamei.marco@unimore.it

Library of Congress Control Number: 2009942250

CR Subject Classification (1998): C.2, G.2.2, G.1.6, I.2.11, J.1, K.6, H.3.4

ISSN   1867-8211
ISBN-10   3-642-11481-4 Springer Berlin Heidelberg New York
ISBN-13   978-3-642-11481-6 Springer Berlin Heidelberg New York

springer.com

© ICST Institute for Computer-Sciences, Social Informatics and Telecommunications Engineering 2010
Printed in Germany

Typesetting: Camera-ready by author, data conversion by Scientific Publishing Services, Chennai, India
Printed on acid-free paper   SPIN: 12828081   06/3180   5 4 3 2 1 0

Preface

These proceedings contain the papers presented at the Third International ICST Conference on Autonomic Computing and Communication Systems, Autonomics 2009, held at the Cyprus University of Technology, Limassol, Cyprus, during September 9–11, 2009.

As for the previous editions of the conference, this year too the primary goal of the event was to allow people working in the areas of communication, design, programming, use and fundamental limits of autonomics pervasive systems to meet and exchange their ideas and experiences in the aforementioned issues.

In maintaining the tradition of excellence of Autonomics, this year we accepted 11 high-quality papers out of 26 submitted and had 5 invited talks, covering various aspects of autonomic computing including applications, middleware, networking protocols, and evaluation.

The wide interest in the autonomic systems is shown by the broad range of topics covered in the papers presented at the conference. All papers presented at the conference are published here and some of them, which are considered particularly interesting, will be considered for publication in a special issue of the *International Journal of Autonomics and Adaptive Communications Systems (IJAACS)*. The conference also hosted the First International Workshop on Agent-Based Social Simulation and Autonomic Systems (ABSS@AS).

Organization

Autonomics 2009

Steering Committee

Roberto Baldoni	University of Rome La Sapienza, Italy
Imrich Chlamtac (Chair)	CREATE-NET, Italy
Daniele Miorandi	CREATE-NET, Italy

General Chair

Thanos Vasilakos	University of Western Macedonia, Greece

Vice Chair

Roberto Beraldi	University of Rome La Sapienza, Italy

Technical Program Chairs

Roy Friedman	Technion, Israel
Marco Mamei	University of Modena, Italy

Conference Coordinator

Gergely Nagy	ICST, Belgium

Local Chair

Rozita Pavlidou	Cyprus University of Technology, Cyprus

Publications and Publicity Chair

Giorgia Lodi	University of Rome La Sapienza, Italy

Web Chair

Marco Platania	University of Rome La Sapienza, Italy

Technical Program Committee

Antonio Manzalini	Telecom Italia, Italy
Aline Carneiro Viana	Inria Saclay, France
Borbala Katalin Benko	Budapest University of Technology and Economics, Hungary
Cristian Borcea	New Jersey Institute of Technology, USA
Christof Fetzer	Dresden University of Technology, Germany
Gregory Chockler	IBM Haifa, Israel
Danny Weyns	Catholic University of Leuven, Belgium
Giovanna Di Marzo Serugendo	Birkbeck University of London, UK
Douglas Blough	Georgia Institute of Technology, USA
Eiko Yoneki	Univeristy of Cambridge Computer Laboratory, UK
Gwendal Simon	Telecom Bretagne, France
Matti Hiltunen	AT&T Labs Research, USA
Hussein Alnuweiri	Texas A&M University of Qatar, Qatar
Luís Rodrigues	INESC-ID/IST, Portugal
Raffaela Mirandola	Politecnico di Milano, Italy
Oriana Riva	ETH Zurich, Switzerland
Manish Parashar	Rutgers, USA
Leonardo Querzoni	University of Rome La Sapienza, Italy
Ravi Prakash	University of Texas at Dallas, USA
Roman Vitenberg	University of Oslo, Norway
Thrasyvoulos Spyropoulos	ETH Zurich, Switzerland
Maarten van Steen	VU University Amsterdam, The Netherlands
Neeraj Suri	Technical University of Darmstadt, Germany
Vinny Cahill	Trinity College Dublin, Ireland
Ted Herman	University of Iowa, USA
Matthias Baumgarten	University of Ulster, UK

ABSS@AS 2009 Workshop

Organizing Committee

Mario Paolucci	Institute for Cognitive Science and Technology (ISTC-CNR) National Research Council
Isaac Pinyol	Artificial Intelligence Research Institute (IIIA-CSIC) Spanish National Research Council

Program Committee

Frederic Amblard	Universite Toulouse 1, France
Luis Antunes	University of Lisbon, Portugal
Cristiano Castelfranchi	ISTC-CNR, Italy
Federico Cecconi	ISTC-CNR, Italy
Helder Coelho	University of Lisbon, Portugal
Rosaria Conte	ISTC-CNR, Italy
Gennaro Di Tosto	ISTC-CNR, Italy
Bruce Edmonds	Centre for Policy Modelling, UK
Boi Faltings	Ecole Polytechnique Federale de Lausane, Switzerland
Nigel Gilbert	University of Surrey, UK
Wander Jager	University of Groningen, The Netherlands
Marco Janssen	Arizona State University, USA
David Hales	University of Bologna, Italy
Jean-Pierre Muller	CIRAD, France
Pablo Noriega	IIIA-CSIC, Spain
Emma Norling	Manchester Metropolitan University, UK
Oswaldo Terán	University of Los Andes, Venezuela
Mario Paolucci	ISTC-CNR, Italy
Juan Pavon Mestras	Universidad Complutense Madrid, Spain
Isaac Pinyol	IIIA-CSIC, Spain
Walter Quattrociocchi	ISTC-CNR, Italy
Jordi Sabater-Mir	IIIA-CSIC, Spain
Jaime Sichman	University of Sao Paulo, Brazil
Carles Sierra	IIIA, Spain
Liz Sonenberg	University Melbourne, Australia
Flaminio Squazzoni	University of Brescia, Italy
Keiki Takadama	Tokyo Institute of Technology, Japan
Klaus Troitzsch	University of Koblenz, Germany
Paolo Turrini	University of Utrecht, The Netherlands
Laurent Vercouter	École Nationale Supérieure des Mines de Saint-Étienne, France
Harko Verhagen	Stockholm University, Sweden

Table of Contents

Autonomics 2009

ABSS@AS 2009 Workshop

A-OSGi: A Framework to Support the Construction of Autonomic OSGi-Based Applications*

João Ferreira, João Leitão, and Luis Rodrigues

IST/INESC-ID
joao.elias.ferreira@ist.utl.pt, jleitao@gsd.inesc-id.pt, ler@ist.utl.pt

Abstract. The OSGi specification is becoming widely adopted to build complex applications. It offers adequate support to build modular applications, where modules can be added and removed at runtime without stopping the entire application. This paper proposes A-OSGi, a framework that leverages on the native features of the OSGi platform to support the construction of autonomic OSGi-based applications. A-OSGi offers a number of complementary mechanisms for that purpose, such as: the ability to extract indicators for the performance of deployed bundles; mechanisms that allow to have a fine grain control of how services bind to each other and to gather this information in runtime; and support for a policy language that allows the administrator to define autonomic behavior of the OSGi application.

Keywords: Autonomic Computing, OSGi, Service Oriented Computing.

1   Introduction

The OSGi specification [1] (initials for the extinct Open Services Gateway initiative) defines a standardized component oriented platform for building Service Oriented Java^{TM} applications. OSGi provides the primitives and runtime support that allows developers to build applications from small, reusable and collaborative components. The OSGi platform also provides the support for dynamically changing such compositions, without requiring restarts. To minimize the level of coupling, the OSGi provides a service-oriented architecture that enables components to dynamically discover each other for collaboration.

OSGi was first developed for embedded systems software and later automotive electronics. However, its advantages also made the technology appealing also to build flexible Desktop Applications [2], Enterprise Applications [3,4], and Web Applications [5,6]. A key issue associated with the deployment and management of complex web applications is to ensure the performance of the application in face of changing workloads. The difficulties in forecasting accurately the demand and in estimating the interference among the deployed applications, makes the

---

* This work was partially supported by FCT, through project Pastramy, PTD-C/EIA/72405/2006.

A.V. Vasilakos et al. (Eds.): AUTONOMICS 2009, LNICST 23, pp. 1–16, 2010.

configuration of web applications a significant challenge [7,8]. The concurrent execution of multiple OSGi bundles, possibly developed by different teams, that invoke each other in patterns which, due to the dynamics of the system evolution, are difficult to predict at design time, makes this challenge even more daunting.

Autonomic computing has emerged as a viable approach to manage complex systems such as the one described above [9]. The idea is that a system must own autonomic management components, able to offer self-configuration, self-optimization, self-healing and self-protection features. The ability to adapt its own behavior in response to changes in the execution environment is the fundamental ability of an autonomic system. The OSGi platform, by allowing components to be removed, added, and replaced at runtime without stopping the system, is particularly appealing for building autonomic web applications.

This paper proposes, describes and evaluates A-OSGi, a framework to support the construction of autonomic OSGi-based applications. A-OSGi offers a number of complementary extensions to the basic OSGi framework that improve its autonomic capabilities. Namely, A-OSGi includes the following features: the ability to extract performance indicators of deployed bundles, mechanisms that allow to have a fine grain control of how services bind to each other and to gather this information at runtime, and support for the interpretation of a policy language, that allows system administrators to define the autonomic behavior of OSGi applications deployed over the A-OSGi framework.

The rest of the paper is organized as follows. Section 2 overviews related work. The design and implementation of A-OSGi is described in Section 3 and Section 4, respectively. The resulting system is illustrated and evaluated in Section 5. Section 6 concludes the paper, providing some pointers for future work.

2   Related Work

In this section we provide a brief description of the OSGi platform architecture. Then we describe the MAPE-K autonomic control loop in the context of the OSGi architecture and, finally, we present some previous works that have explored strategies to enrich the OSGi platform with mechanisms to assist in the creation of autonomic applications, for instance, by proposing adequate monitoring mechanisms.

2.1   OSGi Platform

The OSGi platform [1] is a container supporting the deployment of extensible Java-based applications composed by components, usually named *bundles*. The basic architecture of the platform is depicted in Figure 1. The platform is able to install, update, and remove bundles without stopping or restarting the system. Moreover, the platform supports a Service oriented Architecture (SOA), where bundles interact in a publish/find/bind service model. SOA allow the developing loosely coupled bundles that interact through service interfaces.

In more detail, a bundle can register with the OSGi platform a number of services that it makes available to other bundles; the platform offers a service

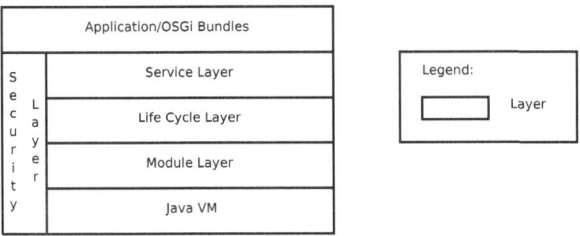

Fig. 1. OSGi Architecture

discovery mechanism that allows a bundle to dynamically find, at runtime, services that it requires to operate.

The platform functionality is divided into the following four layers: i) The *Security Layer* extends the basic Java security architecture specifically the permission model to adapt it to the typical use cases of OSGi deployments; ii) The *Module Layer* defines the modularization model employed by the platform, including the Java packages visibility among bundles(bundle private packages); iii) The *Life Cycle Layer* provides an API to support the mechanisms to install, update, remove, start, and stop individual bundles; iv) The *Service Layer* owns the responsibility of providing the mechanisms to support a service-oriented architecture (SOA) on top of the OSGi platform. This SOA support allows programmers to develop loosely coupled components that can adapt to the changing environment in runtime, without restarting bundles. The SOA becomes even more essential in OSGi due to the platform dynamic nature.

The OSGi platform was initially oriented to embedded systems and network devices, however with its inclusion in the Eclipse IDE, OSGi is now widely used for both desktop and server applications [2,3], and developing web applications [5,6]. OSGi based applications have increased in complexity over the years, however the OSGi platform still lacks support for developing autonomic applications. Namely, the platform does not provide mechanisms to monitor the operation of individual bundles, or to take advantage on distinct service implementations that potentially present different trade-offs between quality of service provided to the clients and resource consumption required to provide that service.

iPOJO. One of the useful properties of OSGi, that can assist in developing autonomic applications, is the Service Oriented Architecture support. However managing the services dynamics in a system like OSGi rises dependencies management issues. For instance a service becomes available or unavailable, as a result of bundle activation or deactivation. This problem is tackled by Service Oriented Component Models that eases the registering of services and dependencies management.

iPOJO is a Service Oriented Component Model that creates a clear separation between the bundle business logic and service oriented mechanisms such as registering a service and binding to other services. This separation allows the bundle to be implemented as simple POJOs[1]. In [10], the authors specifically apply the iPOJO solution over an OSGi platform. Although this approach can ease the management of services binding in runtime, unlike A-OSGi, it lacks the remaining components to build a autonomic system. However we rely in iPOJO to build autonomic bundles on top of A-OSGi.

2.2 MAPE-K Control Loop

Many autonomic systems are modeled through a MAPE-K autonomic management control loop [11]. This loop consists on the following operations: monitoring (M), analysis (A), planning (P), and execution (E). The K stands for a shared knowledge base that supports these operations. We now provide a brief description of each MAPE-K component and discuss how they can be implemented in the context of the OSGi platform.

Monitoring. The monitoring component is responsible for managing the different sensors that provide information regarding the system. In the OSGi context, sensors can capture the current consumption of critical resources (such CPU and memory) but also other performance metrics (such as the number of processed requests per second and the request process latency). The monitoring metrics must be fine grained, i.e. per bundle. Sensors can also raise notifications when changes to the system configuration happen. Such sensors can be implemented using the notifications provided by the OSGi platform during the life cycle of bundles and services, and when bundles bind and unbind to services.

Analysis. The analysis component is responsible for processing the information captured by the monitoring component and to generate high level events. For instance, it may combine the values of CPU and memory utilization to signal an overload condition in the OSGi platform.

Planning. The planning component is responsible for selecting the actions that need to be applied to the system in order to correct some deviation from the desired system state. The planning component relies on a high level policy that describes an adaptation plan for the system. These policies may be described using Event Condition Action (ECA) rules that are defined by a high level language. A ECA rule describes for a specific event and a given condition what action (or actions) should be executed. In the context of OSGi, the actions may affect the deployed bundles, the registered services or the bindings to services.

Execution. The execution component applies the actions selected by the planning component to the target components using the available actuators. In OSGi, we consider three main action types, as follows: *i)* specify rules for service bindings, in such a way that a specific bundle is prohibited, or obliged, to use some specific service implementation; *ii)* change service properties, for instance change a parameter associated with a service implementation; and *iii)* control the life cycle of a bundle, by either starting or stopping bundles.

---

[1] **P**lain **O**ld **J**ava **O**bjects.

Knowledge Base. The knowledge base component maintains information to support the remaining components. In the context of OSGi, it maintains information about managed elements, specifically which services a bundle is using, which services a bundle provides, and other information about the dependencies concerning services.

2.3 OSGi Monitoring

Several previous works have addressed the topic of monitoring OSGi applications [12,13]. Most of these solutions have focused on providing an adequate bundle CPU consumption isolation. The work presented in [12] employs a thread-based approach to monitor each OSGi bundle, by creating threads that are internally associated with an individual bundle. Another approach can be found in [13], where the authors employ Isolates (or other execution environment objects) to achieve the required isolation (unfortunately, this solution only works in specific, modified, JVMs). Other tools could also be applied to monitor the resources, such as bytecode instrumentation for CPU accounting [14].

3 The A-OSGi Framework

The A-OSGi framework offers a number of extensions to the OSGi platform to support the development of autonomic applications. In this section, we provide an overall overview of the A-OSGi architecture followed by a detailed description of each of its components.

The A-OSGi architecture follows the general MAPE-K model (introduced previously in the Section 2.2). More specifically, we have augmented the OSGi platform with functionalities that support monitoring, analysis, planning, execution, and the knowledge aspects of that model. As depicted in Fig. 2 these functionalities are provided by four main components, namely: A-OSGi Monitoring and Analysis component (MAC); A-OSGi Execution component (EC); A-OSGi Knowledge component (KC); and A-OSGi Policy Interpreter and Enforcer (PIE).

A-OSGi Monitoring and Analysis Component (MAC). The MAC component is responsible for retrieving information from sensors; it interacts with the OSGi service and module layers, as well as with the JVM. The MAC component monitors resource consumption, performance metrics, and changes to both bundle and service availability, as well as the binding of services by individual bundles.

Whenever the MAC detects a relevant change in the system, it generates an event to alert any interested component. Such events are routed to all components that have previously subscribed them. In our current architecture, only the PIE component subscribes all provided events. However, by exposing a publish-subscribe interface, we facilitate the extension of our architecture with additional functionalities.

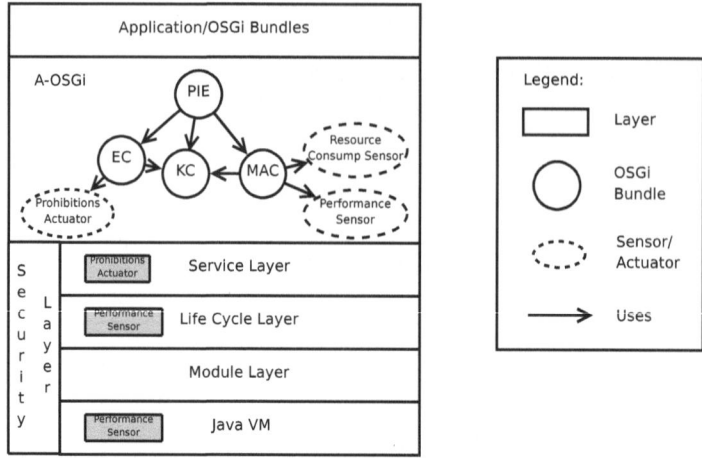

Fig. 2. A-OSGi Architecture

The MAC component is also responsible for generating new events from the composition of other events. In the current prototype, there is no explicit support to specify these using some form of domain specific language constructs: analysis events have to be programmed directly in Java. This pragmatic design choice allowed us to build a running prototype of the A-OSGi architecture that has been used to assess the merits of our approach. As future work we will enrich the analysis component, for instance, integrating previous work by others, such as the Event Distiller described in [15].

A-OSGi Execution Component (EC). The EC component is responsible for executing actions over bundles, individual services, and the OSGi kernel. Its interface exports the primitives that allow to start and stop bundles, change service binding rules in run-time (by adding or removing binding obligations and prohibitions), and also change properties of individual services (for instance by changing parameters associated with the operation of such services). In order to perform these actions, EC interacts with both the service and the life cycles layers of the OSGi architecture. In the current version of the architecture, only the PIE component uses the services of the EC component.

A-OSGi Knowledge Component (KC). The KC component provides a set of mechanisms that allow other components to consult information regarding the state of the A-OSGi execution environment. In more detail, this component maintains, and exports, information concerning the set of installed bundles and registered services, and also on existing dependencies among bundles and services. To maintain such information available, the KC component interacts directly with the module and service layers of the OSGi architecture. In our current architecture the information maintained by the KC is accessed by the PIE component, which uses it to compute adaptation plans.

A-OSGi Policy Interpreter and Enforcer (PIE). The PIE component interprets the system policy, which is described by a set of ECA rules. The activity of PIE is driven by events received from the MAC component, that notify the need to perform adaptations. To select the best course of action, PIE uses the the information about the system provided by the KC component. As a result of its activation, PIE may request to the EC component the execution of one or more actions.

4   Implementation of A-OSGi

In this section we describe in some detail the implementation of A-OSGi. The components of the A-OSGi architecture are implemented, themselves, as OSGi bundles. Naturally, these bundles need to be deployed to support the autonomic behavior of the OSGi system. However, some of the functionality required to implement these bundles requires small changes to the standard OSGi framework. More precisely, we had to augment the life cycle and service layers of the basic OSGi framework. These changes were necessary to support the monitoring and execution components of the MAPE-K cycle.

In the following paragraphs, we first enumerate the technologies that we have used to build our prototype of the A-OSGi framework and, subsequently, describe in more detail the implementation of each component.

4.1   Underlying Technologies

The OSGi specification has several implementations, some of the most well-know are: Eclipse Equinox [16], Apache Felix [17] and Knopflerfish [18]. For the work presented in this paper we have selected the Apache Felix 1.6.0 implementation. Notice however that changes performed over this implementation, and described in this paper, can easily be ported to other existing implementations. Other important component of our architecture is a HTTP server/container that permits the registering of resource and servlets to support the deployment of web applications. In this work we used the Pax Web bundle [19] that implements the OSGi HTTP service specification [6], on top of Jetty HTTP Server [20].

The interfaces of the KC, EC, and MAC components are exported as JMX Managed Beans [21]. Thus, any existing JMX client can use these components, and subscribe the MAC events, or invoke the KC and EC methods. This allows the services provided by these components to be used by third party components and even other applications.

Moreover, the operation of the MAC component requires the inclusion of a JVMTI Agent [22] at the JVM level. Finally, the PIE component is based on the Ponder2 [23] policy interpreter for handling our ECA rules.

4.2   MAC Implementation

The MAC component monitors different aspects of the OSGi execution using the available sensors. Each of these sensors has its own specific requirements in terms of implementation. Namely:

Performance Sensor. A Sensor that monitors the requests received by the HTTP server and stores information concerning the bundle in charge of processing the request. Therefore, this sensor is able to provide information about the absolute number of requests processed by each bundle and the relative distribution of requests among bundles. It also stores the observed latency in the processing of each request. To implement such functionalities, the HTTP server bundle had to be changed in order to monitor the received requests.

Resource Consumption Sensor. A Sensor that monitors CPU usage and memory consumption per bundle. In order to extract this information, some sort of isolation among bundles is necessary. To implement our prototype, we used a thread based approach to achieve the isolation, by creating a hierarchy of *ThreadGroups* that associates a different *ThreadGroup* to each bundle. To create this hierarchy of threads, we have altered the life cycle layer of OSGi such that, whenever a bundle is started, the starting method is executed in a new thread from the *ThreadGroup* of that bundle. As a result, all threads created by the starting thread belong to the *ThreadGroup* associated with the bundle. Furthermore, clients of a service are provided with a proxy that executes the service methods in a thread associated to the bundle that registered the service.

We are aware that the thread based approach used in the current prototype has a number of limitations. In first place, it has a non-negligible overhead as it requires two context switch for each service invocation. Furthermore, it is unable to isolate interactions that do not use the service interfaces (such as when a bundle invokes methods of classes from another bundle). Finally, this approach may cause deadlocks in services with synchronized methods. Therefore, the approach requires a careful configuration of which services need to be isolated. Still, it its able to provide enough feedback to support the required information to implement many relevant autonomic behaviors. Given that the problem of providing isolation among OSGi bundles is a challenging research topic on its own, we expect to incorporate in the future results from complementary on-going research[13].

With thread isolation, CPU usage can be calculated iterating over the threads associated to a bundle ThreadGroup and sum all the threads CPU time. The same approach can be extended to memory since its possible to detect the allocation of objects and assign allocations to the thread that is performing that operation.

Table 1. A-OSGi MAC Events

Event Name	Event Attributes
CPUUsage	BundleID, value, oldvalue
MemoryUsage	BundleID, value, oldvalue
RequestsPerSec	BundleID, value, oldvalue
Latency	BundleID, value, oldvalue
BundleStarted	BundleID
BundleStopped	BundleID
ServiceRegistered	BundleID, ServiceID
ServiceUnregistered	BundleID, ServiceID
ClientRegistered	ClientBundleID, ServiceID
ClientUnregistered	ClientBundleID, ServiceID

OSGi Platform Sensor. This Sensor monitors notifications provided by the OSGi platform concerning the service registration and bundle life cycle. The binding between a bundle and a service is monitored by leveraging on the iPOJO functionalities.

The complete list of events currently provided by the A-OSGi MAC is listed in Table 1.

4.3   EC Implementation

The EC component not only provides an interface to start and stop bundles (something that is directly supported by the standard OSGi implementation) but, more importantly, provides interfaces to control how bundles bind to each other and, as a result, to control which of multiple alternative implementations of a given service can, or should, be used. For that purpose, the EC offers the following mechanisms:

- bindings obligation: a binding obligation specifies that a bundle which operation requires a given service will be obliged to use a specific service implementation. The purpose of this mechanism is to force the use of a service implementation by a bundle.
- binding prohibitions: a binding prohibition specifies that a bundle which operation requires a given service cannot use a specific service implementation. The purpose of this mechanism is to limit the use of service implementations by bundles.
- service property configuration: the EC also provides support to change the value of a property associated to a service implementation. This functionality can be used to alter properties that the developer of the bundle exposed as a service property.

The complete list of actions supported by the EC component is listed in Table 2. In order to implement the EC component we have augmented the OSGi service layer. In A-OSGi, this layer was modified to maintain, for each bundle, the associated obligations and prohibitions. This information is used in run-time to filter the services a bundle can bind, in order to satisfy the constraints defined at each moment. We resort to iPOJO functionality to ensure the correctness of bindings, accordingly to the prohibitions and obligations defined

Table 2. A-OSGi EC Actions

Action Name	Parameters
StartBundle	BundleID
StopBundle	BundleID
SetClientProhibition	BundleID, ServiceID
RemoveClientProhibition	BundleID, ServiceID
RemoveClientProhibitionForServiceName	BundleID, ServiceName
SetClientObligation	BundleID, ServiceID
RemoveClientObligation	BundleID, ServiceID
ChangeServiceProperty	ServiceID, Property, Value

4.4   KC Implementation

The KC provides a set of methods that allow to consult runtime information
about the installed bundles and the registered services, as well as the dependencies between the client bundles and services. To implement these functions,
we use the module layer to extract information about services that a bundle is
using and the service layer to extract information about the bundles being used
by a service. The KC also provides methods to retrieve the current set of service
obligations or prohibitions. The full interface of the KC component is listed in
Table 3.

Table 3. A-OSGi KC functions

A-OSGi Bundle related functions		
Function	Parameters	Returns
getAllBundles		BundleID[ ]
getWebBundles		BundleID[ ]
getBundleName	BundleID	BundleName
getBundleID	BundleName	BundleID
getUsedServiceNames	BundleID	ServiceName[ ]
getUsedServiceIDs	BundleID	ServiceID[ ]
getUsedServiceIDsbyName	BundleID, ServiceName	ServiceID[ ]
getAllUsedServicesIDs	BundleID	ServiceID[ ]
getProvidedServiceIDs	BundleID	ServiceID[ ]
getProvidedServiceNames	BundleID	ServiceName[ ]
getUsingBundles	BundleID	BundleID[ ]
getAllUsingBundles	BundleID	BundleID[ ]
A-OSGi Service related functions		
Function	Parameters	Returns
getAllServices		ServiceID[ ]
getServiceName	ServiceID	ServiceName
getServiceNames	ServiceID	ServiceName[ ]
getServiceBundle	ServiceID	BundleID[ ]
getServiceImplementations	ServiceName	ServiceID[ ]
getUsingBundles	ServiceID	BundleID[ ]
getAllUsingBundles	ServiceID	BundleID[ ]
getAllUsingWebBundles	ServiceID	BundleID[ ]
getClientProhibitions	BundleID	ServiceID[ ]
getClientObligation	BundleID	ServiceID
getServiceProperty	ServiceID, Property	Value

4.5   PEI Implementation

For implementing the PEI component we have used the Ponder2 policy interpreter [23]. With Ponder2 we implemented Managed Objects that we used as
adaptors to interact with the MAC, KC and EC components (using the corresponding JMX MBeans). To describe ECA rules, Ponder provides a language
called PonderTalk. To create an ECA rule we have to specify an event from the
available MAC events, a condition using the KC functions, and actions provided
by EC. The use of Ponder2 also allows the dynamic definition of the policies,
a property very useful in a OSGi system due to the dynamic nature of the
platform.

4.6   Framework Modifications

In order to implement A-OSGi, some modifications to the OSGi Framework were necessary. These modifications can be summarized as follows: i) *JVM level*, a JVMTI agent was implemented to support the monitoring of CPU and memory usage; ii) *Life Cycle Layer*, the execution of the bundle start method was modified in order to execute this method in a new Thread with a corresponding ThreadGroup; iii) *Service Layer*, to implement the prohibitions and obligations mechanism in order to filter services a bundle can find, so the services that a bundle can discover respect the defined constrains.

5   Evaluation

We now illustrate and evaluate the potential of A-OSGi to build autonomic OSGi-based applications. Our case study uses a Web Application that has been implemented using the architecture described in the previous section, and that allows us to demonstrate some of the main features of A-OSGi.

The set of OSGi bundles used by our application is depicted in Figure 3. We consider two web bundles that implement the presentation layer for an on-line store that sells CDs and DVDs. These web bundles are implemented as individual bundles that register with our altered version of the Jetty web server. Both web bundles allow remote clients to: i) list a sub set of products, available in the store and currently in stock, and ii) get details for a specific product. Information about available items in stock is provided by a *stock service* that consults a local database. There are two (independent) bundles that offer this service with distinct trade-offs between quality of service and resource consumption. In more detail, the first implementation of the stock service, simply named *Basic*, only resorts to the internal database to provide information about products. The second implementation of this service, named *Premium*, additionally relies on on a costumer preferences service, to order the product list according to the client preferences. Also, the premium service can offer suggestions about other

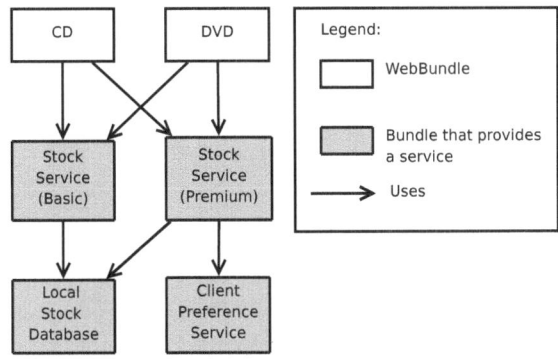

Fig. 3. Case Study Components

products that may be of interest to the user and, therefore, returns additional items when the client searches for either CDs or DVDs.

The functionality provided by the Premium implementation, by offering personalized content, can improve the costumer satisfaction and also generate more revenue to the store. Unfortunately, this additional quality of service comes at the expense of increased resource consumption. In situations where the server becomes overloaded with requests, it may be preferable to satisfy more requests, using the Basic implementation, than to provide the Premium service to a subset of clients and drop the remaining requests. Naturally, when the load allows, one would like to serve all requests using the Premium service. Furthermore, we would like to have the possibility of making these adaptations for each service independently of each other. For instance, if only the CD bundle is overloaded with requests, it may be possible to adapt only the stock implementation used by that service, and continue to use the Premium implementation for DVD buyers. As we will show, the A-OSGi architecture provides support to specify and implement this sort of policies.

5.1   Using A-OSGi

We now describe how A-OSGi can be used to implement the policy described above for our case study. The policy can be described by only two rules, depicted in Listing 1. The first rule simply prohibits any web bundle that is consuming more than 35% of CPU from using the Premium implementation of the stock service. The second rule removes this prohibition when a web bundle uses less than 5% CPU. The adequate thresholds for the CPU usage were determined experimentally. This policy ensures that the most expensive implementation is used, if and only if, the resources are enough to sustain the current load.

Adaptation is performed with bundle-level granularity. The way the rules are specified does not require the CD or DVD web bundles to be named explicitly. Therefore, in run-time, depending on the system load, they may be applied to just the CD service, to just the DVD service, or both. This is possible because the KC component maintains updated information about each bundle, specifically on their bindings. Also, since A-OSGi offers the flexibility to choose which services should be monitored, it is possible to configure the platform in such a way that only the CD and DVD services are monitored, reducing the monitoring overhead to a minimum. Run-time adaptation is performed by restarting the target of the rule. This forces iPOJO to reevaluate the bindings of the target bundle, taking into consideration the new set of rules in the system.

5.2   Performance

To evaluate experimentally A-OSGi we used a workbench composed of two Intel core-2 duo at 2.20 Ghz with 2Gb of memory. Both machines run Linux (Ubuntu 8.10 Desktop Edition) and the Sun Java Virtual Machine 1.6. Both nodes are connected by a 100 Mbit switch. We deployed A-OSGi in one of these machines, and loaded the policy depicted in Listing 1. The other machine is used to generate

Listing 1. Policy

```
newpolicy := root/factory/ecapolicy create.
newpolicy     event: root/event/bundleCPU;
  condition: [: value : bundleID |
    usedstockservice := ((bundles getUsedServiceIDsbyName: \
      bundleID name: "pt.mediaportal.stock.StockService") at: 0).
    usedstockbundle := (services getServiceBundle: usedstockservice).
    stock1bundle := (bundles getBundleID: "pt.mediaportal.stock.Premium").
    (value > 35) & (usedstockbundle == stock1bundle) ];
  action:          [: value : bundleID |
    usedstockservice := ((bundles getUsedServiceIDsbyName: \
      bundleID name: "pt.mediaportal.stock.StockService") at: 0).
    services setClientProhibition: bundleID serviceID: usedstockservice.
    bundles stopBundle: bundleID.
    bundles startBundle: bundleID.
    ];
  active: true.
newpolicy := root/factory/ecapolicy create.
newpolicy     event: root/event/bundleCPU;
  condition: [: value : bundleID |
    usedstockservice := ((bundles getUsedServiceIDsbyName: \
      bundleID name: "pt.mediaportal.stock.StockService") at: 0).
    usedstockbundle := (services getServiceBundle: usedstockservice).
    stock2bundle := (bundles getBundleID: "pt.mediaportal.stock.Basic").
    (value < 5) & (usedstockbundle == stock2bundle) ];
  action:          [: value : bundleID |
    usedstockservice := ((bundles getUsedServiceIDsbyName: \
      bundleID name: "pt.mediaportal.stock.StockService") at: 0).
    services removeClientProhibition: bundleID serviceID: usedstockservice.
    bundles stopBundle: bundleID.
    bundles startBundle: bundleID.
    ];
  active: true.
```

the workload using Apache JMeter 2.3.2 to emulate clients executing requests to the server. Clients operate by requesting a list of either DVDs or CDs from the server, and subsequently requesting details on one of the returned items.

During the experiments the web application is subject to 3 different workloads that we have named, CD/DVD, CD/DVD+, and CD+/DVD+. The CD/DVD workload imposes 50 requests per second to the CD service and another 50 requests per second to the DVD service. This load is low enough such that the Premium implementation of the stock service can be used to answer all requests without overloading the system. The CD/DVD+workload, in addition to the previous requests, imposes an additional load of 1.500 requests per second to the DVD service. To sustain this load, one is required to adapt the implementation of the stock bundle used to process DVD requests (CD requests do not need to be affected by the adaptation at this point). Finally, the CD+/DVD+ workload includes an excess of 700 requests per second to the CD service. At this point, both the DVD and CD requests are required to use the Basic implementation of the stock service to sustain the heavy load.

The system is initiated with the CD/DVD workload. At time 60 the workload is changed to the CD/DVD+ workload. Subsequently, at time 120 the workload is increased again to CD+/DVD+. Finally, at time 180 the workload returns to the baseline CD/DVD workload. Each individual workload was generated by a group of 10 client threads. These workloads are illustrated in Figure 4 (time is measured in seconds).

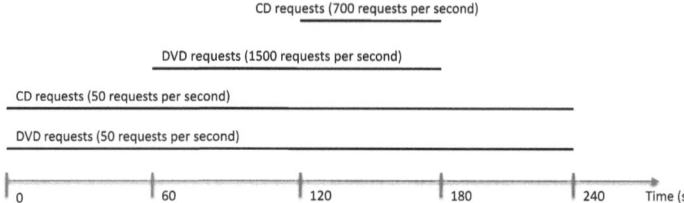

Fig. 4. Workload Description

The results are depicted in Figure 5. The first plot compares the performance of a static configuration (providing the premium service) against the autonomic configuration. The adaptations that result from execution the policy can be inferred by the quality of service provided to the user in plot 5(b). Clearly, the autonomic configuration is able to ensure a much better throughput than the static configuration, by dynamically changing to the less expensive implementation of the stock bundle. Plot 5(c) depicts the total number of requests processed by both configurations. This last plot makes clear that the autonomic version responds better to the increase in the workload.

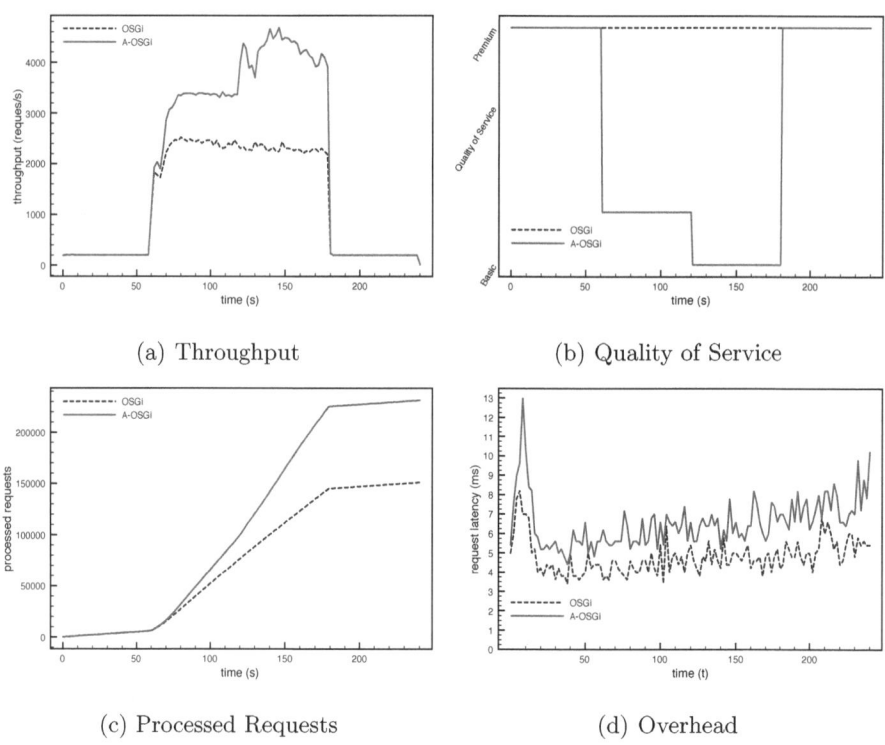

(a) Throughput

(b) Quality of Service

(c) Processed Requests

(d) Overhead

Fig. 5. Performance with and without adaptation

Finally, plot 5(d) compares the average request latency of the application running in the A-OSGi framework against the same application, under the same medium workload, running in a plain OSGi framework. This allows us to assess the overhead induced by the current implementation of the A-OSGi mechanisms. The difference is in the order of 25%, which is not surprising, given that many of the A-OSGi components are not yet fully optimized (in particular the isolation mechanisms required for detailed monitoring).

5.3   Other Policies

Due to lack of space, we have only discussed and evaluated one of several policies that could be applied to the case study. However, we would like to point out some other alternatives that would also be supported by the A-OSGi framework. Alternatively, or in addition to commuting between the Basic and Premium implementation, the policy could also configure the operation of each of these implementations (for instance, by changing the number of recommendations returned to the client by the Premium service). This would require to write rules specific for each bundle implementation, a feature that our simple case-study does not illustrates. Also, instead of setting individual binding constraints, the global behavior of the system could be controlled by simply installing or uninstalling bundles on the fly.

6   Conclusions

In this paper we have proposed A-OSGi, a framework that augments the OSGi platform to support the implementation of autonomic OSGi-based applications. A-OSGi offers a number of complementary mechanisms to this end, including the ability to extract performance indicators about the execution of deployed bundles, mechanisms that allow to have a fine grain control of how services bind to each other, and support to describe the the autonomic behavior of the OSGi application using a policy language.

The architecture has been implemented. Experimental results have illustrated the benefits of the approach: we were able to selectively adapt the implementation of a bundle used by different services, in order to augment the system performance in face of dynamic workloads. As future work, we plan to study ways to optimize the performance of some of the A-OSGi components, such as the MAC (by using more efficient isolation techniques), to reduce the overhead imposed by the autonomic mechanisms.

References

1. The OSGi Alliance: OSGi Service Platform Core Specification, Release 4, Version 4.1 (2007), http://www.osgi.org/Download/Release4V41
2. Gruber, O., Hargrave, B.J., McAffer, J., Rapicault, P., Watson, T.: The eclipse 3.0 platform: Adopting osgi technology. IBM Systems Journal (2005)

3. Sun Microsystems: Sun GlassFish Enterprise Server v3 Prelude Release Notes (2008), http://docs.sun.com/app/docs/coll/1343.7
4. OW2 Consortium: Jonas - White Paper v1.2 (2008), http://wiki.jonas.objectweb.org/xwiki/bin/download/Main/Documentation/JOnAS5_WP.pdf
5. Spring Source: Spring Dynamic Modules for OSGi (2009), http://www.springsource.org/osgi
6. The OSGi Alliance: OSGi Service Platform Service Compendium, Release 4, Version 4.1 (2007), http://www.osgi.org/Download/Release4V41
7. Diao, Y., Gandhi, N., Hellerstein, J., Parekh, S., Tilbury, D.: Using mimo feedback control to enforce policies for interrelated metrics with application to the apache web server. In: Network Operations and Management Symposium, NOMS 2002. 2002 IEEE/IFIP, pp. 219–234 (2002)
8. van der Mei, R., Hariharan, R., Reeser, P.: Web server performance modeling. Telecommunication Systems (2001)
9. IBM: Autonomic computing: Ibm's perspective on the state of information technology. IBM Journal (2001)
10. Escoffier, C., Hall, R., Lalanda, P.: Ipojo: an extensible service-oriented component framework, July 2007, pp. 474–481 (2007)
11. IBM: An architectural blueprint for autonomic computing, fourth edition. Technical report, IBM (2006)
12. Miettinen, T.: Resource monitoring and visualization of OSGi-based software components. PhD thesis, VTT Technical Research Centre of Finland (2008)
13. Geoffray, N., Thomas, G., Clément, C., Folliot, B.: Towards a new Isolation Abstraction for OSGi. In: Proceedings of the First Workshop on Isolation and Integration in Embedded Systems (IIES 2008), Glasgow, Scotland, UK, April 2008, pp. 41–45 (2008)
14. Hulaas, J., Binder, W.: Program transformations for light-weight cpu accounting and control in the java virtual machine. Higher Order Symbol. Comput. 21(1-2), 119–146 (2008)
15. Kaiser, G., Parekh, J., Gross, P., Valetto, G.: Kinesthetics extreme: an external infrastructure for monitoring distributed legacy systems. In: Autonomic Computing Workshop, June 2003, pp. 22–30 (2003)
16. Eclipse Equinox: Homepage, http://www.eclipse.org/equinox/
17. Felix Apache: Homepage, http://felix.apache.org/
18. Knopflerfish: Homepage, http://www.knopflerfish.org/
19. Pax Web: Homepage, http://wiki.ops4j.org/display/paxwev/Pax+Web/
20. Jetty HTTP Server: Homepage, http://www.mortbay.org/jetty/
21. Sun Microsystems: Java Management Extensions, http://java.sun.com/javase/6/docs/technotes/guides/jmx/index.html
22. Sun Microsystems: Java Virtual Machine Tools Interface, http://java.sun.com/javase/6/docs/platform/jvmti/jvmti.html
23. Twidle, K., Lupu, E., Dulay, N., Sloman, M.: Ponder2 - a policy environment for autonomous pervasive systems, June 2008, pp. 245–246 (2008)

A Network-Coding Based Event Diffusion Protocol for Wireless Mesh Networks

Roberto Beraldi[1] and Hussein Alnuweiri[2]

[1] "La Sapienza" University of Rome, Rome, Italy
beraldi@dis.uniroma1.it
[2] Electrical & Computer Engineering Texas A&M University at Qatar
hussein.alnuweiri@qatar.tamu.edu

Abstract. Publish/subscribe is a well know and powerful distributed programming paradigm with many potential applications. In this paper we consider the central problem of any pub/sub implementation, namely the problem of event dissemination, in the case of a Wireless Mesh Network. We propose a protocol based on non-trivial forwarding mechanisms that employ network coding as a central tool for supporting adaptive event dissemination while exploiting the broadcast nature of wireless transmissions. Our results show that network coding provides significant improvements to event diffusion compared to standard blind dissemination solutions, namely flooding and gossiping.

Keywords: Network coding, publish/subscribe, wireless.

1 Introduction

This paper investigates the problem of event diffusion over a wireless mesh network (WMN) by leveraging a recent information dissemination technique called Network Coding; see [8] for a tutorial. The Wireless Mesh Network (WMN) is an emerging communication architecture with many practical applications in such areas as self-organizing community networks, industrial plant automation, wireless sensor networks, etc., [1]. A WMN can be considered as a two-tier architecture. The first tier is a wireless backbone composed of mesh routers capable of packet routing and optionally providing gateway functionality. The second tier is composed of mobile and/or portable wireless devices (e.g. WiFi-enabled smart phones, mobile TV devices, etc.) which can act as clients. A WMN is a self-organizing network with a certain degree of variability in terms of participants and topology. For example, clients can move, new clients can join a network, mesh routers can be occasionally switched off, or some clients can at times act as wireless routers. Having a suitable application level abstraction that can face with such a changes is thus very appealing. In this regards, publish/subscribe (pub/sub) is a mature interaction paradigm that fits such requirements, since it allows for reference-decoupled and asynchronous interactions among the participants [7]. In a pub/sub communication system publishers produce information in form of events and subscribers receive the subset of events that match their interests, expressed as a filter. Pub/sub

---

[1] This paper was supported by the EU STREP SM4ALL FP7-224332.

A.V. Vasilakos et al. (Eds.): AUTONOMICS 2009, LNICST 23, pp. 17–31, 2010.
© Institute for Computer Sciences, Social-Informatics and Telecommunications Engineering 2010

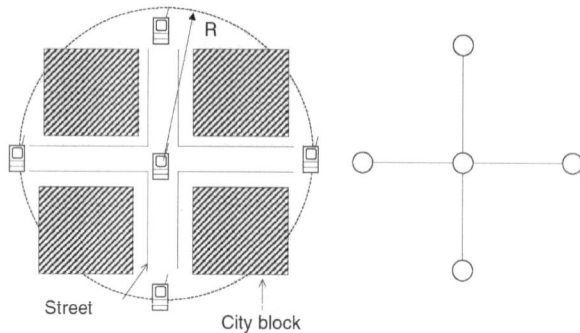

Fig. 1. The grid topology arising from a metropolitan deployment of a WMN

systems have been widely studied in wired a setting, e.g., SIENA [6],Gryphon [11], LeSubscribe [16]. However, while some papers have also focused on pub/sub systems running over networks exploiting wireless technology, e.g. [3], [13], only a very few of them have considered WMNs, [10], [21].

We consider a WMN deployed over a Manhattan like city model, see [1], in which mesh routers can be considered as approximately placed at the intersection of two streets. Since the streets are running est-west and north-south, mesh routers form a regular grid topology, Figure 1. We assume that mesh routers are used as a dispatching structure for supporting event diffusion. This solution is borrowed from the proposal presented in [10]. We assume that each mesh client can communicate with only one mesh router (called its local mesh router), and mesh routers are equipped with additional software appliances that clients interact with. Essentially, when the publisher needs to publish a new event, it contacts its local mesh router and then sends the event to it. The mesh router diffuses the newly event to all the other routers in the network, on behalf of the publisher. A subscriber periodically renews its subscription to its current local mesh router for a specific period of time, thus implementing a lease mechanism. Filtering is done at the mesh router, and filters are not propagated into the network. A router notifies the client as soon as it receives an event matching the filter, given that the client subscription has not expired. In the rest of the paper we refer to a mesh router as a node.

1.1 Contribution of the Work

The contribution of the paper is the proposal of an event dissemination protocol suitable for dynamic environments. The protocol is self tuning in that (i) the behavior of a single node depends on the amount of information is being received as well as on the number of neighbors of a node (node density), (ii) the protocol runs efficiently independently of how many targets there are in the system and where they are located.

The rest of the paper is organized as follows. Section 2 presents a brief tutorial on the main concept of network coding and discusses basic alternatives to implement event diffusion in a wireless mesh setting. Section 3 presents the details of our network-coding based protocol, and Section 4 provides several evaluation results. Finally, conclusions are given in Section 5.

2 Background

Network coding is a relatively recent technique for end-to-end information delivery in communication networks, introduced in the seminal paper of Ahlswede et al., [2] and advanced by others [12] for many applications. Network coding marks a clear departure from the basic network role as a passive relay of data packets or frames, to a more active model in which network nodes can perform algebraic operations on the data before sending it out. With network coding the intermediate nodes between source of information and the destination(s) do not simply relay the received packets. Rather, they are allowed to combine (encode) incoming data in order to generate the data output to be forwarded. The original key advantage of this intermediate combination is for data broadcasting and multicasting. With network coding a source node can always send data at the network's broadcasting rate, while without network coding this is not possible in general. Some concrete examples of network coding based multicast protocols can be found in [14], [15], [5] and [4]. Additional references can be found in [19].

In the following we adopt a linear network coding approach in which operations on packets are confined to algebraic operations over a finite field. More precisely, we confine ourselves to the Galois Field $GF(2^w)$ and interpret each data packet as being composed from a set elements of the field, each of size w bits. We restrict ourselves to apply linear network coding to the problem of broadcasting an original data packet, X, from a source node (e.g. the mesh router on behalf of the publisher) to all the other nodes of a wireless network. The problem solution can be easily generalized to multi-source multicast under reasonable additional constraints. The main symbols used throughout the paper are listed in Table 1.

In linear network coding, the basic operation performed by each network node is generating linear combinations of incoming packets, and transmitting the new "coded" packet. A linear combination is carried over a fixed set of original data *chunks*, called a *generation* of the original packet. More precisely, we assume that special designated nodes split an original data packet X of length l into m chunks, x_i, each of length l/m,

Table 1. Definition of the main symbols

E	Event to be diffused
m	Generation size
x	Original chunk of data
X	Vector of the original m chunks
y	an encoded chunk
Y	Vector of encoded chunks
α	Random coefficient
A	$m \times m$ matrix of random coefficients (decoding matrix)
EV	Encoding vector, coefficients used to create a linear comb.
IV	Information vector, an encoded chunk sent into a packet

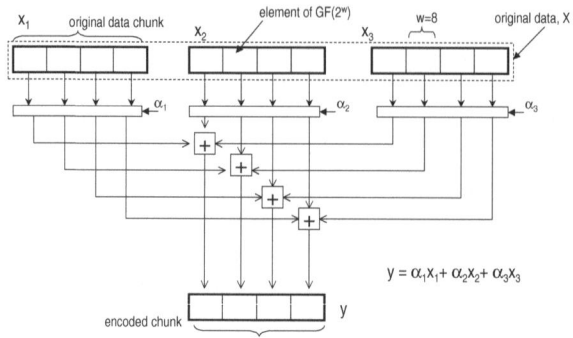

Fig. 2. The operation of linear combination over the original chunks of data

to form the generation $\{x_1, x_2, \ldots x_m\}$. Each chunk of data is composed of k elements of $GF(2^w)$. Hence, $l = k \times m \times w$ (0s are padded if required). The value m is called the *generation size*.

Consider for sake of example only, a data packet X of size 12 bytes, generation size $m = 3$ and element size $w = 8$ bits, or one byte (see Figure 2). The packet is divided into 3 chunks, x_1, x_2, x_3, each composed of 4 elements (bytes). A linear combination is achieved by choosing 3 coefficients of length w bits, $\alpha_1, \alpha_2, \alpha_3$ and computing a new *encoded* chunk

$$y = \alpha_1 x_1 + \alpha_2 x_2 + \alpha_3 x_3$$

Because all operators are defined over the finite field $GF(2^w)$, the above computation is performed element-wise, i.e., if x_j^i is the i-th element of chunk j and y_i is the i-th element of the linear combination we have

$$y_i = \alpha_1 x_1^i + \alpha_2 x_2^i + \alpha_3^i x_3 \quad i = 1, 2, 3$$

However, for the sake of avoiding cumbersome notation, we do not make this replication explicit, and a linear combination is expressed as

$$y = \sum_{i=1}^{m} \alpha_i x_i$$

The result y is called an *encoded chunk* or, when it is sent in a packet, an *Information Vector, IV*. The set $EV = [\alpha_1, \ldots, \alpha_m]$ of coefficients used in the combination is called the *Encoding Vector*. As common with other network coding schemes, we assume that a node sends both the information vector IV and the associated encoding vector EV. Moreover, the coefficients used in a linear combination are generated randomly by the node.

Note that the overhead due to sending the above encoding vector is $m \times w$. Assuming typical values of $m = 16$ and $w = 8$ bits when sending a 1 KB data packet, the overhead is only $16/1024 \approx 1.5\%$.

An important aspect of linear network coding is that encoded chunks can themselves be combined to generate new encoded chunks at an intermediate node. For example, to combine n encoded chunks, $y_1 \ldots y_n$, a node uses n coefficients, say $\alpha'_1 \ldots \alpha'_n$ with which it generates

$$y' = \sum_{j=1}^{n} \alpha'_j y_j$$

Note that the newly encoded chunk, y', also represents a linear combination with respect to the original data chunks. Therefore, since

$$y' = \sum_{j=1}^{n} \alpha'_j (\sum_{i=1}^{m} \alpha_{ji} x_i) = \sum_{i=1}^{m} (\sum_{j=1}^{n} \alpha'_j \alpha_{ji}) x_i$$

where α_{ji} is the i-th coefficient used to calculate y_j, the encoding vector the intermediate nodes is in effect

$$EV = [\sum_{j=1}^{n} \alpha'_j \alpha_{j1}, \ldots, \sum_{j=1}^{n} \alpha'_j \alpha_{jm}]$$

2.1 Event Dissemination with Network Coding

To elucidate the advantage of network coding, we present several alternatives for disseminating (diffusing) an event over a portion of a wireless grid in which each node can reach four neighbor nodes to the north, south, east and west. For this purpose, we consider the 2-dimensional grid topology of Figure 3 and explain several alternatives for disseminating an event E generated by the *source* node S (located at the center of the grid) using the smallest amount of data transmissions. The problem we solve is how to ensure that the event reaches the four *destination* nodes at the four corners of the grid. Nodes 1, 2, 3, 4 function as *relay* nodes in this example.

For the sake of simplicity, we assume here an idealized collision-free broadcast communication channel and that event E fits the size of a single packet (a realistic channel is used in simulations). Our aim is to compare the performance/cost tradeoff of different principle design, where the cost is the total amount of data sent over the network, T, and the performance is measured through the probability P_F that *all* nodes receive E. In the following the cost of sending one packet (containing E) is counted as one.

In general, there are two different approaches for event diffusion: *informed* dissemination and *blind dissemination*. Informed Dissemination requires the source node S to know the topology and coordinate transmissions to the destination nodes. For example, in the case of Figure 3, S may declare, in the packet header, two destinations which have to rebroadcast the packet. For example, S can specify nodes 1 and 3 as destinations. This means relay nodes 2 and 4 discard the packet, while nodes 1 and 3 will rebroadcast their packet to the corner nodes. It is obvious that the cost of this solution is $T = 3$, and $P_F = 1$ (ignoring the small cost of sending the additional destination IDs in the packet header). Clearly, this is the smallest amount of data that must be sent for disseminating the event.

Blind dissemination is an attractive alternative which is more suited to the distributed dynamic nature of wireless mesh networks. In such environment, it may not be easy (or it may be very costly) to maintain the required topology or link-state information for

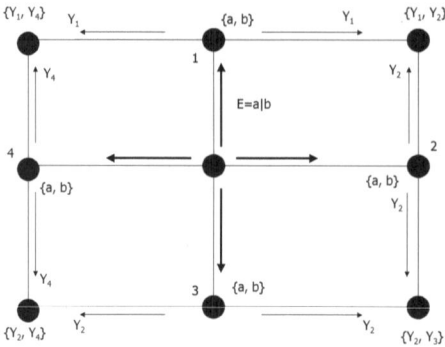

Fig. 3. Event diffusion over a square grid. The source node S is placed at the center.

disseminating events in the presence of node mobility, or some nodes turning off and new nodes turning on arbitrarily.

In this paper, we are interested mainly in blind dissemination. Next, we review several alternatives that avoid the use of coordination among nodes while still requiring the same amount of data transmissions as in informed dissemination. Normally, this comes at the cost of a lower value for P_F. In order to assess the benefit of applying network coding for event diffusion, we will explore and compare the performance of network coding to other blind dissemination techniques, namely flooding, naive dissemination, and gossip. We will briefly explain these techniques with the help of the wireless grid example of Figure 3.

NAIVE RANDOM DISSEMINATION. This technique exploits a very simple random forwarding approach in which coordination is no longer required. Consider the following variation of the approach. Initially, the source S transmits the event to its four neighbors (relay nodes). Then each relay node splits the packet (event E) into two parts, say a and b, of equal size. Now, each relay node chooses only one of these parts, i.e. either a or b, with equal probability, then transmits it to all destination it can reach. Since each part contains only half the information carried in E, the cost of sending one part of the event by a relay node is $\frac{1}{2}$. Thus, the total cost of dissemination is still $T = 3$.

To calculate P_F, let $\{y_1, y_2, y_3, y_4\}$ denote the scheduled parts chosen by the four relay nodes $1, 2, 3$ and 4, respectively. For example, the schedule $\{a, a, a, b\}$ refers to the schedule where relay nodes $1, 2, 3$ each choose to transmit a, and 4 chooses to transmit b. For this example, there are 2^4 different schedules, but only two of them, namely $\{a, b, a, b\}$ and $\{b, a, b, a\}$ allow all the receiver nodes reconstruct the event. Thus, $p_F = 2^{-3} = 1/8 = 0.125$. This result shows the weakness of naive random dissemination.

RANDOM DISSEMINATION WITH NETWORK CODING. The second random solution leverages linear network coding, where we can show that we can achieve forwarding probability $p_F \approx 1$ while maintaining the cost of forwarding practically the same as with the informed dissemination. Instead of just sending one half of the event E randomly, a relay node now sends a random linear combination of the two parts a and b.

To compute the linear combination, the relay node picks two coefficients, α_1 and α_2, uniformly at random from a finite field of size q then broadcasts $y = \alpha_1 a + \alpha_2 b$ together with the coefficient's vector (α_1, α_1) (all Operations are defined over the field F_q). Note that the size of the linear combination $|y| = |a| = |b| = |E/2|$. The transmitted packet will have a size slightly larger than $|y|$ because we must include the coefficients in the packet. However, for most practical cases, the overhead due to including the coefficients in the transmission is small and can be neglected to simplify analysis.

Since a receiver node at one of the corners gets two linear combinations from its neighbors, say $y_1 = \alpha_{11} a + \alpha_{12} b$ and $y_2 = \alpha_{21} a + \alpha_{22} b$ with the corresponding 4 coefficients, it will be able to retrieve (i.e. decode) the original event E if it can solve the following linear system of equations:

$$\alpha_{11} y_1 + \alpha_{12} y_2 = x_1$$

$$\alpha_{21} y_1 + \alpha_{22} y_2 = x_2$$

There are q^4 different possible 2×2 matrices whose coefficients are picked randomly from the field F_q. Considering that the number of linearly independent matrices is $(q^2 - 1)(q^2 - q)$, the probability that the matrix above is non-singular, thus allowing a node to retrieve the event (by inverting the matrix, or using Gaussian elimination), is

$$\frac{(q^2 - 1)(q^2 - q)}{q^4};$$

from which the probability that all of the 4 receivers at the corners of the grid get the event is $P_F = \left[(1 - \frac{1}{q^2})(1 - \frac{1}{q}) \right]^4$. For example, if we choose with $q = 256$, i.e. the field of 8-bit coefficients, then $P_F \approx 0.98$. Note, also that the overhead for sending two coefficients is just two bytes.

GOSSIP. In order to illustrate the powerful utility of network coding in random event dissemination, we compare it against a probabilistic flooding technique commonly referred to as *gossiping*. Using a basic gossip protocol, each relay node transmits the complete event E to its neighbors with probability p. Let

$$[tx_1, tx_1, tx_3, tx_4]$$

be the transmit decision of the four relay nodes such that $tx_i = 1$ means that node i decided to transmit the packet, and $tx_i = 0$ means it decided to discard the packet. There are two sets of decisions that allow for all four receivers at the corners of the grid to get the event. They are

$$TX = [1, *, 1, *] \quad TX' = [*, 1, *, 1]$$

where $*$ means any decision. In other words, all receivers will get E when either (at least) both relay nodes 1 and 3 transmit E, or both nodes 2 and 4 transmit E. A decision belongs to one of these 2 sets with probability p^2. Note that the decision $1, 1, 1, 1$ is common to both sets and occurs with probability p^4. Therefore, the probability that all receivers get the event is the probability to observe a retransmission pattern of type TX or TX' on the relay nodes, or

$$P_F = 2p^2 - p^4$$

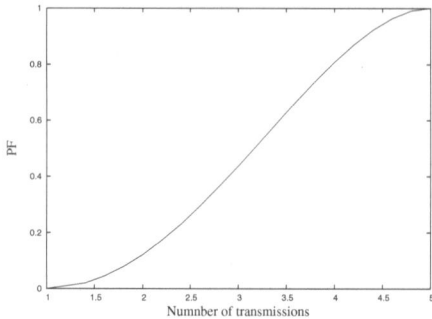

Fig. 4. Probability that all nodes receive the event, P_F vs total number of transmissions, T

To compare gossiping to RDNC, we first observe that in gossiping, the average number of transmissions is $T = 1 + 4p$, giving us $p = (T - 1)/4$, where T is the cost of dissemination. Figure 4 shows the complete reception probability P_F as a function of T for the gossip technique. The above result shows that if we are to maintain the cost at $T = 3$ transmissions, then we get $p = 0.5$; from which we determine $P_F \approx 0.43$. On the other hand, in order for gossip to achieve the same dissemination reliability as network coding, i.e. $P_F \approx 0.98$, we need to set $p \approx 0.9$, which means the cost will increase to $T \approx 4.6$ transmissions.

3   Proposed Protocol

The prosed protocol utilizes a push-pull method whereby the event is diffused in two phases. During the first phase, the information content of the event is partially pushed throughput all the network. During the second phase the fraction of nodes that need to fully decode the event pull the missed information from their neighbor nodes.

   The key idea of the protocol is to assure that although a single node doesn't have the whole information required to detect the event, the remaining part can be retrieved from nearby nodes. Roughly speaking, the pushing phase is such that *any* group of nodes, which is composed by a node and its neighbors, has the full information required to decode the event.

   In the first phase of the protocol a node adapts its behavior according to the amount of information being received. A kind of self adaptation is also incorporated in the second phase, where a feedback mechanism is used to adapt the reaction of each node according to the current node density.

3.1   Basic Data Structure and Assumption

Each node manages the following data structures

- $m \times m$ matrix, A, called Decoding Matrix containing the elements of the encoding vector
- $1 \times m$ vector, Y, called the local encoded data vector, which contains the encoded chunks

- Operation bit mode, *op* (normal or collecting)
- Transmission counter, C

A node has direct access to the local broadcast link layer primitive, `bcast(P)`, with which it can send a packet P to all its neighbor nodes, namely, the ones that are within its transmission range. A packet P contains the following fields

- Information Vector, $P.IV$
- Encoding Vector, $P.EV$
- Generation ID, $P.gen$
- Topic ID, $P.topic$

We assume that an event E fits the size of a packet (a simple variation allows to overcome this limitation). Each event E is uniquely identified through the concatenation of a generation number, managed by the publisher, and the publisher's ID.

3.2 Protocol Description

In our algorithm the nodes are classified according to their roles as:

- Source node (the node that sends the packet carrying the event to be diffused)
- Bootstrap nodes (the neighbors of the source)
- Intermediate nodes (all the of other nodes)

An intermediate node may operate into two different modes: *normal* ($op = 0$) mode and *collecting* mode ($op = 1$). The nodes initially operate in the normal mode.

The proposed protocol uses the following four key parameters to define a flexible network-coding specific forwarding policy. These parameters can be tuned to optimize the performance of the forwarding policy on a wireless mesh. The parameters are:

- BF, Bootstrap Factor
- ΔT_C, Collecting Time
- FF, Forwarding Factor
- $MaxTx$, Maximum number of allowed transmissions

Push phase. The actions performed by each node in the first push phase are the following ones.

PUBLISHER. When the publisher needs to send a new event E, it issues the publish primitive on the local node (router), say node S. Node S acts as a source of the event on behalf of the publisher, and sends E by a local broadcast indicating the generation ID.

BOOTSTRAP NODES. When a bootstrap node, B, receives event E from S, it splits the event into m chunks, $x_1, \ldots, x_m$. Then, B executes BF times the following four steps:

1. generate m random coefficients, $EV = [\alpha_1, \ldots, \alpha_m]$
2. compute the linear combination $y = \alpha_1 x_1 + \ldots + \alpha_m x_m$
3. prepare a new packet with $P.IV = y$, $P.EV = EV$
4. send P via the local broadcast primitive.

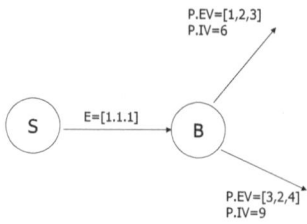

Fig. 5. The initial phase of event diffusion

The Bootstrap Factor parameter, BF, varies in the range $[1..m]$ and determines the number of linear combinations sent by a bootstrap node.

Figure 5 shows an example for $m = 3$. For simplicity, we have assumed that E is composed of 3 bytes, each of value 1. The bootstrap node B generates two linear combinations with the coefficients $[1, 2, 3]$ and $[3, 2, 4]$. Thus, the bootstrap factor is 2.

INTERMEDIATE NODES. When an intermediate node, say I, receives a packet P containing a chunk for a new event, it creates a new decoding matrix, A, associated with the event. The matrix contains all 0s, except for the first row which contains the coefficients carried in the EV. The node also creates a new local encoded data vector, Y, containing all 0s in all entries except the first one, which stores the IV of packet P. Finally, it sets the transmissions counter C to $MaxTx$.

After these steps, I enters into the collecting operation mode ($op = 1$) and starts a new *collecting phase*. A new collecting phase is initiated each time the node enters this mode. During a collecting phase I waits for other possible new innovative chunks if order to decide how many linear combinations to send.

The Collecting time parameter, ΔT_C, determines the minimum duration of the collecting phase. If a new innovative packet is received while the node operates in the collecting mode, say at time t, then the duration of collecting phase will be postponed until time $t + \Delta T_C$. Should a new innovative packet arrive before the new deadline, the collecting phase will be again deferred by ΔT_C.

The application of collecting-time is sketched in Figure 6. Node I has received two packets, $P1$ and $P2$, carrying two linearly independent combinations over the original chunks. The decoding matrix thus contains the corresponding encoding vectors. Packet $P1$ triggers the collecting phase, while $P2$ prolonged the phase.

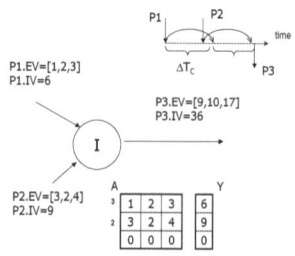

Fig. 6. Generating linear combination at an intermediate node

After ΔT_C from when $P2$ was received, the node does the following: create a new linear combination using the coefficients 3 and 2 (they are shown to the left of the matrix in figure 6), create a new information vector whose value is $3 \times 6 + 2 \times 9 = 36$, create the new encoding vector, $3[1, 2, 3] + 2[3, 2, 4] = [9, 10, 17]$, then send the new packet P3.

The reason for not immediately sending new linear combinations is that we have empirically seen by simulations that innovative packets tend to arrive as burst[1]. As the generation of a new linear combination is more useful if it is done over a wider number of chunks, deferring the generation time - in the hope of increasing the number of chunks - is a simple pragmatic way for improving the benefit of network coding. In fact, the newly generated combination is more likely to be independent from the ones stored at a larger number of neighbors.

The prolongation of the collecting phase thus acts a very simple adaption mechanism that uses the reception of a new innovative chunk as an indicator that further useful chunks are likely to be received in the near future. Please note that if the rank of the decoding matrix at the beginning of a collecting phase is r, then the node can receive no more than $m - r$ new linear combinations; thus, the overall duration of the collecting phase cannot exceed $(m - r) \times \Delta T_C$.

At the end of the collecting phase the node sends out $k = min\{\lceil FF \times r \rceil, C\}$ new combinations, where r is the total number innovative packets collected during the collecting phase. The parameter FF, called the Forwarding Factor is a real number in the range $(0..1]$. The forwarding factor regulates the "verbosity" of a node, and can be useful in reducing the amount of data flooded in the network, or for conserving energy by reducing the amount of transmissions from a node. Finally the node I decreases the transmission counter C of k. The push phase ends when $C = 0$.

3.3 Pull Phase

The pull phase is initiated by a node A after ΔT_P time unit from when $C = 0$ (recall that each packet carries the topic identification so A triggers this phase only for chunks belonging to events of interest).

To pull the missed chunks, node A broadcasts a requesting message to its neighbors. This message contains the event ID and the number of missed chunks, say c, required for full decoding. Upon receiving such a request, say at time t, a A's neighbor, say B, schedules the transmissions of c linear combinations at time $t + \Delta T$, where ΔT is a random jitter picked in the interval $(0, T_J)$ (T_J is maximum jitter). During ΔT, B listen for possible updates sent by A. A sends an update message each time it receives some innovative chunks from any of its neighbors. An update message contains the number $c' < c$ of innovative packets that A now wishes to receive. In such an update message is received, B adjusts the scheduled number of transmissions to c'. Clearly, the transmissions are cancelled when $c' = 0$. Thus for example, a request message with $c = 3$ triggers the scheduling of 3 linear combinations at each A's neighbors (suppone they are they B and D). Let assume that a neighbor D sends 3 chunks. If after the reception of the new chunks A needs, say $c' = 1$ more new chunks, A will send a new requesting message containing this new value. In this way B decreases its contribution from 3 to 1 chunk.

---

[1] We have used the nam animation tool shipped with ns-2.

4   Evaluation

This section reports an in-depth performance evaluation study of the proposed protocols carried out by extensive simulations using ns-2.31 simulation tools [18]. We used a 914 MHz Lucent WaveLAN DSSS radio interface model available in the simulator. Each node has a transmission range R. We used a two-ray ground refection model as the wireless propagation model. Each reading of the simulation was taken after 100 independent runs. We have simulated 400 nodes arranged on a grid at distance of 250 meters from one other. The transmission range is either fixed to $R = 250m$ (with connectivity degree = 4 nodes), or to $R = 750m$ (connectivity degree = 20 nodes). The duration of each simulation was 500 seconds. The source of the event is placed at the center of the grid and transmits a new packet of $1KB$ in size, every 1 s. The default collecting time is $\Delta T_C = 50\ ms$ while the maximum jitter is $T_J = 10\ ms$. The target nodes are placed *at random*.

Arithmetic operations are performed on the Galois field 2^8. We have used the library available at [9]. To speed up the simulation, the decoding matrix is managed in the Gaussian triangular form. We say that a node decodes the event when the associated encoding matrix has full rank. Using this simulation environment, we provide estimates of the following performance metrics,

- Percentage of Decoding: percentage of nodes that successfully decode the event.
- Decoding Delay: the amount of time elapsed from when the event is generated until it is decoded
- Cost of Diffusion: total number of bytes sent transmitted in the network for disseminating the event

4.1   Protocol Tuning

Our first set of experiments concerns the effect of collecting time and the generation size on the protocol performance. In these preliminary tests we have considered a push-only protocol, i.e., nodes have no limit on the number of transmissions they are allowed to perform.

The left plot in Figure 7 shows the full decoding probability as a function of the generation size and the collecting time given as a parameter. The decoding probability is highly affected by the collecting time. When the decoding phase is not applied, $\Delta T_C = 0$, the decoding probability is very low, especially for small generation size. The decoding probability increases with the collecting time, meaning that waiting allows a node for collecting a larger amount of innovative information and then generating linear combinations that are useful for a larger number of neighbors.

The right side plot in Figure 7 shows the cost as function of the generation size and the collecting time as a parameter. The cost increases with the collecting time since a longer collecting time allows for more node to full decoding and then sending data.

4.2   Performance on a Grid

The left side plot in Figure 8 shows the total per event diffusion cost as a function of the number of subscribers for a 20×20 grid with connectivity 4. The cost of the proposed protocol has been compared against three not adaptive protocols: flooding, gossip

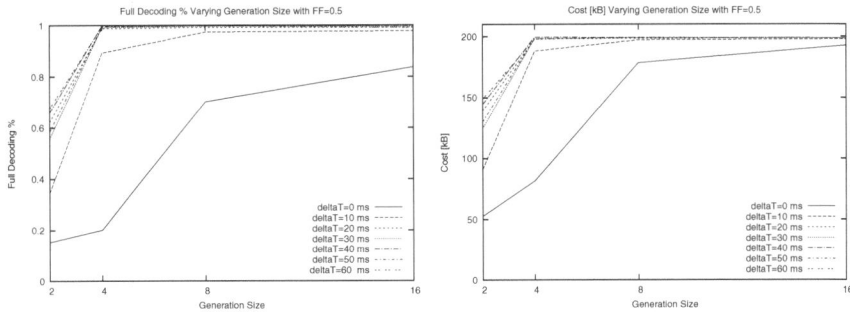

Fig. 7. Full decoding and cost vs generation size, collecting time is a parameter

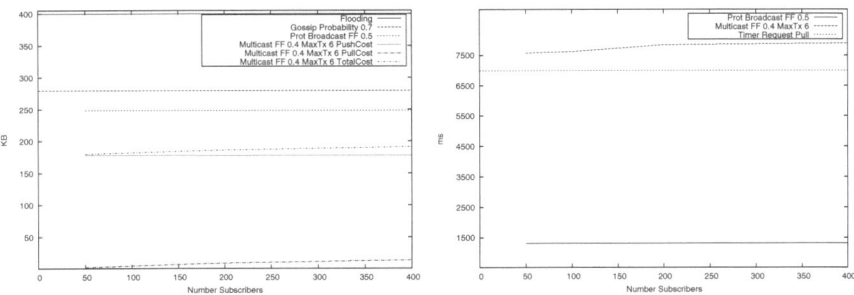

Fig. 8. Cost performance in a 20×20 grid, connectivity 4. Number of bytes sent (left), decoding delay (right).

and a push-only protocol with forwarding factor FF=5 and no limit on the number of transmissions (labelled Prot Broadcast $FF = 0.5$). In the gossip protocol a node sends the packet with probability $p_c = 0.7$. The proposed *multicast* push-pull runs with $FF = 0.4, MaxTx = 6$.

The cost of flooding is always 400 KB as all nodes send the packet whereas the cost of gossip is 280 KB due to the probabilistic transmission. The cost of our protocol is composed of a fixed part related to the pushing phase, and a variable part due to the pulling phase, which increases linearly with the number of subscribers. The cost is always below the gossip's one.

The right side plot in Figure 8 shows the total decoding delay. The push-only protocol is used to estimate an upper bound on the delay of the push-phase. The total decoding delay is dominated by the time after which a node initiates the pulling phase. In our simulations the pull phase starts after $\Delta T_P = 7$ s. This high value was selected for clearly distinguish the pull phase component of the delay. From the plot we can see that the pull phase requires approximatively 1 s to terminate. As the push phase requires no more than 1.5 s., the total decoding delay can be as low as 2.5 sec.

To test the self-adapting mechanism of the pull phase, we have conducted a set of experiments considering the higher connectivity degree of 20 neighbors. The parameters used in this new setting are $FF = 0.3, MaxTX = 3$. These values are determined

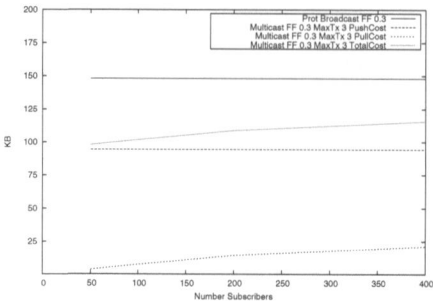

Fig. 9. Cost performance in a 20×20 grid, connectivity 20

empirically from the simulation and provide near optimal performance. The cost, see Figure 9, shows a similar behavior and it always is lower than the cost of a push-only protocol.

5   Conclusions

In this paper we have proposed an adaptive network-coding based event-diffusion protocol for wireless mesh networks. The proposed partial diffusion protocol uses a push-pull method to reduce the initial dissemination cost (amount of data sent), but adds a recovery cost incurred by the subscriber nodes. Compared to other blind dissemination protocols, most notably probabilistic flooding or gossip, our protocol has the advantage that it requires significantly smaller amount of data to be transmitted into the network. This advantage can be especially important for situations when nodes have limited energy and/or bandwidth, such as sensor networks, or battery-powered wireless nodes. Even in wireless meshes harnessed with higher-bandwidth IEEE 802.11 wireless routers, reducing the amount of data transmitted is necessary in broadcast environments to reduce contention on channels, since collisions can lead to excessive waiting delays or dropping connections.

Wireless mesh networks can especially benefit from our protocols because the nodes are not only transmitting and receiving their own data, but are also involved in relaying data between close and distant neighbors.

References

1. Akyildiz, I.F., Wang, X., Wang, W.: Wireless Mesh Networks: A Survey. Computer Networks Journal (March 2005)
2. Ahlswede, R., Cai, N., Li, S.-Y.R., Yeung, R.W.: Network Information Flow. IEEE Transactions on Information Theory, IT-46, 1204–1216 (2000)
3. Baldoni, R., Beraldi, R., Cugola, G., Migliavacca, M., Querzoni, L.: Structure-less content-based routing in mobile ad hoc networks. In: Costa, P., Picco, G. (eds.) IEEE International Conference on Pervasive Services, ICPS (2005)

4. Deb, S., Medard, M., Chote, C.: Algebraic Gossip: A network coding approach to optimal multiple rumor mongering. IEEE/ACM Transactions on Networking, 2486–2507 (June 2006)
5. Mosk-Aoyama, D., Shah, D.: Information Dissemination via Network Coding. In: IEEE Intl. Symp. on Info. Theory (2006)
6. Carzaniga, A., Rosenblum, D., Wolf, A.: Design and evaluation of a wide-area event notification service. ACM Trans. Comput. 19(3), 332–383 (2001)
7. Eugster, P.T., Felber, P.A., Guerraoui, R., Kermarrec, A.M.: The many faces of publish/ subscribe. ACM Computing Surveys 35(2), 114–131 (2003)
8. Fragouli, C., Le Boudec, J.-Y., Widmer, J.: Network Coding: An Instant Primer. In: ACM SIGCOMM 2006 (2006)
9. http://www.partow.net/projects/galois/
10. Gavidia, D., Voulgaris, S., van Steen, M.: A Gossip-based Distributed News Service for Wireless Mesh Networks. In: Proc. Third Int'l Conf. Wireless On-demand Network Systems and Services (WONS) (January 2006)
11. Gryphon Web Site, http://www.research.ibm.com/gryphon/
12. Ho, T., Mdard, M., Effros, M., Karger, D.: The benefits of coding over routing in a randomized setting. In: Proc. IEEE Symp. Information Theory, Yokohama, Japan (June/July 2003)
13. Ionescu, M., Marsic, I.: Stateful Publish-Subscribe for Mobile Environments. In: ACM International Workshop on Wireless Mobile Applications and Services on WLAN Hotspots, WMASH (2004)
14. Katti, S., et al.: XORs in The Air: PracticalWireless Network Coding. In: ACM Sigcom (2006)
15. Lin, Y., Li, B., Liang, B.: Stochastic Analysis of Network Coding in Epidemic Routing. JSAC 26(5) (June 2008)
16. Preotiuc-Pietro, R., et al.: Publish/subscribe on the web at extreme speed. In: Proc. of ACM SIGMOD Conf. on Management of Data, Cairo, Egypt (2000)
17. Mansouri, M., Shah Pakravan, M.R.: Network Coding Based Reliable Broadcasting in Wireless Ad-hoc Networks. In: 15th IEEE International Conference on Publication Networks (ICON 2007), November 2007, pp. 525–530 (2007)
18. http://www.isi.edu/nsnam/ns/
19. http://tesla.csl.uiuc.edu/koetter/NWC/
20. Wu, Y., Chou, P.A., Kung, S.-Y.: Minimum-Energy Multicast in Mobile Ad Hoc Networks Using Network Coding. IEEE Transaction on Communications 53(11) (November 2005)
21. Zheng, Y., Cao, J., Liu, M., Wang, J.: Dept., Efficient Event Delivery Publish/Subscribe Systems forWireless Mesh Networks. In: Proc. of IEEE Wireless Communications and Networking Conference, WCNC (2007)

Expressing Adaptivity and Context Awareness in the ASSISTANT Programming Model

Carlo Bertolli, Daniele Buono, Gabriele Mencagli, and Marco Vanneschi

Dept. of Computer Science, University of Pisa, Largo Pontecorvo 3, Pisa I-56127 Italy
bertolli@di.unipi.it
http://www.di.unipi.it/~bertolli

Abstract. Pervasive Grid computing platforms are composed of a variety of fixed and mobile nodes, interconnected through multiple wireless and wired network technologies. Pervasive Grid Applications must adapt themselves to the state of their surrounding environment *(context)*, which includes the state of the resources on which they are executed. By focusing on a specific instance of emergency management application, we show how a complex high-performance problem can be solved according to multiple parallelization methodologies. We introduce the ASSISTANT programming model which allows programmers to express multiple versions of a same parallel module, each of them suitable for particular context situations. We show how the exemplified programs can be included in a single ASSISTANT parallel module and how their dynamic switching can be expressed. We provide experimental results demonstrating the effectiveness of the approach.

Keywords: Adaptivity, Context Awareness, Parallel Programming, High-Performance Computing.

1 Introduction

Pervasive Grid computing platforms [15] are composed of a variety of fixed and mobile nodes, interconnected through multiple wireless and wired network technologies. In these platforms the term *context* represents the state of logical and physical resources and of the surrounding environment (e.g. acquired by sensor data). An example of Pervasive Grid application is risk and emergency management [4]. These applications include data- and compute-intensive processing (e.g. forecasting models) not only for off-line centralized activities, but also for on-line, real-time and decentralized ones: these computations must be able to provide prompt and best-effort information to mobile users. In general these applications are composed of multiple software modules interconnected in some graph structure (e.g. work- or data-flow). In abstract terms each module is responsible for solving a specific sub-problem. Clearly, each problem can be solved according to different methods featuring different characteristics. They are suitable for different parallelization techniques and optimized for being mapped onto different resources. For instance, a method can be optimized for the parallelization according to task farm instead of data parallel. These computations can also

A.V. Vasilakos et al. (Eds.): AUTONOMICS 2009, LNICST 23, pp. 32–47, 2010.

be different in the provided Quality of Service (QoS). In this paper we consider the term QoS as a set of metrics, which reflect the run-time behavior of a computation w.r.t. factors such as its memory occupation, its estimated performance (e.g. the average service time for a stream computation or the completion time of a single task) and the the quality of computed results. In this paper we show how multiple *versions* of a same parallel module can be introduced to target performance issues for different computing nodes, to face with the dynamic nature of pervasive grids and to meet dynamic user requests. We show that each version is best suited to be selected depending on conditions which are verified only at run-time (e.g. failures and user requests). This contribution is synthesized in the novel ASSISTANT programming model (ASSIST [18] with Adaptivity and Context-Awareness) which allows programmers to:

- express multiple versions of a same parallel module, exploiting the structured parallelism paradigm [6] (e.g. skeletons). This feature is inherited from the previous ASSIST parallel programming model [18];
- dynamically select which parallel version must be performed, in response to user-defined events or context changes (e.g. related to resource availability and sensor data). This feature can be expressed by exploiting performance models of structured parallel computations [19].

Thus, an ASSISTANT parallel module can be used to implement a fully autonomic computation.

We show a prototype implementation of ASSISTANT and we use it on a specific problem which is part of flood management application. A main module is involved in computing a flood forecast and it includes the resolution of a large number of tridiagonal linear systems. We show some different methods to solve tridiagonal systems, we discuss their properties and how these influence their parallelization scheme. We show experimental results of the execution of two parallel programs on best suited computing platforms. Thus, in this paper we show an example focusing on self-healing, self-configuring and self-optimization properties, leaving to future work the description of the remaining self- properties.

The paper is organized as follows: Sect. 2 discusses related works. Sect. 3 introduces the flood management application. Sect. 4 describes the different versions solving tridiagonal systems. Sect. 5 introduces the ASSISTANT programming model. Sect. 6 describes the implementation of the tridiagonal solver methods in the ASSISTANT model and it shows experimental results.

2   Related Work

Adaptivity has been introduced for mobile and pervasive applications by exploiting the concept of context [3]. Context definition includes environmental data, such as air temperature, the state of network links and processing nodes, and high-level information. Smart Space systems [16] mainly consist in providing context information to applications, which possibly operate on controllers to meet some user defined requirements. Some works focus on abstracting useful information from raw sensor data for adaptivity purposes. For instance, [5] exploits ontologies to model various context information, to reason, share and disseminate them.

General mobile applications must adapt themselves to the state of the context. For instance a mobile application can exploit optimized algorithms [12], protocols [7] or systems [2]. In this vision, it is the run-time support (e.g. the used protocol) which is in charge of adapting its behavior to the context. In a more advanced vision adaptivity can be defined as part of the application logic itself [4]. For instance, in Odyssey [13] an operating system is responsible of monitoring resources. Significant changes in resource status are notified to applications, which adapt themselves to meet a *fidelity* degree. Adaptivity is expressed in terms of the choice of the used services.

High-performance for context-aware applications is introduced in [11]. Computations are defined as data stream flows of transformations, data fusions and feature extractions. They are executed on centralized servers, while mobile nodes are only demanded to result collection and presentation. We go beyond this vision by: (i) allowing programmers to express multiple versions of a same program with different QoS; (ii) allowing programmers to execute proper versions also on mobile nodes.

Independently of pervasive environments, several research works are focused on adaptivity for high-performance programs [19]. In [1] it is shown how hierarchical management can be defined for structured parallel component-based applications [6]. Adaptivity for service-oriented applications is also targeted in [14], but application adaptivity is only discussed for large-scale simulations of nonlinear systems. We inherit and extend these research works in our programming model. In this paper we mainly focus on the programming model mechanisms to express adaptivity between multiple versions of a same computation, and on their performances according to a known cost model.

3   A Flood Management Application

We consider a schematic view of an application for fluvial flood management (see Fig. 1). During the "normal" situation several parameters are periodically monitored and acquired through sensors and possibly by other services (meteo and GIS). For instance sensors can monitor the current value and the variation of flow level and surface height. A forecasting model is periodically applied for specific geographical areas and for widest combinations of these areas. An example

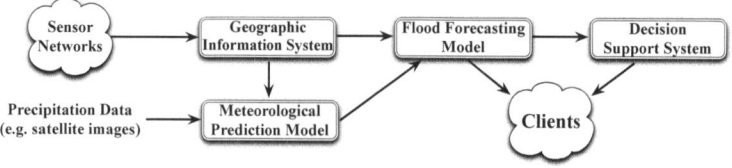

Fig. 1. Scheme of the flood management application

is the TUFLOW [17] hydrodynamic model, which is based on mass and momentum partial differential equations to describe the flow variation at the surface. Their discrete resolution requires, for each time slice, the resolution of a very large number of linear (tri-diagonal) systems. The quality of the forecasts also depends on the size of the tridiagonal systems. There exist parallel techniques which allow us to obtain reasonable response times in scalable manner.

During the execution, the forecasting model may signal abnormal situations which could lead to a flood. Thus, *performance is a critical parameter* concerning the response time of the forecasting model *per se* and also concerning all graphic and visualization activities.

Consider an example of a critical situation: the network connection of the human operator(s) with the central servers is down or unreliable. This is possible because we are making use of a (large) set of mobile interconnection links which are geographically mapped onto a critical area. To manage the potential crisis in real time, *we can think to execute the forecasting model and visualization tools on a set of decentralized resources* whose interconnections are currently reliable.

Just limiting to this scenario, *it is clear that there is a complex problem in dynamic allocation of software components to processing and communication resources.* Some resources may have specific constraints in terms of storage, processing power, power consumption: the same version of the software components may be not suitable for them, or even may be impossible to run it. Thus, the application must be designed with several levels of adaptivity in order to be able to cover different resource availability situations and dynamic QoS requirements. In this paper we show how multiple versions can be introduced for the flood emergency management application.

4   Defining Parallel Versions

We focus on the problem of solving tridiagonal linear systems of equations and we show how multiple parallel versions can be introduced each with different characteristics. There exist several resolution methods for *generic* linear systems: if the system is tridiagonal specialized techniques can be employed [9]. For performance modeling purposes, in this paper we focus on direct methods, which attempt to find an exact solution in a fixed, statically known, number of steps. Examples of direct approaches for tridiagonal systems are twisted factorization and cyclic reduction [9]. In this paper we are interested in defining multiple versions by exploiting different parallelization schemas of the same method: we focus on cyclic reduction methods because they can be easily generalized to banded and block tridiagonal systems [9]. In [10] two algorithms are introduced for solving tridiagonal systems of generic size N. For brevity, we avoid to introduce the mathematical formulations of these algorithms: interested readers can refer to [10].

First Algorithm. This algorithm includes two main parts. The first part (denoted by *transformation*) transforms in $q - 1$ steps ($q = log_2(N + 1)$) the input system. At each step l we consider all rows i such as $i \bmod 2^l = 0$. We solve:

$$a_i^l = \alpha_i a_{i-2^{l-1}}^{l-1} \qquad\qquad c_i^l = \gamma_i c_{i+2^{l-1}}^{l-1}$$
$$b_i^l = b_i^{l-1} + \alpha_i c_{i-2^{l-1}}^{l-1} + \gamma_i \qquad k_i^l = k_i^{l-1} + \alpha_i k_{i-2^{l-1}}^{l-1} + \gamma_i k_{i+2^{l-1}}^{l-1} \qquad (1)$$
$$\alpha_i = -a_i^{l-1}/b_{i-2^{l-1}}^{l-1} \qquad\qquad \gamma_i = -c_i^{l-1}/b_{i+2^{l-1}}^{l-1}$$

where a_i, b_i, c_i and k_i are the diagonal coefficients and the constant term of the i-th row. The superscripts denote the computational step at which their values are taken. α and γ are used in this notation to make equation reading easier. The stencil, i.e. the functional dependencies between successively computed values, refers the same element i and two neighbors: rows $i - 2^{l-1}$ and $i + 2^{l-1}$.

The second part of this algorithm is denoted *resolution*. We compute the solutions of the system, according to a fill-in procedure. It includes q steps for $l = q, q-1, \ldots, 1$. At each step l we consider all rows i for which $i \bmod 2^l = 0$

$$x_i = (k_i^{l-1} - a_i^{l-1} x_{i-2^{l-1}} - c_i^{l-1} x_{i+2^{l-1}})/b_i^{l-1} \qquad (2)$$

In this case we do not need multiple x values for each computation step. The stencil is the same of the first part of the algorithm.

Second Algorithm. The second algorithm includes two parts as the previous one. The first part includes q steps. Unlike the first algorithm, we solve the same equations (1) but for *all* rows at each step. The second part includes only a single step in which we directly get all the solutions of the system. These are computed in the following way: $x_i = k_i^q/b_i^q$. Notice that we only need the last values of the transformed system, instead of all the ones computed in the first part.

Discussion. We discuss the features of each algorithm to define the best parallelization schemas. *In this paper we focus on context events including the state of the used computing resources and their associated performance.* We avoid to consider environmental data (e.g. sensor data) influencing the version selection policy (demanded to future work).

The performance features of the described algorithms can be characterized as following:

- **Number of steps:** The first algorithm performs $q-1 = log_2(N-1)-1$ steps during the transformation part and $q = log_2(N-1)$ in the resolution one. The second algorithm performs less steps: $q = log_2(N-1)$ transformation steps and only one resolution step.
- **Number of Operations:** The first algorithm performs a lower number of operations in the first part w.r.t. the second algorithm. This because the second algorithm applies, at each step, the equations (1) to all system elements, instead of only a subset of them. The second part of both algorithms involves the same number of operations.
- **Number of Functional Dependencies:** The first algorithm includes a lower number of functional dependencies because, at each step of the transformation part, equations 1 are solved only for a subset of elements.

In our application we need to solve a stream of tridiagonal systems, i.e. a possibly unlimited sequence of systems. We need to consider the parallel efficiency on the single system and on the stream of systems. Looking at the algorithms above we can think to use two kinds of parallel structures:

- **Task Farm:** The systems (tasks) belonging to the input stream are scheduled w.r.t. several replicated workers according to a load balancing strategy, each worker executing the sequential algorithm. An output stream of results is produced. As known, this parallelism paradigm does not decrease the processing latency of a single element of the stream, but it decreases the service time (increases the throughput), provided that the stream interarrival time is sensibly less than the sequential processing time (stream processing situation in the true meaning).
- **Data Parallel:** Each tridiagonal system is partitioned (scattered) onto several replicated workers, each one performing the sequential algorithm for its respective partition. In the considered algorithms, workers cooperate during each step according to a proper communication stencil. The whole result is obtained by gathering the partial results. With respect to the farm structure, this parallelism paradigm works both in a stream processing situation, and when only a single system has to be processed (i.e. equivalently, when the stream interarrival time is greater than the sequential processing time for a single task). Moreover, it is able to decrease the processing latency of a single tridiagonal system and the memory size per node. In a stream situation, the disadvantage of a stencil-based data parallel structure, w.r.t. the farm paradigm, is a potential load unbalance and a more critical impact of the communication/computation time ratio, thus in general a greater service time.

Two structuring modalities of the whole computation have to be distinguished: an acyclic graph structure or a cyclic one. In the former case, a pipeline-like effect is present, provided that a real stream processing situation occurs. In the latter, the overall computation is a client-server schema. Each client sends the input data to the tridiagonal solver module (i.e. the server) and it waits for the corresponding results. To increase the performance of each client we can parallelize the server with a proper parallelism degree:

- in a task farm structure it is equal to the number of clients;
- in a data parallel structure it is independent of the number of clients and can be obtained by the proper cost model of the parallel structure. Moreover, the reduced latency time contributes to decrease the server response time, thus the client service time.

All the described situations (stream vs single element processing, acyclic graph vs client-server structure) can be taken into account in an adaptive and context-aware computation. The farm and the data- parallel paradigms are able to optimize specific situations. In general it may be convenient, or necessary, to switch from one structure to another one dynamically, thus to switch from a version of the computation implemented according to a parallelism paradigm to another version, implemented according to the other parallelism paradigm: this feature characterizes our approach to high-performance adaptive and context-aware application design.

We have seen that the second algorithm performs more operations than the first one (but less steps), but also more communications. These can be buffered and, provided that their support is efficient, the second algorithm can be parallelized according to the data parallel structure. Communication efficiency characterizes multicores: communications between cores are implemented as accesses to shared variables and the computation can take advantage of the hardware caching support. Thus, we implement this version on a multicore architecture (see below).

The first algorithm minimizes the number of operations performed in the whole computation. Thus, it seems reasonable to: (a) parallelize it according to the task farm model, which has not the strong requirements, in terms of communication efficiency, of the data parallel; (b) implement it for both cluster and multicore architectures. We show experimental results for both versions and we discuss how the best version is dynamically selected according to specific context situations.

We show the data parallel program for the second algorithm in the ASSIST syntax. For brevity, we avoid to show the program of the task farm version.

4.1   The ASSIST Model

ASSIST [18] is a programming environment for expressing parallel and distributed computations according to the structured parallel paradigm. In ASSIST it is not possible to natively express an adaptive application, which is one of the intended goals of ASSISTANT. An ASSIST application is expressed in terms of a set of ParMods (i.e. Parallel Modules) interconnected by means of typed streams. The ParMod construct includes three sections:

- **input_section:** It is used to express the distribution of received data from input streams to the parallel activities performing the computation, according to primitive constructs (e.g. *on-demand* and *scatter*) or user-programmed ones;
- **virtual_processors:** They are used to express the parallel computation applied to each input data, possibly producing an output result. Virtual processors are the abstract units of parallelism in ASSIST, which are mapped onto a set of implementation processes;
- **output_section:** In this section we express the collection of virtual_processors results and their delivery to output streams, by means of primitive strategies (e.g. *gather*) or user-programmed ones.

4.2   Parallel Programs in ASSIST

We implement an ASSIST parmod for the data parallel version using the second algorithm (see Fig. 2). The parmod *TSM-DP* receives a stream of tridiagonal systems (*syst_row* data structure) as input tasks (line 1). For each system

```
1 parmod TSM–DP(input_stream syst_row input_syst[N] output_stream
     solutions sols) {
2   topology array [i:N] vp;
3   attribute syst_row computing_syst[2][N] scatter S[][*i] onto VP[i
     ];
4
5   do input_section {
6     guard1: on , , input_syst {
7       distribution input_syst[*s] scatter to comput_syst[0][s];
8     }
9   } while (true)
10
11  virtual_processors {
12    solve_system (in guard1 out sols) {
13      VP i {
14        for(l = 1; l <= q; l++)
15          transform(i, computing_syst[i][l-1], computing_syst[i-pow
             (2,l-1)],computing_syst[i+pow(2,l-1)],sols[i]);
16
17          solve(i, comput_syst[i][], sols[i]);
18        }
19      }
20    }
21
22    output_section {
23      collects sols from ALL vp[i];
24    }
25 }
```

Fig. 2. Data parallel program based on the second cyclic reduction algorithm

it computes the correct solution (*solutions* data structure). The topology command (line 2) gives integer numbers (from 1 to N) as names of virtual processors. Virtual processors are assigned a single system row on which they apply the algorithm. In the implementation, multiple virtual processors are mapped onto a set of implementation processes. At line 3 an attribute (a ParMod variable) is used to store two successive system values during its transformation and it is scattered amongst the virtual processors. The input section (line 5) is fired when a system is received (line 6) on the input stream. The system is scattered onto the first position of the attribute (line 7). In the virtual_processors section (line 11) each VP_i in parallel: (a) transforms the input system in q steps (line 15); (b) computes the results (line 17). In the output_section we gather all computed results (*from ALL* keyword), which are automatically delivered onto the output stream.

4.3 Experiments

We have tested the parallel efficiency of the different tridiagonal solver versions on a emulation of a pervasive grid. The aim of this section is to experimentally show that different versions can be used to target different context situations, related to the state of the used computing resources (e.g. their availability) and on their performance. The experiments are performed on a prototype of ASSISTANT: we avoid to show the reconfiguration costs (i.e. version selection) because it is out of the scope of this paper. The pervasive grid is emulated by the following nodes:

- a centralized server, emulated with a cluster architecture. The cluster is composed by 30 nodes Pentium III 800 MHz with 512 KB of cache, 1 GB of main memory and interconnected with a 100 Mbit/s Fast Ethernet. We map the task farm version onto this platform;
- an interface node between the cluster and the mobile distributed platform of PDA nodes and sensor devices. The interface node is emulated with an Intel E5420 Dual Quad Core multicore processor, featuring 8 cores of 2.50 GHz, 12 MB L2 Cache and 8 GB of main memory. Both data parallel and farm versions are mapped onto this architecture.

Fig. 3, 4 and 5 show the results in terms of service time and scalability. We define the service time as the time passing between the consuming of two successive systems from the input stream (not their complete resolution but only their consuming). Scalability can be defined as the ratio between the sequential computation time (parallelism equal to 1) and the parallel one: it represents the quality of parallelization of the module. Notice that the cluster service time is higher than the multicore one because of the sensible difference between the processing power of their single nodes (800 MHz versus 2.5 GHz). For comparable processing powers, the cluster would provide higher scalability and parallelism degrees. As discussed at the end of Sect. 4, for acyclic graph application structures:

- the farm version is effective only when the computation operates on a stream processing situation: in this case it performs better than the data-parallel solution;
- the data parallel version has to be adopted when the computation operates on a single tridiagonal system (i.e. too large interarrival time).

For client-server cyclic graph application structures, both versions are potentially feasible: one or the other will be selected dynamically according the performance comparison between the farm and the data parallel version, i.e. according to the optimal parallelism degree of the two versions.

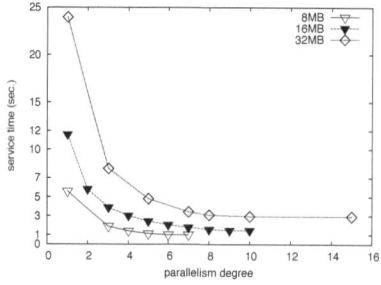 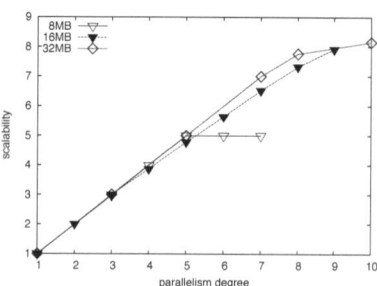

Fig. 3. Experimental results of the cluster task farm version (first algorithm): service time (left) and scalability (right)

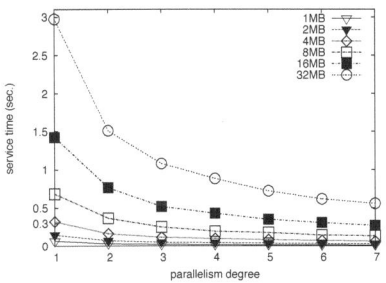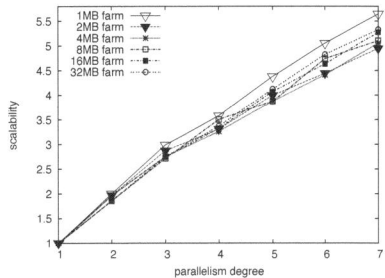

Fig. 4. Experimental results of the multicore task farm version (first algorithm): service time (left) and scalability (right)

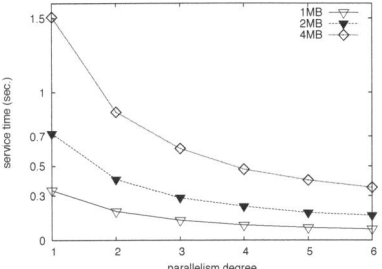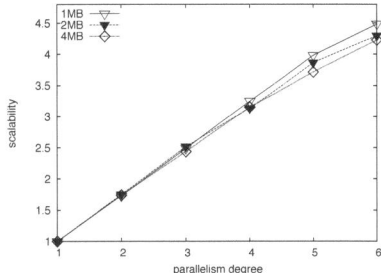

Fig. 5. Experimental results of the multicore data parallel version (second algorithm): service time (left) and scalability (right)

From this discussion we can conclude that pervasive grid applications must be programmed as multiple parallel modules, each provided in multiple versions. Moreover, *the programmer needs to express the conditions (or events) under which each version is dynamically selected,* according to the available resources and user needs. This can be done by specifying different parallel programs and by introducing a policy for the dynamic selection of the best version. The actual implementation of version selection can be automatized in the following points:

- low-level version switching: this includes the re-direction of input data streams between the versions;
- data consistency: the channel re-direction must guarantee that no input elements are lost or re-ordered.

In the next section we introduce an high-level programming model in which all these actions are automatized, while the application programmer just focuses on the abstract events inducing a version switching.

5   The ASSISTANT Programming Model

We introduce the novel ASSISTANT programming model for high performance pervasive applications with adaptive and context-aware behaviors.

With ASSISTANT we target application adaptivity by allowing programmers to express how the computation evolves reacting to specified events. We enable this kind of expressivity with a new ParMod construct. We can characterize three main logics (Fig.6 (left)) that can be used to describe the semantic and the behavior of a ParMod:

- **Functional** logic: This includes all the versions solving the same problem in the ASSIST syntax. Functional logics of different ASSISTANT ParMods communicate by means of typed data streams.
- **Control** logic: This includes the adaptivity strategies, i.e. the *reconfiguration* actions performed to adapt the ParMod behavior in response to specified events. For instance, the control logic can select the best version between multiple ones, according to specific cost models. The programmer is provided with high-level constructs to directly express the control logic with the corresponding adaptivity strategies. Control logics of different application ParMods can interact by means of *control events*.
- **Context** logic: This includes all the aspects which link the ParMod behavior with the surrounding context. The programmer can specify events which correspond to sensor data, monitoring the environmental and resource state (e.g. air temperature and network bandwidth). It can also specify events related to the dynamic state of the computation (e.g. the service time of a ParMod). These *context events* can be provided by the run-time support of the programming model, or in other cases primitive *context interfaces* (e.g. failure detectors) which communicate with the application modules by context events.

The different versions of a same ParMod are expressed by means of the *operation* construct. Each ParMod can include multiple operations (see Fig. 7), all solving the same problem according to different algorithms and parallel structures. All operations of the same ParMod must feature the same input and output interfaces, in terms of streams. Each operation includes its own part of functional, control and context logic of the ParMod in which it is defined. That is, each operation features its own parallel algorithm, but also its own control and context logics. Notice that operations are not merely alternative sections of code inside a module definition: they are the adaptation and deployment units of the

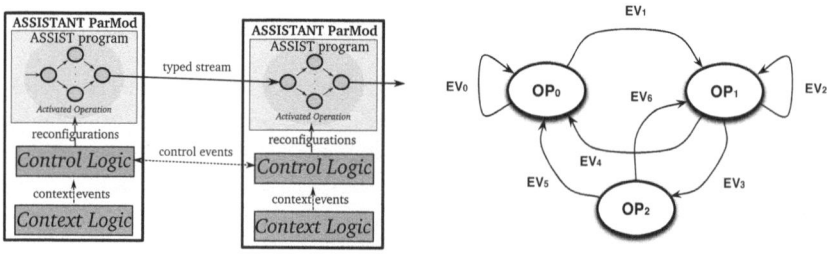

Fig. 6. Example of ASSISTANT ParMods (left) and of event-operation graph (right)

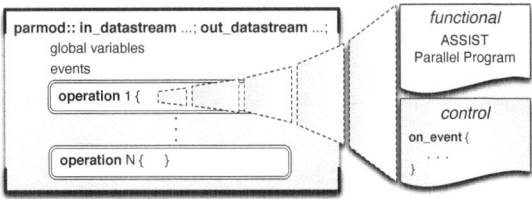

Fig. 7. Syntactic view of a ParMod

versions of the same Assistant module. For this reason, a new suitable construct is required to achieve our goals.

Syntactically, the ParMod has a *name* and a set of *input* and *output streams*. It can feature a *global state* shared between operations and it can define *events* which it is interested to sense. Events may be context ones, whose monitoring can be provided by context interfaces, or control events obtained from the control logic of other ParMods. Semantically, only one operation for each ParMod can be currently activated by its control logic. When a ParMod is started a user-specified *initial* operation is performed, possibly deploying it on dynamically discovered resources. During the execution the context logic of a ParMod, or the control logic of other modules, can notify one or more events. The control logic exploits a mapping between these events and reconfiguration actions, defined by the programmer, to either select a new operation to be executed, or modify the run-time support of the current operation (e.g the parallelism degree of a parallel computation as described in [19]). The control logic of an ASSISTANT ParMod can be described as a graph (see Fig. 6(b)(right)): nodes are operations of the ParMod and arcs are events (or their combinations with logical expressions on the actual ParMod state). Semantically a ParMod control logic is a sequential automaton: this is done to avoid nondeterministic behaviors. In the case of concurrent events we serialize and manage them according to a priority defined in the control logic itself (i.e. their definition sequence). In the example, the initial operation is OP_0. If the event EV_0 occurs, we continue executing OP_0 but we modify some aspects of its implementation (e.g. its parallelism degree). That is, self-arcs, starting and ending in the same node, correspond to run-time system reconfigurations. Consider now the arc from OP_0 to OP_1 fired by event EV_1. In this case the programmer specifies that if we are executing OP_0 and event EV_1 occurs, we stop executing OP_0 and we start OP_1. This switching can include pre- and post- elaborations: for instance, we can reach some consistent state before moving from OP_0 to OP_1 in order to allow the former operation to start from a partially computed result, instead of from the beginning.

Reconfigurations can be performed in the case: (a) some pre-determined events happen and/or (b) some predicates on the module state are satisfied. That is, the control logic of a ParMod is stateful. This behavior is expressed in each operation of a ParMod by means of the *on_event* construct. Syntactically, the programmer makes use of nondeterministic clauses which general structure is described as shown in Fig. 8:

```
event_combination:
   do    <reconfiguration code>
   enddo
```

Fig. 8. General structure of the on_event construct

If the *event_combination* logical expression is satisfied, the corresponding re-configuration code is executed. Programmers are also provided with a *parallelism* construct, to specify a modification of the parallelism degree of the current operation (i.e. a run-time system reconfiguration).

6   Programming Adaptivity for the Flood Application in ASSISTANT

We show how to encapsulate the different versions, to solve tridiagonal systems, in a single ParMod.The flood management application is composed of the following ParMods:

- **Generator:** This module emulates all the application phases preceding the flood forecasting model. It generates a stream of double precision floating point data related to the conditions of each point in the river [17].
- **Tridiagonal Solver Module** (TSM): This module implements the forecasting model (see Sect. 4), which is applied to each input stream element received by the Generator. For each input data, it generates and solves four tridiagonal systems (e.g. see [17]). The TSM includes three different operations.
- **Visualization:** This module implements the post-processing activities, visualizing forecast results on the user's display.

In this program we consider three different operations for the TSM: the first one is *clusterOperation* which is the task-farm mapped onto the cluster architecture as described in Sect. 4. The second one is *interfaceNodeFarm*) which is the task farm version executed on interface multicore nodes. The third operation is *interfaceNodeD.-P-* which is the data parallel version executed on interface multicore nodes. As the TSM functional logic has been described in Sect. 4, we are interested in the control logic. This is responsible of deciding which context changes have to be monitored and which ones cause a reconfiguration. The considered context changes are:

- a network event from the TSM context logic related to the current status of network connections (e.g. their availability or presence of high-latency links). We have considered only two disconnection events: *mainNetFail* and *mobileNetFail* which provide a boolean information related to the connection capability (based on current latency and connection status) between the cluster and the considered interface node (the former) and the interface node and the user's PDAs (the latter);

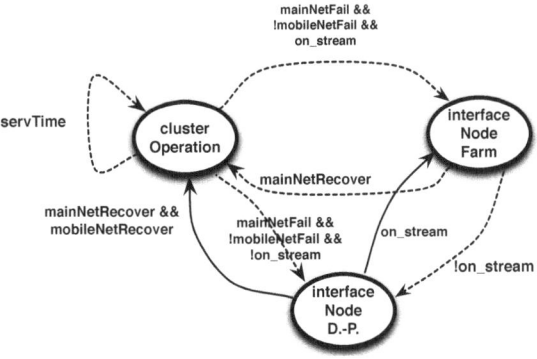

Fig. 9. Event-Operation graph of the parmod TSM. Bold arrows are implemented in Fig. 6.

```
parmod TSM(..) {
  operation interfaceNodeD.-P. {
    //Parallel Computation in an ASSIST-like fashion:
    see Section 3..

    //Management section of this operation:
    on_event {
      mainNetRecover && mobileNetRecover:
        do
          notify(Generator_Module, con_fail);
          notify(Client_Module, con_fail);
          //Stop the operation consistently:
          this.stop();
          //Activation of the new operation:
          clusterOperation.start();
        enddo
      on_stream():
        do
          ...
          interfaceNodeFarm.start();
        enddo
    }
  }
  ...
```

Fig. 10. Control part of *interfaceNodeD.-P.* operation inside the parmod TSM definition

– the average interarrival time to the ParMod TSM is lower than a maximum threshold. This event is denoted with *on_stream*.

Fig. 9 shows the event-operation graph of the TSM: bold arrows are those expressed in the *interfaceNodeD.-P.* control part. Dotted arrows are expressed in the other two operations. Fig. 6 shows the corresponding *on_event* section of the operation inside the TSM definition, implementing its adaptive behavior when the *interfaceNodeD.-P.* is executed. This *on_event* instance describes two different conditions: while executing *interfaceNodeD.-P.* the *mainNetRecover* and *mobileNetRecover* events can be received. We choose to execute the forecasting model on the cluster, because it enables higher parallelism degree than the multicore and, consequently, lower service times. If the *on_stream* event is received, the input stream interarrival time is now lower than the TSM service time. In this case we can switch to the *interfaceNodeFarm* operation, which provides better on stream scalability (see Sect. 4). We also need to notify the generator and client modules of this re-configuration (i.e. the *notify* function).

7   Conclusions

In this paper we have shown how adaptivity for pervasive grid applications can be defined by exploiting multiple versions for the same application module. We have exemplified our approach on the specific problem of solving tridiagonal systems of linear equations, introducing two resolution algorithms and parallelizing them. The two algorithms are shown to be best suited for being executed on a cluster architecture and on a multicore one. The experimental results show that: the cluster version and the interface node version, as well as the respective farm and data parallel schemas, have clear pros and cons that can drive the selection of the best adaptation strategy at run time. As an example, the cluster version has to be preferred to the interface node one if the network status provides a reasonable communication latency between the cluster and the mobile users. We have introduced the novel ASSISTANT programming model, providing constructs to express multiple versions of a same parallel module and to adapt its execution by dynamically selecting the best one. We have implemented a flood forecasting module exploiting the ParMod construct, including the two resolution algorithms and their dynamic selection policy.

References

1. Aldinucci, M., Danelutto, M., Kilpatrick, P.: Co-design of distributed systems using skeletons and autonomic management abstractions. In: César, E., et al. (eds.) Euro-Par 2008 Workshops. LNCS, vol. 5415, pp. 403–414. Springer, Heidelberg (2009)
2. Balasubramanian, A., Levine, B.N., Venkataramani, A.: Enhancing interactive web applications in hybrid networks. In: 14th ACM International Conference on Mobile Computing and Networking, pp. 70–80. ACM, New York (2008)
3. Baldauf, M., Dustdar, S., Rosenberg, F.: A survey on context-aware systems. Int. J. Ad Hoc Ubiquitous Computing 2, 263–277 (2007)
4. Bertolli, C., Fantacci, R., Mencagli, G., Tarchi, D., Vanneschi, M.: Next generation grids and wireless communication networks: towards a novel integrated approach. Wireless Comm. and Mobile Computing 9, 445–467 (2009)
5. Chaari, T., Ejigu, D., Laforest, F., Scuturici, V.M.: A comprehensive approach to model and use context for adapting applications in pervasive environments. Journal of Syst. Softw. 80, 1973–1992 (2007)
6. Cole, M.: Bringing skeletons out of the closet: a pragmatic manifesto for skeletal parallel programming. Par. Comp. 30, 389–406 (2004)

7. Curtmola, R., Rotaru, C.N.: BSMR: Byzantine-Resilient Secure Multicast Routing in Multi-hop Wireless Networks. IEEE Trans. on Mobile Comp. 8, 263–272 (2009)
8. Danelutto, M.: QoS in Parallel Programming through Application Managers. In: 13th Euromicro Conf. on Parallel, Distributed and Network-Based Processing, pp. 282–289. IEEE Press, Washington (2005)
9. Duff, I.S., Van der Vorst, H.A.: Developments and trends in the parallel solution of linear systems. Par. Comp. 25, 1931–1970 (1999)
10. Hockney, R.W., Jesshope, C.R.: Parallel Computers: Architecture, Programming and Algorithms. Institute of Physics Publishing, Bristol (1981)
11. Lillethun, D.J., Hilley, D., Horrigan, S., Ramachandran, U.: MB++: An Integrated Architecture for Pervasive Computing and High-Performance Computing. In: 13th IEEE Intl. Conf. on Embedded and Real-Time Computing Systems and Applications, pp. 241–248. IEEE Press, Washington (2007)
12. Mishra, A., Shrivastava, V., Agrawal, D., Banerjee, S., Ganguly, S.: Distributed channel management in uncoordinated wireless environments. In: 12th Intl. Conf. on Mobile Computing and Networking, pp. 170–181. ACM, Los Angeles (2006)
13. Noble, B.D., Satyanarayanan, M.: Experience with adaptive mobile applications in Odyssey. Mob. Netw. Appl. 4, 245–254 (1999)
14. Plale, B., Gannon, D., Brotzge, J., Droegemeier, K., Kurose, J., McLaughlin, D., Wilhelmson, R., Graves, S., Ramamurthy, M., Clark, R.D., Yalda, S., Reed, D.A., Joseph, E., Chandrasekar, V.: CASA and LEAD: Adaptive Cyberinfrastructure for Real-Time Multiscale Weather Forecasting. Computer 39, 56–64 (2006)
15. Priol, T., Vanneschi, M.: From Grids To Service and Pervasive Computing. Springer, Heidelberg (2008)
16. Román, M., Hess, C., Cerqueira, R., Ranganathan, A., Campbell, R.H., Nahrstedt, K.: A Middleware Infrastructure for Active Spaces. IEEE Perv. Comp. 1, 74–83 (2002)
17. Syme, B.: Dynamically Linked Two-Dimensional/One-Dimensional Hydrodynamic Modelling Program for Rivers, Estuaries and Coastal Waters. WBM Oceanics, Aus (1991)
18. Vanneschi, M.: The programming model of ASSIST, an environment for parallel and distributed portable applications. Par. Comp. 28, 1709–1732 (2002)
19. Vanneschi, M., Veraldi, L.: Dynamicity in distributed applications: issues, problems and the ASSIST approach. Par. Comp. 33, 822–845 (2007)

Experiences in Benchmarking of Autonomic Systems*

Xavier Etchevers, Thierry Coupaye, and Guy Vachet

Orange Labs, France Télécom Group
28 chemin du Vieux Chêne, F-38240 Meylan, France
{xavier.etchevers,thierry.coupaye,guy.vachet}@orange-ftgroup.com

Abstract. Autonomic computing promises improvements of systems quality of service in terms of availability, reliability, performance, security, etc. However, little research and experimental results have so far demonstrated this assertion, nor provided proof of the return on investment stemming from the efforts that introducing autonomic features requires. Existing works in the area of benchmarking of autonomic systems can be characterized by their qualitative and fragmented approaches. Still a crucial need is to provide generic (i.e. independent from business, technology, architecture and implementation choices) autonomic computing benchmarking tools for evaluating and/or comparing autonomic systems from a technical and, ultimately, an economical point of view. This article introduces a methodology and a process for defining and evaluating factors, criteria and metrics in order to qualitatively and quantitatively assess autonomic features in computing systems. It also discusses associated experimental results on three different autonomic systems.

Keywords: Autonomic computing, benchmark, metrics, criteria, evaluation, comparison, return on investment, ROI.

1 Introduction

From an industrial perspective, the overall motivation underlying the emergence of autonomic computing is based on the observation that the costs related to the IT infrastructures are quickly and massively migrating from new investments (development and licensing costs) to maintenance expenses (deployment and exploitation costs). [10] This evolution illustrates the transformation of computing systems in terms of size, distribution, sophistication, dynamism, heterogeneity and interoperability that results in even more complex –and thus expensive– management tasks. In this context, autonomic computing aims basically insofaras possible at automating the deployment and management (administration) of computing systems in order to reduce human interventions and all associated costs.

---

* This work is partially funded by European IST FWP6 Selfman project [5].

A.V. Vasilakos et al. (Eds.): AUTONOMICS 2009, LNICST 23, pp. 48–63, 2010.

Bit by bit research in autonomic computing is starting to generate industrial solutions. This is the case, for example, of workload management in J2EE web applications servers (e.g. JOnAS, JBoss, WebSphere, WebLogic). However, even in such a well-known area, the autonomic features embedded in different products do not necessarily address the same problems and/or can have different maturity levels. Moreover, even if some experimental results like [2], [14] or [17] tend to demonstrate that autonomic behaviors[1] (ABs) improve systems efficiency (by comparing it when enabling and disabling the autonomic features), there are still no models and no tools for formally measuring and comparing the technical and economical benefits these autonomic capabilities are supposed to offer.

Nowadays industrial companies continuously strive to improve and simplify their process for increased customer satisfaction, operational efficiency and cost reduction. This implies improved efficiency of computing systems in terms of productivity, performances, quality of service (QoS), trust, etc. Autonomic computing appears to be an appealing solution. However this quite recent and disruptive approach still raises a number of issues regarding:

(technical) efficiency: can the impact of autonomic computing on the quality of service (QoS)[2] be exhibited or even "proved" - especially when ABs are built into the system? Is this impact measurable generically with respect to various business fields, architecture or implementation of different systems?

profitability (i.e. economical efficiency): when comparing all the costs linked to an autonomic system versus a non-autonomic equivalent one, does autonomic computing come out as a cheaper or a more expensive technology? How long will it take to become profitable?

applicability: considering the previous questions, are there some more or less pertinent areas to which autonomic computing should or should not be applied?

Such questions raise the need for some autonomic computing benchmark(s) that would allow for evaluation and comparison of autonomic systems, both from a technical and an economical perspective. This article hypothesizes that, even if developing benchmarks are business-specific (i.e. specific to web applications servers or to P2P systems for the benchmarks considered in this article), some generic benchmarking elements could be shared among all specific domains and therefore could allow for inter-area comparison. The contribution of this paper is twofold. First, it introduces a methodology and a process for defining and evaluating factors, criteria and metrics in order to qualitatively and quantitatively assess autonomic features in computing systems. Second, it analyzes associated experimental results on three different autonomic systems. The article is organized as follows. Section 2 discusses existing works on autonomics benchmarking. Section 3 defines a methodology and a process, based on criteria and metrics, in order to benchmark autonomics both qualitatively and quantitatively. Section 4 describes and analyzes experimental results. Section 5 concludes.

---

[1] *Autonomic behavior* (AB) designates the implementation of a particular control loop or MAPE-loop.

[2] In this article, the term 'QoS' is used in its most general meaning.

2 Background

2.1 Models and Metrics

Since autonomic computing aims essentially at improving the QoS of systems, [15] and [19] try to define an autonomic computing evaluation model based on the ISO/IEC 9126 standard[3] [13]. They describe the qualitative binding between autonomic characteristics[4] and the ISO/IEC 9126 factors (see figure 1). At the same time, [11] provides a qualitative hierarchy between the eight autonomic characteristics (the four main ones and the four secondary ones) (see figure 2). Finally some works focus on the definition of metrics in order to evaluate autonomic capabilities. For example, [12], [3] and [4] propose a non-exhaustive set of criteria and metrics that are not related to any evaluation standard such as ISO/IEC 9126.

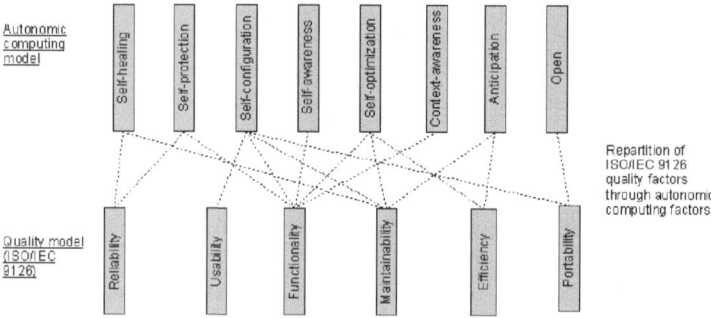

Fig. 1. Organization of autonomic computing characteristics based on ISO/IEC 9126 standard quality factors [15] [19]

These works constitute indispensable steps toward providing a complete autonomics benchmark, but they can be characterized by a qualitative and fragmented approach. [15] and [19] do not provide a quantitative composition of the factors for evaluating experimentally the autonomic characteristics, nor an adaptation of ISO/IEC 9126 model to the autonomic computing field. [11] is not based on any Factors-Criteria-Metrics model and still has to be quantitatively validated. Finally, [12], [3] and [4] list some criteria and metrics. However, for most of criteria (i.e. granularity, flexibility, robustness, adaptability), no metrics

---

[3] ISO/IEC 9126 is an adaptation of a generic Factors-Criteria-Metrics model [6] applied to the software quality field. However it defines only factors and criteria because software quality metrics are specific to the business area of the considered system.

[4] [7] and [9] define autonomic computing thanks to four main characteristics (self-configuration, self-healing, self-protection and self-optimization) and four secondary characteristics (self-awareness, context-awareness, openness and anticipation).

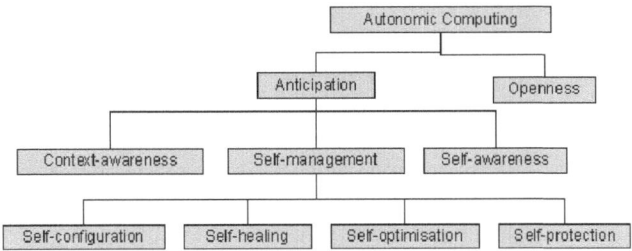

Fig. 2. Hierarchy between autonomic computing characteristics [11]

have been defined. Moreover the proposed metrics are either qualitative (i.e. degree of autonomy) and therefore their evaluation is subjective, or quantitative (i.e. adaptation time, reaction time, stabilization duration, latency) but they remain difficult to measure because of their abstraction level.

2.2   Benchmarking Methodologies and Tools

[1] lists the specificities associated to the benchmarking of autonomic features. A "classical" performance benchmark and an autonomic computing benchmark differ along three main axes: 1) environment stability that is questioned for autonomic computing by the injection of disturbances, 2) management interactions that must not occur in a classical performance benchmark, but that constitute the AB as well, 3) the antagonism between test realism (regarding the representativeness of the workload or the disturbances load the system under test will have to face) and the benchmark requirements, especially in terms of reproducibility, cost and legality. Among the works dealing with the experimental evaluation of autonomics efficiency, [2] seems to be one of the most advanced. It describes one of the first autonomic computing benchmark dealing with self-healing evaluation. It details its experimental protocol for validating the tool. This consists in measuring the impact of thirty different classes of disturbances on two metrics. The disturbances are sequentially injected.

The three main specificities of an autonomic computing benchmark (highlighted by [1]) illustrate the addition of a second dimension into an autonomic computing benchmark compared to a classical QoS benchmark. Each benchmark can indeed be associated with a function that get injection profiles as inputs and that returns a vector of evaluated metrics. Thus a classical QoS benchmark is a function with a single input (i.e. the injected workload profile) whereas an autonomic computing benchmark consists in a function getting two inputs (i.e. the injected workload profile and the injected disturbance profile). Concerning this last type of benchmark, constraints relative to costs, reproducibility and legality, which are antagonistic to test realism, can be declined into constraints on the injection profiles (synchronization of workload and disturbances injections, stability of workload injection, etc.). In other words an autonomic computing benchmark consists of two coupled evaluation tools. The first one is dedicated to the QoS measurement: it is business specific and is essentially made of quantitative metrics. The second one focuses on the evaluation of self-management behaviors. It should include only generic (i.e. domain independent) aspects. Concerning the experimental assessment, works like [2], [14], [17], although they tend to demonstrate that autonomic features improve systems efficiency, do not achieve to define a scale offering a synthesized and absolute[5] view. Thus [2] and [3] define a value ranging between 0 (non autonomic) and 1 (fully autonomic) for assessing self-* features. However this indicator is obtained by restraining the number of different.disturbances.

3 Assessment Methodology and Process

This section introduces the first contribution of this article, namely the proposed generic benchmarking methodology and process.

3.1 Methodology

The first step of the proposed approach is to define the constituents, i.e. *factors*, *criteria* and *metrics*, of an hybrid (refined) ISO/IEC 9126 model for the autonomic computing area, similarly to the refinement of ISO/IEC 9126 model proposed in [18] and addressing the concrete case of test specification. Then high-level indicators are defined for qualifying autonomic features. A bottom-up approach integrating, supplementing and adapting some existing works (see section 2) is adopted.

Metrics. An exhaustive set of metrics participating in the empirical evaluation of the ISO/IEC 9126 model has been identified (by using and enriching the list defined by [12], [3] and [4]). These metrics have to be:

Generic, i.e. independent from the field of application, in order to be applied to evaluating various business areas or to inter-domain comparisons;

Measurable, i.e. these metrics can be assessed independently from architecture, design, implementation or technologies choices;

Quantifiable for metrics processed as inputs of composition functions (see below) in order to calculate quantitative higher-level indicators. Quantitative metrics are intrinsically quantifiable whereas qualitative ones are characterized by the subjectivity of their evaluation. However some of qualitative metrics can be composed of discrete but ordered values (e.g. the level of maturity of an AB) whereas others are made of non ordered values (e.g. the list of standards or technologies with which a component complies). Only the first ones are quantifiable.

---

[5] *Absolute* is relative to a scale that could classify the results from a system without any autonomic features to an idealistic system that could autonomously anticipate or deal with any disturbances (without any delay and any impact on the quality of service).

Composition Functions. After having qualitatively defined a hybrid ISO/IEC 9126 model for autonomic computing (see figure 3)[6] and tied it to the proposal of [15] and [19] concerning the coupling between autonomic characteristics and software quality factors, the next step consists in defining composition functions for enabling the computation of higher-level quantitative indicators. Among these indicators some have to be compared –and be thus quantifiable– whereas others only relate interesting properties. As for an autonomic computing benchmark, the four main self-* characteristics, i.e. *self-configuration, self-optimization, self-healing, self-protection*, have to be quantifiable so as to be compared, whereas *self-awareness, context-awareness, openness* do not directly impact ABs efficiency [7] . In order to get the property of *absoluteness* of the comparison scale (i.e. to avoid a restriction of the events set that can occur, see section 2.2), the value I of each quantifiable high-level indicators is ranged between 0 (non autonomic) and $+\infty$ (idealistic fully autonomic). This value I will be obtained by adding the result of the corresponding composition function f_I applied on each event e handled by the system and weighted with the occurrence probability p_e of this event:

$$I = \sum_e p_e f_I(e) \qquad (1)$$

The indicator value highlights the wideness of the events spectrum the system is able to compute and the utility of dealing with each event.

3.2 Qualitative Assessment

Table 1 summarizes criteria and metrics defined for qualitative assessment of ABs.

3.3 Quantitative Assessment

Table 2 summarizes the criteria and metrics defined for quantitative assessment of ABs.

Four separate time measurements –monitorability, analyzability, planning capability and changeability– that are mapped on each phase of the widely accepted concept of MAPE-loop (see figure 4) are defined. This decomposition offers two main advantages. On the one hand, these quantitative (i.e. per se quantifiable) metrics are generic and measurable because they are based on the control loop that is a fundamental concept of ABs implementation. On the other hand, they can be separately measured, according to the system maturity, because of the use of a common concept (i.e. the stages of the MAPE-loop) for defining these metrics and the maturity levels.

---

[6] Efficiency factor and associated criteria and metrics are not, strictly speaking, constituents of this model but they appear on this figure in order to highlight the relationship between this hybrid model for autonomic computing and business specific metrics.

[7] Whether *anticipation* is or is not a quantifiable high-level indicator is still to be determined.

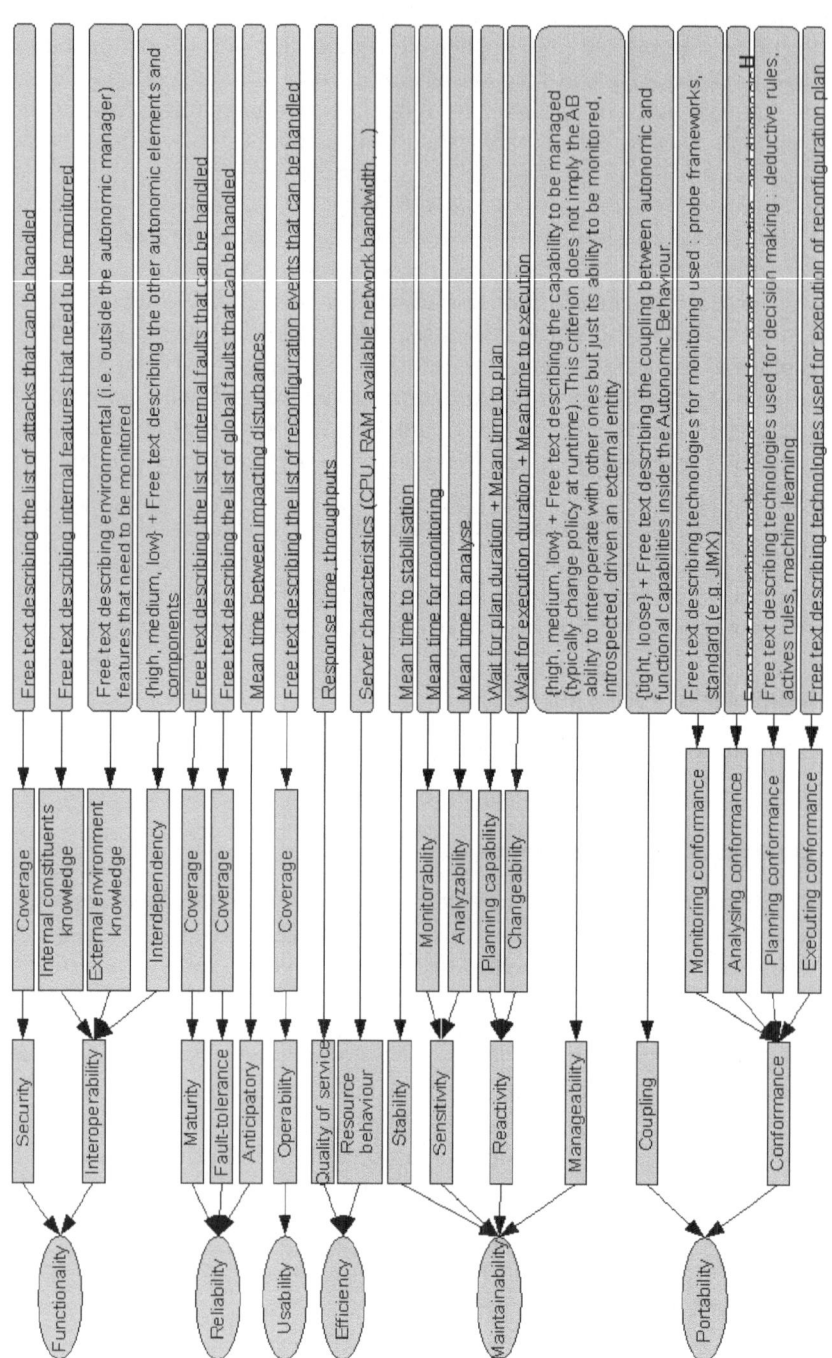

Fig. 3. Quality model for autonomic computing assessment

Table 1. Qualitative criteria for ABs assessment

Criterion	Criterion type
Description of AB	Free text. Short description of the AB.
Related to self-* characteristic	{self-configuration, self-healing, self-optimization, self-protection}
Coverage	Free text describing the list of disturbances the AB is able to deal with
Interdependency	high, medium, low + Free text. Description of the other ABs and components the current AB is depending on.
Internal constituents knowledge	Internal features that need to be monitored for this AB
External environment knowledge	Environmental (i.e. outside the AB) features that need to be monitored for this AB
Level of automation	{-, M, MA, MAP, MAPE}
Monitoring compliance	Free text. Technologies used for monitoring: probe frameworks, standard (e.g. JMX)
Analyzing compliance	Free text. Technologies used for event correlation and diagnosis
Planning compliance	Free text. Technologies used for decision making: deductive rules, actives rules, machine learning
Executing compliance	Free text. Technologies used for execution of reconfiguration plan
Coupling	tight, loose + Free text. Description of coupling between autonomic and functional capabilities inside the AB.
Manageability	high, medium, low + Free text. Capability to be managed (typically change policy at runtime). This criterion does not imply the AB ability to interoperate with other ones but just its ability to be monitored, introspected, driven an external entity.

Table 2. Metrics for quantitative assessment of ABs

Criterion	Sub-criterion	Metric
Sensitivity	Monitorability	Mean time for monitoring
	Analyzability	Mean time to analyze
Reactivity	Planning capability	Wait time for plan duration + Mean time to plan
	Changeability	Wait time for execution duration + Mean time to execution
Anticipation		Mean time between impacting disturbances
Stability		Mean time to stabilization

Anticipation measures the system ability to deal with events while maintaining its QoS at a satisfying level: the human administrator defines an interval of satisfaction for QoS values. Thus, among the overall disturbances the system is

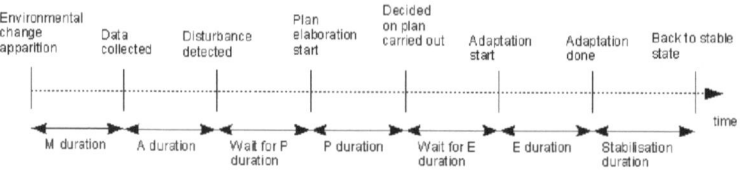

Fig. 4. Mapping between MAPE-loop stages and their duration

able to compute, the impacting ones are those which modify the system efficiency (QoS) outside this interval.

Stability measures the time the system needs to return to a stable state, i.e. a state for which QoS values are stable.

Test Process. [2] proposes a three-step test process for evaluating ABs.

1. First a workload, that will be maintained constant during all the three steps, is injected in the system under test (SUT). This first step lasts until the system reaches a stable state regarding its QoS metrics.
2. Then a single disturbance is injected in order to trigger an autonomic reaction of the SUT.
3. The last step consists in observing metrics (related to QoS and autonomic computing) until the SUT returns to a stable state (possibly different from the initial one) regarding the QoS metrics.

This test process measures autonomic features efficiency. It exhibits some kind of reproducibility property (see section 2.2) coming from the stability of the workload injection during all the test duration. However due to this stability and to the uniqueness of the injected disturbance, this test process does not fit the realism property (see the third difference between a "classical" benchmark and an autonomic computing benchmark in section 2.2). Notwithstanding, this process is adopted for quantitative experimentations. The injected disturbance is considered as a default for evaluating self-healing, an attack for evaluating self-protection, an extra constant workload for evaluating self-optimization and an event triggering a reconfiguration for evaluating self-configuration (e.g. *churn* in P2P systems).

3.4 Economical Assessment (Return on Investment)

As mentioned in section 1, the ultimate goal of autonomic benchmarking, from an industrial point of view, is to define means for evaluating the return on investment (ROI) of developing and deploying autonomic computing technologies. Once the technical evaluation stage is over, available data relative to the economical cost management (i.e. cost of licenses, disturbance rate and frequency, mean time to resolve a disturbance, human administrator salary, energy consumption, etc.) can be collected. Then these economical metrics will be placed in the autonomic computing model in order to obtain an estimation of the economical return on investment. The economical evaluation could be carried out by using utility functions as proposed in [16]. Utility functions can be seen as composition functions whose result is a financial value expressed in a given currency. Utility functions might allow for comparison of the economical efficiency of different autonomic features in the system under consideration or in different autonomic systems (e.g. "is there more ROI to be expected by introducing autonomic features in web applications servers or machine-to-machine platforms?").

4   Experimental Results

This section introduces and discusses the experimental results that have been obtained by applying the methodology described to eleven ABs in three autonomic systems. The three systems considered cover the architectural classification of autonomic systems proposed by [12]:

- wide distributed systems composed of a large number of independent collaborative intelligent nodes. In our case, a peer-to-peer (P2P) transactional storage system, namely Scalaris from ZIB (Zuse Institute Berlin) designated as SUT1, and a P2P system on mobile phones, namely the gPhone application from UCL (Université Catholique de Louvain) designated as SUT2. Both are based on a common structured overlay network (SON). SUT1 and SUT2 implement eight different ABs (some are common, some are specific). SUT1 and SUT2 are research-oriented and have been developed in the EU funded IST Selfman project [5].
- more centralized systems composed of a hierarchy of components. In our case, an industrial workload management application, designated as SUT3[8] based on a centralized manager and a cluster of JOnAS [8] J2EE application servers. SUT3 implements three ABs.

4.1   Qualitative Assessment

The qualitative assessment consisted in filling out each line (i.e. computing each qualitative metric) of the table 1 in the context of each AB. Altogether, eleven tables have been obtained (eight for SUT1 and SUT2 and three for SUT3) that make up dozens of pages in Selfman project deliverables concerning SUT1 and SUT2. Due to space limitation in this article, results have been summarized in table 3 that reports the major tendencies highlighted by these individual assessments.

Among these major tendencies, some are common to all evaluated ABs. The lack of self-protection ABs is explained by the systems editors as follows: some of them assume that their solution runs in a safe environment whereas others estimate that security has to be delegated to another system contributor. However a further explanation is that self-protection ABs have to deal with more complex events and thus cannot be elementary behaviors, exclusively focusing

---

[8] For motives of confidentiality, the name of this solution has not been mentioned.

Table 3. Synthesis of qualitative assessment on eleven ABs coming from SUT1, SUT2 and SUT3

Criterion	Tendency
Related to self-* characteristic	None of the assessed ABs is related to self-protection. Moreover, the distinction between self-optimization and self-configuration is not self-evident.
Coverage	All of these ABs can be considered elementary: they only deal with one or two low-level events ('peer joining' or 'excessive response time to a request' for instance)
Interdependency	Only one of the ABs interoperates with another one but this interaction is limited to a single query/reply and does not consist in a structured dialog
Internal constituents knowledge	All these ABs run according to the monitoring data they get from the resources they manage
External environment knowledge	All these ABs have very little (or even no) knowledge of their external environment
Level of automation	However all are fully autonomous regarding their level of maturation: none of them claims to be autonomic without fully implementing the four stages of the MAPE-loop
Monitoring compliance	Concerning their compliance with a standard and/or a wide spread / opened technology, all ABs propose highly proprietary implementation for this MAPE-loop stage
Analyzing compliance	Concerning their compliance with a standard and/or a wide spread / opened technology, all ABs propose highly proprietary implementation for this MAPE-loop stage
Planning compliance	Concerning their compliance with a standard and/or a wide spread / opened technology, all ABs propose highly proprietary implementation for this MAPE-loop stage
Executing compliance	Concerning their compliance with a standard and/or a wide spread / opened technology, all ABs propose highly proprietary implementation for this MAPE-loop stage
Coupling	All the ABs of SUT1 and SUT2 are tightly coupled with the business functionalities in term of implementation whereas this coupling is loose concerning the three ABs of SUT3
Manageability	Three of the eleven ABs include manageability capabilities. However the management policy elements they can get are quite rudimentary (like a timer value or a combination of simple conditions). This illustrates the difficulty in converting high-level management policies into simpler rules understandable by the autonomic managers

on their internal resources. They might result from the composition (implying sophisticated interoperability) between different lower-level ABs. Thus the composition of ABs becomes an important research domain. However the lack of compliance with open standards, which characterizes elementary ABs, prevents any progress on their interoperability.

The main difference resulting from these qualitative assessments concerns the coupling between autonomic features and application specific (business) functionalities. This is due to:

systems architecture: SUT1 and SUT2 lie on a wide distributed multi-peers architecture whereas SUT3 is based on a centralized and hierarchical architecture;
development history: SUT1 and SUT2 have been developed from scratch whereas SUT3 consists in a clustered workload management layer added to an existing of J2EE application server.

4.2 Quantitative Assessment

Again, due to space limitation, this section only focuses on a single AB that illustrates the kind of results obtained by applying the quantitative metrics defined in table 2 on all implemented ABs. This AB concerns self-optimization in SUT3. It aims at increasing or reducing the number of physical machines (nodes of a cluster of JOnAS application servers) according to a level of workload addressed to a given web application. The workload manager is centralized and, in this example, is deployed on a dedicated machine. All requests addressed to the given web application are routed through the workload manager. The experimentation consists in injecting an extra workload causing a violation of the management policy concerning the response time limitation.

An evaluation methodology close to the one described in [2] (see section 3.3) has been used for this benchmark. Preliminarily the workload threshold at which the manager decides to release the AB, has been determined. This depends mainly on the material configuration of the test platform. A value of 50 requests per second (req/s) has been obtained. Then, the same three-step approach has been applied to each different experimentation.

1. The first stage consists in submitting SUT3 to the maximum workload a single J2EE application server can handle without breaching the response time limitation (set to 100 ms).
2. Then, after having achieved the stabilization of response time, an extra workload is injected.
3. The last step focuses on the observation of response time (QoS metric) and of the different autonomic durations until the system returns to a stable state (with a satisfying response time).

For any SUT, it is possible to draw a figure showing the impact of the injected disturbance on the QoS. According to the autonomic characteristic –and thus to the type of disturbance– different injection profiles can be defined. In the field of self-optimization for instance, the injection profile of extra workload consists in setting its rate and its emerging speed. Figure 5 illustrates the three-step test protocol and shows the impact of an extra workload on SUT3 response time. In this example, the injection profile is characterized by a rate of 20% (+10 requests per second) and an arising speed of 1 second (i.e. a step profile).

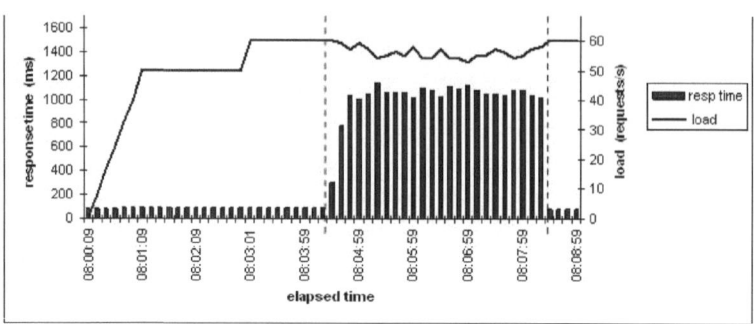

Fig. 5. Impact of an extra workload on SUT3 response time

Table 4. Quantitative evaluation of resource optimization of SUT3

	Workload profile 1	Workload profile 2	Workload profile 3
Extra workload rate	20% - 10 req/s	20% - 10 req/s	40% - 20 req/s
Extra workload arising speed	1 s - 10 req/s^2	60 s - 0.17 req/s^2	60 s - 0.33 req/s^2
M + A durations	61 s.	64 s.	66 s.
Wait for P duration	17 ms.	24 ms.	27 ms.
P duration	485 ms.	475 ms.	475 ms.
Wait for E duration	2 ms.	4 ms.	15 ms.
E duration	136 s.	136 s.	131 s.
Stabilization duration	46 s.	53 s.	39 s.

Table 4 synthesizes the results obtained by applying three different extra workload profiles. Autonomic computing durations were not obtained graphically but through the analysis of workload manager log files.

These results lead to three major observations.

1. Monitoring and analysis durations[9] are amazingly high whereas the time lag for planning is extremely short. A first explanation might conclude that monitoring and analysis are really inefficient. However this is not the case. It results from two causes. On the one hand SUT3 provides a mechanism for avoiding hyper-sensitivity (i.e. reacting too frequently): the violation of response time limitation has to last at least one minute before triggering a reaction plan. SUT3 is effectively an industrial solution whose objectives are an hourly optimization. It does not find any interest in a finer grained precision for resource optimization. On the other hand, SUT3 considers that the disturbance consists of the extra workload injection and its continuation during a configured duration (at least one minute) whereas in the approach proposed in this article, the disturbance consists only in the extra workload

---

[9] Data collected from the SUT3 benchmark did not make possible the distinction between monitoring and analysis durations.

(the configured duration defining the minimal length of the wait for planning duration).

2. Another observation is that the self-optimization behavior is independent from the injected disturbances profiles. SUT3 does not seem to include anticipation features because it remains insensitive to changes of arising speed or rate of extra workload (see table 4).

3. Stabilization duration is quite long. It confirms that this measure remains essential for evaluating the risk the system tends to remain a long time in an unstable state.

4.3 Discussion

This section discusses the proposed methodology for evaluating autonomic computing whereas section 4.1 and section 4.2 focus on presenting and analyzing the results coming from the technical benchmark.

On the one hand, the methodology used for qualitative benchmark is relatively generic: it has been applied successfully, especially as far as qualitative assessment is concerned, to eleven ABs coming from three applications that differ in business areas, architectures and implementations.

On the other hand, our experiments question the genericity of the approach, and in particular of the test process (see section 3.3), with respect to system architecture. Indeed, although the process is adapted for evaluating systems for which QoS and autonomic metrics are evaluated at the same abstraction level (this is the case of autonomic systems with a hierarchical architecture of components, like SUT3), it is not so adequate for systems where QoS and autonomic metrics are measured at different levels of abstraction, like in a P2P systems (such as SUT1 and SUT2) where ABs are local and QoS is evaluated at a macroscopic level. In the latter situation, an injected disturbance will trigger an autonomic adaptation that could have no measurable effect on the QoS (due to the important number of peers contributing to the QoS). A solution could be to define a methodology for composing the results obtained at the peer (local) level in order to extrapolate the impact of ABs at a global level. This is one of the subjects that needs further investigations.

Finally the definition of anticipation and stabilization durations independently from the domain of application and the architecture is also a challenge for a quantitative autonomics benchmark. It has been experimented indeed that for now, in some cases, the stabilization duration can be difficult or even impossible to measure (for example in an AB running periodically without any "real" triggering event).

5   Conclusion

This article has introduced a methodology and a process for qualitative and quantitative assessment of autonomic features in computing systems together with associated experimental results on eleven autonomic behaviors in three different autonomic systems (two P2P systems and one n-tiers system). Our experiments provide an uneven feedback that altogether questions the reachable genericity of autonomic benchmarking and the possibility of evaluating autonomic systems in which autonomic features are deeply built-in (e.g. in the P2P systems considered in this article).

Further investigations related to this work are needed so as to validate or to invalidate these first feedbacks and:

- to improve the ways of measuring some criteria, such as anticipatory or stabilization;
- to work out a methodology for composing elementary ABs in order to obtain a higher level behavior and allow the comparison of QoS metrics and autonomic computing indicators at the same level of abstraction;
- to define ways of coupling cost models with a technical benchmark. This would enable to evaluate the return on investment of autonomic computing and to determine the pertinence of applying such a disruptive technology to different fields of application.

Since benchmarking tools have already managed to provide important improvements on the efficiency of microprocessors or middleware as mentioned in [1], autonomics benchmarking is expected to constitute a decision-making tool for IT managers to assess and hopefully foster the adoption of autonomics in industry - but this is not an easy road.

References

1. Brown, A.B., Hellerstein, J.L., Hogstrom, M., Lau, T., Lightstone, S., Shum, P., Yost, M.P.: Benchmarking Autonomic Capabilities: Promises and Pitfalls. In: 1st International Conference on Autonomic Computing, pp. 266–267. IEEE Computer Society, New York (2004)
2. Brown, A.B., Redlin, C.: Measuring the Effectiveness of Self-Healing Autonomic Systems. In: 2nd International Conference on Autonomic Computing, pp. 328–329. IEEE Computer Society, New York (2005)
3. Chen, H., Hariri, S.: An Evaluation Scheme of Adaptive Configuration Techniques. In: 22nd IEEE/ACM International Conference on Automated Software Engineering, pp. 493–496. ACM Press, New York (2007)
4. De Wolf, T., Holvoet, T.: Evaluation and Comparison of Decentralized Autonomic Computing Systems. Technical report. Department of Computer Science, K.U. Leuven, Leuven, Belgium (2006)
5. European IST 6th FWP Selman project (self management for large-scale distributed systems based on structured overlay networks and components), http://www.ist-selfman.org/wiki/index.php/Selfman_project
6. Forse, T.: Qualimétrie des Systèmes Complexes - Mesure de la Qualité du Logiciel (Qualimetry of Complex Systems - Measurement of Software Quality). Editions d'organization (1989)
7. Horn, P.: Autonomic Computing: IBM's Perspective on the State of Information Technology, http://www.research.ibm.com/autonomic/manifesto/ autonomic_computing.pdf

8. JOnAS OpenSource Java EE Application Server, http://jonas.ow2.org/
9. Kephart, J.O., Chess, D.M.: The Vision of Autonomic Computing. IEEE Computer 1, 41–50 (2003)
10. Kluth, A.: Make It Simple. The Economist (2004-10-28)
11. Lin, P., MacArthur, A., Leaney, J.: Defining Autonomic Computing: A Software Engineering Perspective. In: 16th Australian Software Engineering Conference, pp. 88–97. IEEE Computer Society, New York (2005)
12. McCann, J.A., Huebscher, M.C.: Evaluation Issues in Autonomic Computing. In: Grid and Cooperative Computing - GCC 2004 Workshops: GCC 2004 International Workshops, IGKG, SGT, GISS, AAC-GEVO, and VVS, pp. 597–608. Springer, Heidelberg (2004)
13. Milicic, D.: Software Quality Models and Philosophies. In: Lundberg, L., Mattsson, M., Wohlin, C. (eds.) Software Quality Attributes and Trade-offs, p. 100. Blekinge Institute of Technology (2005)
14. Oyenan, W.H., DeLoach, S.A.: Design and Evaluation of a Multiagent Autonomic Information System. In: 2007 IEEE/WIC/ACM International Conference on Intelligent Agent Technology, pp. 182–188. IEEE Computer Society, New York (2007)
15. Salehie, M., Tahvildari, L.: Autonomic Computing: Emerging Trends and Open Problems. ACM SIGSOFT Software Engineering Notes 4, 1–4 (2005)
16. Walsh, W.E., Tesauro, G., Kephart, J.O., Das, R.: Utility Functions in Autonomic Systems. In: 1st International Conference on Autonomic Computing, pp. 70–77. IEEE Computer Society, New York (2004)
17. Wildstrom, J., Stone, P., Witchel, E.: Autonomous Return on Investment analyzis of Additional Processing Resources. In: 4th International Conference on Autonomic Computing, p. 15. IEEE Computer Society, New York (2007)
18. Zeiss, B., Vega, D., Schieferdecker, I., Neukirchen, H., Grabowski, J.: Applying the ISO 9126 Quality Model to Test Specifications - Exemplified for TTCN-3 Test Specifications. In: Software Engineering 2007, Fachtagung des GI-Fachbereichs Softwaretechnik, pp. 231–244. GI (2007)
19. Zhang, H., Whang, H., Zheng, R.: An Autonomic Evaluation Model of Complex Software. In: International Conference on Internet Computing in Science and Engineering, pp. 343–348 (2008)

An Online Adaptive Model for Location Prediction

Theodoros Anagnostopoulos, Christos Anagnostopoulos,
and Stathes Hadjiefthymiades

Pervasive Computing Research Group, Communication Networks Laboratory,
Department of Informatics and Telecommunications, University of Athens,
Panepistimiopolis, Ilissia, Athens 15784, Greece
Tel.: +302107275127
{thanag,bleu,shadj}@di.uoa.gr

Abstract. Context-awareness is viewed as one of the most important aspects in the emerging pervasive computing paradigm. Mobile context-aware applications are required to sense and react to changing environment conditions. Such applications, usually, need to recognize, classify and predict context in order to act efficiently, beforehand, for the benefit of the user. In this paper, we propose a mobility prediction model, which deals with context representation and location prediction of moving users. Machine Learning (ML) techniques are used for trajectory classification. Spatial and temporal on-line clustering is adopted. We rely on Adaptive Resonance Theory (ART) for location prediction. Location prediction is treated as a context classification problem. We introduce a novel classifier that applies a Hausdorff-like distance over the extracted trajectories handling location prediction. Since our approach is time-sensitive, the Hausdorff distance is considered more advantageous than a simple Euclidean norm. A learning method is presented and evaluated. We compare ART with Offline kMeans and Online kMeans algorithms. Our findings are very promising for the use of the proposed model in mobile context aware applications.

Keywords: Context-awareness, location prediction, Machine Learning, online clustering, classification, Adaptive Resonance Theory.

1 Introduction

In order to render mobile context-aware applications intelligent enough to support users everywhere / anytime and materialize the so-called ambient intelligence, information on the present *context* of the user has to be captured and processed accordingly. A well-known definition of context is the following: *"context is any information that can be used to characterize the situation of an entity. An entity is a person, place or object that is considered relevant to the integration between a user and an application, including the user and the application themselves"* [1]. Context refers to the current values of specific ingredients that represent the activity of an entity / situation and environmental state (e.g., attendance of a meeting, location, temperature).

One of the more intuitive capabilities of the mobile context-aware applications is their *proactivity*. Predicting user actions and contextual ingredients enables a new class of applications to be developed along with the improvement of existing ones.

A.V. Vasilakos et al. (Eds.): AUTONOMICS 2009, LNICST 23, pp. 64–78, 2010.
© Institute for Computer Sciences, Social-Informatics and Telecommunications Engineering 2010

One very important ingredient is location. Estimating and predicting the future location of a mobile user enables the development of innovative, location-based services/applications [2], [12]. For instance, location prediction can be used to improve resource reservation in wireless networks and facilitate the provision of location-based services by preparing and feeding them with the appropriate information well in advance. The accurate determination of the context of users and devices is the basis for context-aware applications. In order to adapt to changing demands, such applications need to reason based on basic context ingredients (e.g., time, location) to determine knowledge of higher-level situation.

Prediction of context is quite similar to information classification / prediction (*offline* and *online*). In this paper, we adopt ML techniques for predicting location through an adaptive model. ML is *the study of algorithms that improve automatically through experience*. ML provides algorithms for learning a system to cluster preexisting knowledge, classify observations, predict unknown situations based on a history of patterns and adapt to situation changes. Therefore, ML can provide solutions that are suitable for the location prediction problem. Context-aware applications have a set of pivotal requirements (e.g., flexibility and adaptation), which would strongly benefit if the learning and prediction process could be performed in real time. We argue that the most appropriate solutions for location prediction are offline and online clustering and classification. Offline clustering is performed through the Offline *k*Means algorithm while online clustering is accomplished through the Online *k*Means and Adaptive Resonance Theory (ART). Offline learners typically perform complete model building, which can be very costly, if the amount of samples rises. Online learning algorithms are able to detect changes and adapt / update only parts of the model thus providing for fast adaptation of the model. Both forms of algorithms extract a subset of patterns / clusters (i.e., a knowledge base) from an initial dataset (i.e., a database of user itineraries). Moreover, online learning is more suited for the task of classification / prediction of the user mobility behavior as in the real life user movement data often needs to be processed in an online manner, each time after a new portion of the data arrives. This is caused by the fact that such data is organized in the form of a data stream (e.g., a sequence of time-stamped visited locations) rather than a static data repository, reflecting the natural flow of data. Classification involves the matching of an unseen pattern with existing clusters in the knowledge base. We rely on a Hausdorff-like distance [5] for matching unseen itineraries to clusters (such metric applies to convex patterns and is considered ideal for user itineraries). Finally, location prediction boils down to location classification w.r.t. Hausdorff-like distance.

We assess two training methods for training an algorithm: (i) the "nearly" *zero-knowledge* method in which an algorithm is incrementally trained starting with a little knowledge on the user mobility behavior and the (ii) *supervised* method in which sets of known itineraries are fed to the classifier. Moreover, we assess a learning method for the online algorithms regarding the success of location prediction, in which a misclassified instance is introduced into the knowledge base updating appropriately the model.

We evaluate the performance of our models against the movement of mobile users. Our objective is to predict the users' future location (their next move) through an on-line adaptive classifier. We establish some important metrics for the performance assessment process taking into account low system-requirements (storage capacity) and effort for model building (processing power). Specifically, besides the prediction accuracy, i.e., the precision of location predictions, we are also interested in the size of the derived knowledge base; that is the produced clusters out of the volume of the training patterns, and the capability of the classifier to adapt the derived model to unseen patterns. Surely, we need to keep storage capacity as low as possible while maintaining good prediction accuracy. Lastly, our objective is to assess the *adaptivity* of the proposed schemes, i.e., the capability of the predictor to detect and update appropriately the specific part of the trained model. The classifier (through the location prediction process) should rapidly detect changes in the behavior of the mobile user and adapt accordingly through model updates, however, often at the expense of classification accuracy (note that an ambient environment implies high dynamicity). We show that increased adaptivity leads to high accuracy and dependability.

The rest of the paper is structured as follows. In Section 2 we present the considered ML models by introducing the Offline kMeans, Online kMeans and ART algorithms. In Section 3 we elaborate on the proposed model with context representation. Section 4 presents the proposed mobility prediction model based on the ART algorithm. The performance assessment of the considered model is presented in Section 5, where different versions of that model are evaluated. Moreover, in Section 6, we compare the ART models with the Offline / Online kMeans algorithms. Prior work is discussed in Section 7 and we conclude the paper in Section 8.

2   Machine Learning Models

In this section we briefly discuss the clustering algorithms used throughout the paper. Specifically, we distinguish between offline and online clustering and elaborate on the Offline/Online kMeans and ART.

2.1   Offline kMeans

In Offline kMeans [3] we assume that there are $k > 1$ initial clusters (groups) of data. The objective of this algorithm is to minimize the reconstruction error, which is the total Euclidean distance between the instances (patterns), $\mathbf{u}_i$, and their representation, i.e., the cluster centers (clusters), $\mathbf{c}_i$. We define the reconstruction error as follows:

$$E\left(\{\mathbf{c}_i\}_{i=1}^k \mid U\right) = \frac{1}{2}\sum_t \sum_i b_{i,t} \parallel \mathbf{u}_t - \mathbf{c}_i \parallel^2 \tag{1}$$

where

$$b_{i,t} = \begin{cases} 1, & if \parallel \mathbf{u}_t - \mathbf{c}_i \parallel = \min_l \parallel \mathbf{u}_t - \mathbf{c}_l \parallel, \\ 0, & otherwise \end{cases}$$

$U = \{\mathbf{u}_t\}$ is the total set of patterns and $C = \{\mathbf{c}_i\}$, $i = 1,\ldots, k$ is the set of clusters. $b_{i,t}$ is 1 if $\mathbf{c}_i$ is the closest center to $\mathbf{u}_t$ in Euclidean distance. For each incoming $\mathbf{u}_t$ each $\mathbf{c}_i$ is updated as follows:

$$\mathbf{c}_i = \frac{\sum_t b_{i,t} \mathbf{u}_t}{\sum_t b_{i,t}} \tag{2}$$

Since the algorithm operates in offline mode, the initial clusters can be set during the training phase and cannot be changed (increased or relocated) during the testing phase.

2.2 Online kMeans

In Online kMeans [3] we assume that there are $k > 1$ initial clusters that split the data. Such algorithm processes unseen patterns one by one and performs *small* updates in the position of the appropriate cluster ($\mathbf{c}_i$) at each step. The algorithm does not require a training phase. The update for each new (unseen) pattern $\mathbf{u}_t$ is the following:

$$\mathbf{c}_i = \mathbf{c}_i + \eta \cdot b_{i,t} \cdot \left(\mathbf{u}_t - \mathbf{c}_i \right)$$

This update moves the *closest* cluster (for which $b_{i,t} = 1$) toward the input pattern $\mathbf{u}_t$ by a factor of η. The other clusters (found at bigger distances from the considered pattern) are not updated. The semantics of $b_{i,t}$, η and ($\mathbf{u}_t - \mathbf{c}_i$) are:

> $b_{i,t} \in \{0, 1\}$ denotes which cluster is being modified,
> $\eta \in [0, 1]$ denotes how much is the cluster shifted toward the new pattern, and,
> ($\mathbf{u}_t - \mathbf{c}_i$) denotes the distance to be learned.

Since the algorithm is online, the initial clusters should be known beforehand[1] and can only be relocated during the testing phase. The number of clusters remains constant. Therefore, the algorithm exhibits limited flexibility.

2.3 Adaptive Resonance Theory

The ART approach [4] is an online learning scheme in which the set of patterns U is not available during training. Instead patterns are received one by one and the model is updated progressively. The term *competitive learning* is used for ART denoting that the (local) clusters *compete* among themselves to assume the "responsibility" for representing an unseen pattern. The model is also called *winner-takes-all* because one cluster "wins the competition" and gets updated, and the others are not updated at all.

The ART approach is *incremental*, meaning that one starts with one cluster and adds a new one, if needed. Given an input $\mathbf{u}_t$, the distance b_t is calculated for all clusters $\mathbf{c}_i$, $i = 1, .., k$, and the closest (e.g., minimum Euclidean distance) to $\mathbf{u}_t$ is updated. Specifically, if the minimum distance b_t is smaller than a certain threshold value, named the *vigilance*, ρ, the update is performed as in Online kMeans (see Eq.(3)). Otherwise, a new center $\mathbf{c}_{k+1}$ representing the corresponding input $\mathbf{u}_t$ is added in the model (see Eq.(3)). It is worth noting that the vigilance threshold refers to the criterion of considering two patterns equivalent or not during the learning phase of a classifier. As it will be shown,

---

[1] One possible approach to determine the initial k clusters is to select the first k distinct instances of the input sample U.

the value of vigilance is considered essential in obtaining high values of corrected classified patterns. The following equations are adopted in each update step of ART:

$$b_t = \| \mathbf{c}_i - \mathbf{u}_t \| = \min_{l=1}^{k} \| \mathbf{c}_l - \mathbf{u}_t \|$$

$$\begin{cases} \mathbf{c}_{k+1} \leftarrow \mathbf{u}_t & \text{if } b_t > \rho \\ \mathbf{c}_i = \mathbf{c}_i + \eta(\mathbf{u}_t - \mathbf{c}_i) & \text{otherwise} \end{cases} \qquad (3)$$

3 Context Representation

Several approaches have been proposed in order to represent the movement history (or history) of a mobile user [15]. We adopt a spatiotemporal history model in which the movement history is represented as the sequence of 3-D points (3DPs) visited by the moving user, i.e., time-stamped trajectory points in a 2D surface. The spatial attributes in that model denote latitude and longitude.

Let $\mathbf{e} = (x, y, t)$ be a 3DP. The *user trajectory* $\mathbf{u}$ consists of several time-ordered 3DPs, $\mathbf{u} = [\mathbf{e}_i] = [\mathbf{e}_1, ..., \mathbf{e}_N]$, $i = 1, ..., N$ and is stored in the system's database. It holds that $t(\mathbf{e}_1) < t(\mathbf{e}_2) < ... < t(\mathbf{e}_N)$, i.e., time-stamped coordinates. The x and y dimensions denote the latitude and the longitude while t denotes the time dimension (and $t(\cdot)$ returns the time coordinate of $\mathbf{e}$). Time assumes values between 00:00 and 23:59. To avoid state information explosion, trajectories contain time-stamped points sampled at specific time instances. Specifically, we sample the movement of each user at $1.66 \cdot 10^{-3}$ Hertz (i.e., once every 10 minutes). Sampling at very high rates (e.g., in the order of a Hertz) is meaningless, as the derived points will be highly correlated. In our model, $\mathbf{u}$ is a finite sequence of N 3DPs, i.e., $\mathbf{u}$ is a $3 \cdot N$ dimension vector. We have adopted a value of $N = 6$ for our experiments meaning that we estimate the future position of a mobile terminal from a movement history of 50 minutes (i.e., 5 samples). Specifically, we aim to query the system with a N-1 3DP sequence so that our classifier / predictor returns a 3DP, which is the predicted location of the mobile terminal.

A *cluster trajectory* $\mathbf{c}$ consists of a finite number of 3DPs, $\mathbf{c} = [\mathbf{e}_i]$, $i = 1, ..., N$ stored in the knowledge base. Note that a cluster trajectory $\mathbf{c}$ and a user trajectory $\mathbf{u}$ are vectors of the same length N. This is because $\mathbf{c}$, which is created from ART based on unseen user trajectories, is a representative itinerary of the user movements. In addition, the *query trajectory* $\mathbf{q}$ consists of a number of 3DPs, $\mathbf{q} = [\mathbf{e}_j]$, $j = 1, ..., N$-1. It is worth noting that $\mathbf{q}$ is a sequence of N-1 3DPs. Given a $\mathbf{q}$ with a N-1 history of 3DPs we predict the $\mathbf{e}_N$ of the closest $\mathbf{c}$ as the next user movement.

4 Mobility Prediction Model

From the ML perspective the discussed location prediction problem refers to an $m+l$ model [13]. In $m+l$ models we have m steps of user movement history and we want to predict the future user movement after l steps (the steps have time-stamped coordinates). In our case, $m = N$-1, i.e., the query trajectory $\mathbf{q}$, while $l = 1$, i.e., the predicted $\mathbf{e}_N$. We develop a new spatiotemporal classifier (C) which given $\mathbf{q}$ can predict $\mathbf{e}_N$. Specifically, $\mathbf{q}$ and $\mathbf{c}$ are trajectories of different length thus we use a Hausdorff-like measure for calculating the $\|\mathbf{q} - \mathbf{c}\|$ distance. Given query $\mathbf{q}$, the proposed classifier C attempts to find the nearest cluster $\mathbf{c}$ in the knowledge base and, then, take $\mathbf{e}_N$ as the *predicted* 3DP. For evaluating C, we compute the Euclidean distance between the predicted 3DP and the *actual* 3DP (i.e., the real user movement). If such distance is greater than a preset error threshold θ then prediction is not successful. After predicting the future location of a mobile terminal, the C classifier receives feedback from the environment considering whether the prediction was successful or not, and reorganize the knowledge base accordingly [14]. In our case, the feedback is the actual 3DP observed in the terminal's movement. Thus the C classifier reacts with the environment and learns new patterns once an unsuccessful prediction takes place.

Specifically,

> ➤ in case of an unsuccessful prediction, the C appends the actual 3DP to $\mathbf{q}$ and updates (i.e., learns) such extended sequence in the model considering as new knowledge, i.e., an unseen user movement behavior.
> ➤ in the case of a successful prediction, C dos not need to learn. A successful prediction refers to a well-established prediction model for handling unseen user trajectories.

The heart of the proposed C classifier is the ART algorithm. ART clusters unseen user trajectories to existing cluster trajectories or creating new cluster trajectories depending on the vigilance value. ART is taking the $\mathbf{u}_1$ pattern from the incoming set U of patterns and stores it as the $\mathbf{c}_1$ cluster in the knowledge base. For the t-th unseen user trajectory the following procedure is followed (see Table 1): The algorithm computes the Euclidean distance b_t between $\mathbf{u}_t$ and the closest $\mathbf{c}_i$. If b_t is smaller than the vigilance ρ then $\mathbf{c}_i$ is updated from $\mathbf{u}_t$ by the η factor. Otherwise, a new cluster $\mathbf{c}_j \equiv \mathbf{u}_t$ is inserted into the knowledge base. The ART algorithm is presented in Table 1.

Table 1. The ART Algorithm for the C classifier

1.	$j \leftarrow 1$
2.	$\mathbf{c}_j \leftarrow \mathbf{u}_j$
3.	**For** $(\mathbf{u}_t \in U)$ **Do**
4.	$b_t = \|\mathbf{c}_j - \mathbf{u}_t\| = \min_{l=1,\dots,j}\|\mathbf{c}_l - \mathbf{u}_t\|$
5.	**If** $(b_t > \rho)$ **Then** /*expand knowledge*/
	$j \leftarrow j + 1$
6.	$\mathbf{c}_j \leftarrow \mathbf{u}_t$
7.	**Else**
8.	$\mathbf{c}_j \leftarrow \mathbf{c}_j + \eta(\mathbf{u}_t - \mathbf{c}_j)$ /*update model locally*/
9.	**End If**
10.	**End for**

Let T, P be subsets of U for which it holds that $T \subseteq P \subseteq U$. The T set of patterns is used for training the C classifier, that is, C develops a knowledge base corresponding to the supervised training method. The P set is used for performing on-line predictions. We introduce the C-T classifier version, which is the C classifier trained with the T set. In addition, once the T set is null then the C classifier is not trained beforehand corresponding to the zero-knowledge training method and performs on-line prediction with the set P. In this case, we get the C-nT classifier corresponding to the C classifier, when the training phase is foreseen.

Moreover, in order for the C classifier to achieve prediction, an approximate Hausdorff-like metric [5] is adopted to estimate the distance between $\mathbf{q}$ and $\mathbf{c}$. Specifically, the adopted formula calculates the point-to-vector distance between $\mathbf{e}_j \in \mathbf{q}$ and $\mathbf{c}$, $\delta'(\mathbf{e}_j, \mathbf{c})$, as follows:

$$\delta'(\mathbf{e}_j,\mathbf{c})=\parallel \mathbf{f}_i -\mathbf{e}_j \parallel_{\min_{\mathbf{f}_i} |t(\mathbf{f}_i)-t(\mathbf{e}_j)|}$$

where $\parallel \cdot \parallel$ is the Euclidean norm for $\mathbf{f}_i \in \mathbf{c}$ and $\mathbf{e}_j$. The $\delta'(\mathbf{e}_j, \mathbf{c})$ value indicates the minimum distance between $\mathbf{e}_j$ and $\mathbf{f}_i$ w.r.t. the time stamped information of the user itinerary, that is the Euclidean distance of the closest 3DPs in time. Hence, the overall distance between the N-1 in length $\mathbf{q}$ and the N in length $\mathbf{c}$ is calculated as

$$\delta_{N-1}(\mathbf{q},\mathbf{c}) = \frac{1}{N-1}\sum_{\mathbf{e}_j \in \mathbf{q}} \delta'(\mathbf{e}_j,\mathbf{c}) \qquad (4)$$

Figure 1 depicts the process of predicting the next user movement considering the proposed C classifier. Specifically, once a query trajectory $\mathbf{q}$ arrives, then C attempts to classify $\mathbf{q}$ into a known $\mathbf{c}_i$ in the knowledge base w.r.t. Hausdorff metric. The C classifier returns the predicted $\mathbf{e}_N \in \mathbf{c}_i$ of the closest $\mathbf{c}_i$ to $\mathbf{q}$. Once such result refers to an unsuccessful prediction w.r.t. a preset error threshold θ then the C-T (or the C-nT) extend the $\mathbf{q}$ vector with the actual 3DP and insert $\mathbf{q}$ into the knowledge base for further learning according to the algorithm in Table 1 (feedback).

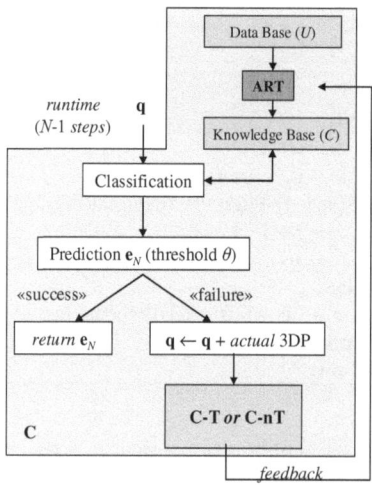

Fig. 1. The proposed adaptive classifier for location prediction

5 Prediction Evaluation

We evaluated our adaptive model in order to assess its performance. In our experiments, the overall user movement space has a surface of 540 km². Such space derives from real GPS trace captured in Denmark [6]. The GPS trace was fed into our model and the performance of the C system w.r.t. predefined metrics was monitored. Table 2 indicates the parameters used in our experiments.

Table 2. Experimental Parameters

Parameter	Value	Comment
Learning rate (n)	0.5	In case of a new pattern $\mathbf{u}_t$, the closest cluster $\mathbf{c}_i$ is moved toward $\mathbf{u}_t$ by half the spatial and temporal distance.
Spatial coefficient of vigilance (ρ_s)	100 m	Two 2D points are considered different if their spatial distance exceeds 100 meters.
Temporal coefficient of vigilance (ρ_t)	10 min	Two time-stamps are considered different if their temporal distance exceeds 10 minutes.
Precision threshold / location accuracy (θ)	10 m	The predicted location falls within a circle of radius 10 meters from the actual location.[2]

The GPS traces including 1200 patterns were preprocessed and we produced two training files and two test files as depicted in Figure 2. The first training file, *TrainA*, is produced from the first half of the GPS trace records. The second training file, *TrainB*, consists of a single trace record. The first test file, *TestA*, is produced from the entire set of trace records, including -in ascending order- the first half of the GPS traces and the other half of unseen traces. Finally the second test file, *TestB*, is produced from the entire set of the GPS trace records, including -in ascending order- the second half of unseen traces and the first half of the GPS traces. During the generation of the training/test files, white noise was artificially induced into the trace records.

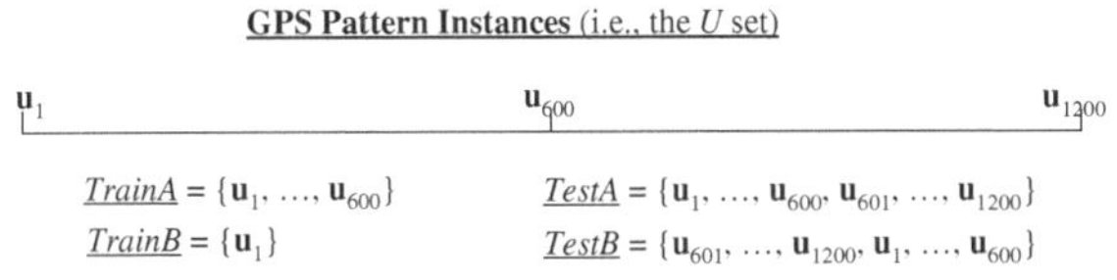

GPS Pattern Instances (i.e., the U set)

$\mathbf{u}_1$ $\mathbf{u}_{600}$ $\mathbf{u}_{1200}$

TrainA = {$\mathbf{u}_1$, ..., $\mathbf{u}_{600}$}   *TestA* = {$\mathbf{u}_1$, ..., $\mathbf{u}_{600}$, $\mathbf{u}_{601}$, ..., $\mathbf{u}_{1200}$}
TrainB = {$\mathbf{u}_1$}   *TestB* = {$\mathbf{u}_{601}$, ..., $\mathbf{u}_{1200}$, $\mathbf{u}_1$, ..., $\mathbf{u}_{600}$}

Fig. 2. The generated GPS trace files for experimentation

---

[2] Such accuracy level is considered appropriate for the kind of applications where location prediction can be applied (see the Introduction section or [12]).

We have to quantitatively and qualitatively evaluate the proposed model. For that reason, we introduce the following quantitative and qualitative parameters: (a) the precision achieved by the prediction scheme –the higher the precision the more accurate the decisions on the future user location- (b) the size of the underlying knowledge base –we should adopt solutions with the lowest possible knowledge base size (such solutions are far more efficient and feasible in terms of implementation) and (c) the capability of the model to rapidly react to changes in the movement pattern of the user/mobile terminal and re-adapt. We define *precision*, *p*, as the fraction of the correctly predicted locations, p_+, against the total number of predictions made by the C system, p_{total}, that is,

$$p = \frac{p_+}{p_{total}}$$

In the following sub-sections, we evaluate the diverse versions of the C classifier w.r.t. training methods by examining the classifier convergence (speed of learning and adaptation) and the derived precision on prediction future locations.

5.1 Convergence of *C-T* and *C-nT*

The C classifier converges once the knowledge base does not expand with unseen patterns, i.e., the set U does not evolve. In Figure 3, we plot the number of the clusters, $|U|$, that are generated from the C-T/-nT models during the testing phase. The horizontal axis denotes the incoming (time-ordered) GPS patterns. The point (.) marked line depicts the behavior of the C-T-1 model trained with *TrainA* and tested with *TestA*. In the training phase, the first 600 patterns of *TrainA* have gradually generated 70 clusters in U. In the testing phase, the first 600 patterns are known to the classifier so there is no new cluster creation. On the other hand, in the rest 600 unseen patterns, the number of clusters scales up to 110 indicating that the ART algorithm learns such new patterns.

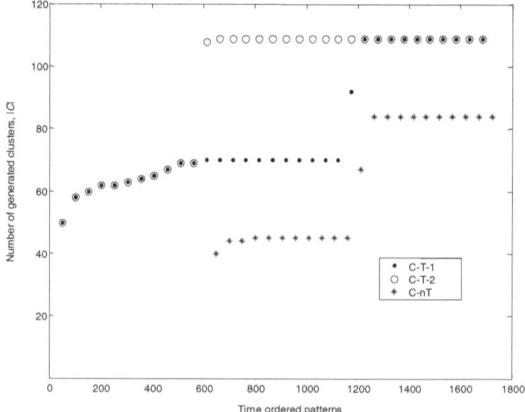

Fig. 3. Convergence of C-T/-nT

The circle (o) marked line depicts the C-T-2 model, which is trained with *TrainA* and tested with *TestB*. Since the train file is the same as in the C-T-1 model, the first generated clusters are the same in number ($|U| = 70$). In the testing phase, we observe a significant difference. ART does not know the second 600 unseen patterns, thus, it learns new patterns up to 110 clusters. In the next 600 known patters, C-T-2 does not need to learn additional clusters thus it settles at 110 clusters.

We now examine the behavior of the C-nT model corresponding to the zero-knowledge training method. The asterisk (*) marked line depicts the training phase (with *TrainB*) followed by the testing phase (with *TestA*) of C-nT. In this case, we have an incremental ART that does not need to be trained. For technical consistency reasons, it only requires a single pattern, which is the unique cluster in the knowledge base at the beginning. In the testing phase, for the first 600 unseen patterns of *TestA* we observe a progressive cluster creation (up to 45 clusters). For the next 600 unseen patterns, we also observe a gradual cluster creation (up to 85 clusters) followed by convergence. Comparing the C-T-1/2 and C-nT models, the latter one achieves the minimum number of clusters (22.72% less storage cost). This is due to the fact that C-nT starts learning only from unsuccessful predictions in an incremental way by adapting pre-existing knowledge base to new instances. Nevertheless, we also have to take into account the prediction accuracy in order to reach safe conclusions about the efficiency and effectiveness of the proposed models.

5.2   Precision of *C-T* and *C-nT*

In Figure 4 we examine the precision achieved by the algorithms. The vertical axis depicts the precision value p achieved during the testing phase. The point (.) marked line depicts the precision of the C-T-1 model trained with *TrainA* and tested with *TestA*. During the test phase, for the first 600 known patterns C-T-1 achieves precision value ranging from 97% to 100%. In the next 600 unseen patterns, we observe that for the first instances the precision drops smoothly to 95% and as C-T-1 learns, i.e., learn new clusters and optimize the old ones, the precision converges to 96%.

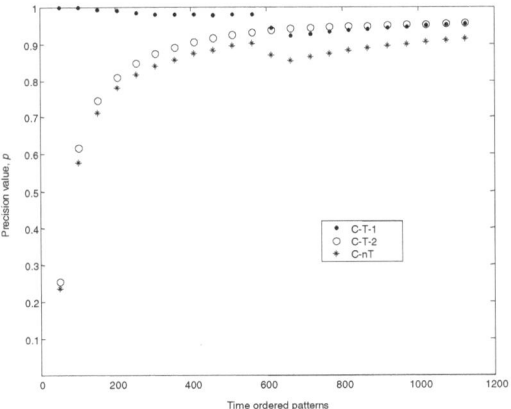

Fig. 4. Precision of C-T/-nT

The circle (o) marked line depicts the precision behavior for the C-T-2 model tested with *TestB* and trained with *TrainA*. With the first 600 totally unseen patterns during the test phase, C-T-2 achieves precision from 26% to 96%. This indicates that the model is still learning during the test phase increasing the precision value. In the next 600 known patterns, the model has nothing to learn and the precision value converges to 96%.

The asterisk (*) marked line depicts the precision behavior of the C-nT model tested with *TestA* and trained with *TrainB*. In this case, C-nT is trained with only one pattern instance, i.e., the algorithm is fully incremental, thus, all the instances are treated as unseen. In the test phase, for the first 600 patterns, the model achieves precision, which ranges from 25% to 91% In the next 600 patterns, we can notice that for the first instances the precision drops smoothly to 88% and as the model learns, precision gradually converges to 93%.

Evidently, the adoption of the training method, i.e., the C-T-1/-2 models, yields better precision. However, if we correlate our findings with the results shown in Figure 3, we infer that a small improvement in precision has an obvious storage cost. Specifically, we need to store 110 clusters, in the case of C-T, compared to 85 clusters in the case of C-nT (22.72% less storage cost). Furthermore, the user movement patterns can be changed repeatedly over time. Hence, by adopting the training method, one has to regularly train and rebuild the model. If the mobile context-aware application aims at maximizing the supported quality of service w.r.t. precision, while keeping the storage cost stable, the C-nT model should be adopted.

6   Comparison with Other Models

We compare the C-nT model with other known models that can be used for location prediction. Such models implement the Offline kMeans and Online kMeans algorithms. Such models require a predefined number of $k > 1$ initial clusters for constructing the corresponding knowledge base. We should stress here that, the greater the k the greater the precision value achieved by Offline/Online kMeans. In our case, we could set $k = 110$, which is the convergence cluster-count for the C models (Section 5). For C-nT, we use *TrainB* for the training and *TestA* for the testing phase (such model adopts the zero-knowledge training method). Moreover, for the Offline/Online kMeans models we use *TrainA* for the training and *TestA* for the testing phase because both models require $k > 1$ initial clusters.

Figure 5 depicts the precision achieved by the C-nT (the point (.) marked line), Offline kMeans (the asterisk (*) marked line) and Online kMeans (the circle (o) marked line) models. The horizontal axis represents the ordered instances and the vertical axis represents the achieved precision. We can observe in the first 600 patterns C-nT achieves precision levels ranging from 25% to 91% indicating adaptation to new knowledge. This is attributed to the learning mechanism (C-nT recognizes and learns new user movements). In the next 600 patterns we notice that for the first instances, the precision drops smoothly to 88% and as the knowledge base adapts to new movements and optimizes the existing ones, precision converges to 93%.

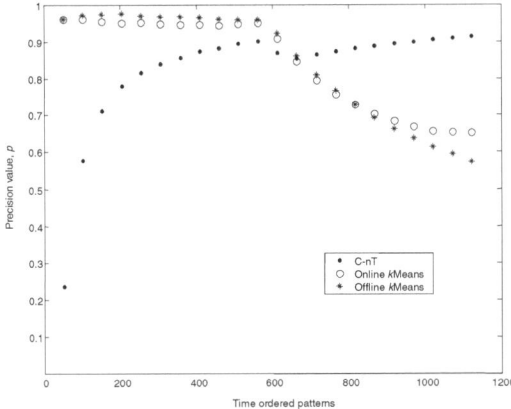

Fig. 5. Comparison of C-nT with the Offline/Online *k*Means models

In the case of Offline *k*Means, we observe that for the first 600 patterns, it achieves precision levels ranging from 96% to 98% once the initial clusters are set to $k = 110$. In the next 600 patterns we notice that the precision drops sharply and converges to 57% as the knowledge base is not updated by unseen user movements. By adopting Online *k*Means we observe that for the testing phase (the first 600 patterns) it achieves precision levels ranging from 94% to 97% given the train file *TrainA*. In the next 600 patterns we notice that for the first instances the precision drops rather smoothly to 86% and, as the knowledge base is incrementally adapting to new patterns, the precision value converges to 65%. Evidently, by comparing such three models, the most suitable model for location prediction is the C-nT since (i) it achieves greater precision through model adaptation and (ii) requires a smaller size of the underlying knowledge base (i.e., less clusters) than the Offline/Online *k*Means models.

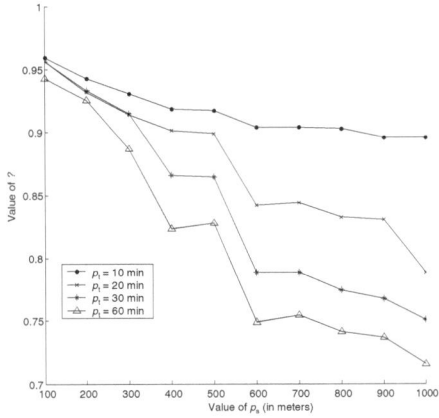

Fig. 6. The behavior of the γ parameter vs. temporal and spatial coefficients of the vigilance threshold

Up to this point we have concluded that the C-nT model achieves good precision with limited memory requirements, which are very important parameters for mobile context-aware systems. However, we need to perform some tests with C-nT in order to determine the best value for the spatiotemporal parameter vigilance ρ. In other words, we aim to determine the best values for both spatial ρ_s and temporal ρ_t vigilance coefficients in order to obtain the highest precision with low memory requirements. We introduce the weighted sum γ as follows:

$$\gamma = w \cdot p + (1 - w) \cdot (1 - a)$$

where a is the proportion of the generated clusters by the classifier (i.e., the size of the knowledge base in clusters) out of the total movement patterns (i.e., the size of the database in patterns), that is: $a = |C|/|U|$; $|C|$ is the cardinality of the set C. The weight value $w \in [0, 1]$ indicates the importance of precision and memory requirements; a value of $w = 0.5$ assigns equal importance to a and p. In our assessment, we set $w = 0.7$. We require that a assumes low values minimizing the storage cost of the classifier. A low value of a indicates that the applied classifier appropriately adopts and learns the user movements without retaining redundant information. The value of γ indicates which values of ρ_s and ρ_t maximize the precision while, on the same time, minimize the memory requirements. Hence, our aim is to achieve a high value of γ indicating an adaptive classifier with high value of precision along with low storage cost. As illustrated in Figure 6, we obtain a global maximum value for γ once $\rho_s = 100m$ and $\rho_t = 10min$ (which are the setting values during the experiments – see Table 2).

7   Prior Work

Previous work in the area of mobility prediction includes the model in [7], which uses Naïve Bayes classification over the user movement history. Such model does not deal with fully / semi- random mobility patterns and assumes a normal density distribution for the underlying data. However, such assumptions are not adopted in our model as long as mobility patterns refer to real human traces with unknown distribution. Moreover, the learning automaton in [8] follows a linear reward-penalty reinforcement learning method for location prediction. However, such model does not provide satisfactory prediction accuracy, as reported in [8]. The authors in [9] apply evidential reasoning in mobility prediction when knowledge on the mobility patterns is not available (i.e., similarly to this paper). However, such model assumes large computational complexity (due to the adopted Dempster-Schafer algorithm) once the count of possible user locations increases and requires detailed user information (e.g., daily profile, preferences, favorite meeting places). Other methods for predicting trajectory have also been proposed in the literature [10] but these have generally been limited in scope since they consider rectilinear movement patterns only (e.g., highways) and not unknown patterns. A closely related work to ours has been reported in [11], where a GPS system is used to collect location information. The proposed system then automatically clusters GPS data taken into meaningful locations at multiple scales. These locations are then incorporated into a similar Markov model to predict the user's future location. The authors in [16] adopt a data mining approach (i.e., rule extraction) for predicting user locations in mobile environments. This approach achieves prediction accuracy lower than ours (i.e., in the order of 80% for deterministic movement). In [17], the authors adopt a clustering method for the location prediction problem. Prediction accuracy is still low (in the order of 66% for deterministic movement). The authors in [18] introduce a framework where for each user an individual function is computed in order to capture its movement. This approach achieves prediction accuracy lower than ours (i.e., in the order of 70% for deterministic movement). In [19], the authors apply movement rules in mobility prediction given the user's past movement patterns. Prediction accuracy is still low (i.e., in the order of 65% for deterministic movement). The authors in [20] introduce a prediction model that uses grey theory (i.e., a theory used to study uncertainty). This approach achieves prediction accuracy lower than ours (i.e., in the order of 82% for deterministic movement).

8 Conclusions

We presented how ML techniques can be applied to the engineering of mobile context-aware applications for location prediction. Specifically, we use ART (a special Neural Network Local Model) and introduce a learning method. Furthermore, we deal with two training methods for each learning method: in the supervised method the model uses training data in order to make classification and in the zero-knowledge method the model incrementally learns from unsuccessful predictions. We evaluated our models with different spatial and temporal parameters. We examine the knowledge bases storage cost (i.e., emerged clusters) and the precision measures (prediction accuracy). Our findings indicate that the C-nT model suits better to context-aware systems. The advantage of C-nT model is that (1) it does not require pre-existing knowledge in the user movement behavior in order to predict future movements, (2) it adapts its on-line knowledge base to unseen patterns and (3) it does not consumes much memory to store the emerged clusters. For this reason, C-nT is quite useful in context-aware applications where no prior knowledge about the user context is available. Furthermore, through experiments, we decide on which vigilance value achieves the appropriate precision w.r.t. memory limitations and prediction error. Finally, in the Neural Networks Local Models literature there are other models (e.g., Self-Organizing Maps) that we have not examined in this paper. We intent to implement and evaluate them with C-nT by means of knowledge base requirements, precision of the location prediction and adaptation.

References

1. Dey, A.: Understanding and using context. Personal and Ubiquitous Computing 5(1), 4–7 (2001)
2. Hightower, J., Borriello, G.: Location Systems for Ubiquitous Computing. IEEE Computer 34(8) (August 2001)
3. Alpaydin, E.: Introduction to Machine Learning. MIT Press, Cambridge (2004)
4. Duda, R., Hart, P., Stork, D.: Pattern Classification. Wiley-Interscience, Hoboken (2001)

5. Belogay, E., Cabrelli, C., Molter, U., Shonkwiler, R.: Calculating the Hausdorff Distance between Curves. Information Processing Letters 64(1), 17–22 (1997)
6. Site, http://www.openstreetmap.org/traces/tag/Denmark
7. Choi, S., Shin, K.G.: Predictive and adaptive bandwidth reservation for hand-offs in QoS-sensitive cellular networks. In: ACM SIGCOMM (1998)
8. Hadjiefthymiades, S., Merakos, L.: Proxies+Path Prediction: Improving Web Service Provision in Wireless-Mobile Communications. ACM/Kluwer Mobile Networks and Applications, Special Issue on Mobile and Wireless Data Management 8(4) (2003)
9. Karmouch, A., Samaan, N.: A Mobility Prediction Architecture Based on Contextual Knowledge and Spatial Conceptual Maps. IEEE Trans. on Mobile Computing 4(6) (2005)
10. Viayan, R., Holtman, J.: A model for analyzing handoff algorithms. IEEE Trans. on Veh. Technol. 42(3) (August 1993)
11. Ashbrook, D., Starner, T.: Learning Significant Locations and Predicting User Movement with GPS. In: Proc. Sixth Int'l Symp. Wearable Computes (ISWC 2002), October 2002, pp. 101–108 (2002)
12. Priggouris, I., Zervas, E., Hadjiefthymiades, S.: Location Based Network Resource Management. In: Ibrahim, I.K. (ed.) Handbook of Research on Mobile Multimedia. Idea Group Inc. (May 2006)
13. Curewitz, K.M., Krishnan, P., Vitter, J.S.: Practical Prefetching via Data Compression. In: Proceedings of ACM SIGMOD, pp. 257–266 (1993)
14. Narendra, K., Thathachar, M.A.L.: Learning Automata – An Introduction. Prentice Hall, Englewood Cliffs (1989)
15. Cheng, Jain, R., van den Berg, E.: Location prediction algorithms for mobile wireless systems. In: Wireless Internet handbook: technologies, standards, and application, pp. 245–263. CRC Press, Boca Raton (2003)
16. Yavas, G., Katsaros, D., Ulusoy, O., Manolopoulos, Y.: A data mining approach for location prediction in mobile environments. Data and Knowledge Engineering 54(2) (2005)
17. Katsaros, D., Nanopoulos, A., Karakaya, M., Yavas, G., Ulusoy, O., Manolopoulos, Y.: Clustering Mobile Trajectories for Resource Allocation in Mobile Environments. In: Berthold, M.R., Lenz, H.-J., Bradley, E., Kruse, R., Borgelt, C. (eds.) IDA 2003. LNCS, vol. 2810, pp. 319–329. Springer, Heidelberg (2003)
18. Tao, Y., Faloutsos, C., Papadias, D., Liu, B.: Prediction and Indexing of Moving Objects with Unknown Motion Patterns. In: ACM SIGMOD (2004)
19. Nhan, V.T.H., Ryu, K.H.: Future Location Prediction of Moving Objects Based on Movement Rules. In: ICIC 2006. LNCIS, vol. 344, pp. 875–881. Springer, Heidelberg (2006)
20. Xiao, Y., Zhang, H., Wang, H.: Location Prediction for Tracking Moving Objects Based on Grey Theory. In: IEEE FSKD 2007 (2007)

MPM: Map Based Predictive Monitoring for Wireless Sensor Networks[*]

Azad Ali, Abdelmajid Khelil, Faisal Karim Shaikh, and Neeraj Suri

Technische Universität Darmstadt, Hochschulstr. 10, 64289 Darmstadt, Germany
{azad,khelil,fkarim,suri}@informatik.tu-darmstadt.de

Abstract. We present the design of a Wireless Sensor Networks (WSN) level event prediction framework to monitor the network and its operational environment to support proactive self* actions. For example, by monitoring and subsequently predicting trends on network load or sensor nodes energy levels, the WSN can proactively initiate self-reconfiguration. We propose a Map based Predictive Monitoring (MPM) approach where a selected WSN attribute is first profiled as WSN maps, and based on the maps history, predicts future maps using time series modeling. The "attribute" maps are created using a gridding technique and predicted maps are used to detect events using our regioning algorithm. The proposed approach is also a general framework to cover multiple application domains. For proof of concept, we show MPM's enhanced ability to also accurately "predict" the network partitioning, accommodating parameters such as shape and location of the partition with a very high accuracy and efficiency.

Keywords: Predicitve Monitoring, Time Series Analysis, Wireless Sensor Networks, Event Prediction.

1 Introduction

Wireless Sensor Networks (WSN) typically entail an aggregation of sensing/ communicating sensor nodes to result in an ad hoc network linking them to the base station or sink. The sensor nodes typically possess limited storage and computational capabilities and require low-energy operations to provide longevity of operational time.

WSN's are often used for monitoring of spatially distributed attributes such as detection of physical events, e.g., fire, temperature gradients and high/low pressure. To maintain the required WSN dependability, network events such as partitioning of the network are also to be detected. The reporting of such events beyond simple monitoring becomes highly useful if these events can be predicted in advance. Consequently, appropriate autonomic actions could be taken either to avoid or delay events from happening by triggering self* actions.

Varied works exist for event detection [1] such as fire detection [2] and network partitioning [3, 4, 5]. Most of these efforts are specialized for specific scenarios.

---

[*] Research supported in part by HEC, MUET, EC INSPIRE, EC CoMiFiN, and DFG GRK 1362 (TUD GKMM).

A.V. Vasilakos et al. (Eds.): AUTONOMICS 2009, LNICST 23, pp. 79–95, 2010.

Some consider generic scenarios [6], but they suppose the event to take certain shapes and patterns. Also most efforts focus on detecting the events after they have already happened. In the perspective of reacting to the event, it may be too late to detect the *events* after it has already taken place. It would be far more practical if we could predict the event occurrence.

Multiple efforts exists for predictions [7,12,8,11,13]. However, most are either limited to predicting specifically a certain attribute like energy or provide only node level short-term prediction for data compression to minimize data to be reported from the network.

It is natural to combine prediction of physical and network events with generalized event detection to take proactive actions. To the best of our knowledge there exists no work that proposes generalized event prediction. In this paper we develop a framework to predict the future states of the network for the attribute of interest such as temperature or residual energy. Based on the predicted states of the network we develop a generalized event detection technique. In this work we target long-term predictions that require more computational resources than a sensor node has and sufficient history of the attribute that contains all variation patterns. Hence, we collect many profiles of the network and perform the modeling operations on the sink. Typically, the sink uses optimized data collection techniques to collect such a history of the interested attributes from the network [8]. We refer to the collection of attribute values or samples from network as *profiling*.

In the WSN environment, events usually are defined as spatially correlated attribute distributions [6]. Thus, the spatial distribution of attribute of interest needs to be quantified in order to detect events. One such natural spatial quantification is the *Map*. For a WSN an eMap is an energy map that represents the current residual energy of the network [9], or tMap for temperature etc. Maps can generally be created for any attribute and provide a basic utility to detect events using pattern matching [6]. The use of maps in our work emphasizes the fact that there is a wide class of events in which a discrete view of a network on node level is either not necessary or not efficient. Rather, a view at a higher abstraction level of regions and maps, that represent a group of nodes, is needed. For example in temperature, pressure, humidity and residual energy monitoring a map is far more meaningful than discrete node values. This also relates to the inherent WSN node redundancy that leads to spatially correlated node states. On this background this paper presents three specific contributions, namely

- Design of a generalized framework for sink aided profile prediction. The framework is adaptable to different simple as well as complex physical and network events.
- Development of a regioning technique to detect events from predicted profiles to support autonomous actions in the specified regions of the network.
- As validation for our Map based Predictive Monitoring (MPM) approach, we propose a solution of network partition prediction as a case study.

The remainder of the paper is organized as follows. Section 2 discusses the related work. Section 3 gives the preliminaries. Section 4 details our MPM approach for predictive monitoring in WSN. In Section 5, the case study is given. In Section 6, we evaluate our approach for the given case study through simulations. Section 7 concludes our work and outlines future directions.

2 Related Work

A variety of work is available for event detection [1]. The most relevant work to our event detection strategy is [6] that investigates map based event detection and requires the user to (a) specify the distribution of an attribute over space and (b) the variation of distribution over time incurred by the event. They further define three common types of events namely Pyramid, Fault and Island. Our work is independent of event shape due to our regioning technique. Furthermore, our framework predicts the events rather than just detecting it. In [10], Banerjee et al. present a technique to detect multiple events simultaneously. They employ a polynomial based scheme to detect event regions with boundaries and propose a data aggregation scheme to perform function approximation of events using multivariate polynomial regression. Our work in addition to the capability of detecting multiple events, can predict events beforehand. Various other works exist that address specific event scenarios like partition detection [3], fire detection [2] and others [1]. These specific solutions do not feature portability to adapt to different application scenarios.

There is a variety of work to monitor WSN's and to minimize the overhead of data collection for the monitoring. Strategies in [7] and [11] predict the power consumption in WSN. In [12] Mini et al. propose a network state model and use it to predict the energy consumption rate and construct the energy map accordingly. In [13], the authors focus on predicting energy efficiency of multimedia networks. These works concentrate on predicting specifically the energy, also they do not provide any extension of their work to attributes other than energy. Authors in [14] give a theoretical framework for abstracting the world as WSN maps. Our work in this paper however presents a practical approach not only for abstraction but also facilitates to predicts the events to take place in future.

As we present a case study for partition prediction, hence here we discuss the related work in this respect. In [3], this problem has been addressed for a sub-class of linearly separable partitions, i.e., cuts. Memento [4] is a health monitoring system that suggests to continuously collect connectivity information at the sink to be able to detect network partitioning. The Partition Avoidance Lazy Movement protocol for mobile sensor networks [5] is a decentralized approach, where a sensor node can locally suspect network partitioning and move to avoid it. A node periodically collects the position of all its neighbors and checks if at leat one neighbor is located in a small angle towards the sink. If no neighbor is located in this "promising zone", the node suspects network partitioning. Based on our event prediction framework as an example we propose a solution that is generalized and is not dependent on the shape, size or location of the partition. Moreover, we provide prediction of the time, when network partitioning is going to happen.

3 Preliminaries

We now describe the system model, the requirements driving our approach and give basic definitions.

3.1 System Model

We consider a WSN composed of N static sensor nodes and one static sink. Sensor nodes are battery powered and usually entail limited processing and storage capabilities. Sensor nodes are assumed to know their geographic position either using distributed localization methods [15] or GPS. A typical WSN deployment may contain hundreds of sensor nodes with varying densities according to the coverage requirements. In this work, we do not consider a particular node distribution. We assume all sensor nodes to be homogeneous. Hence the nodes have the same transmission range R and same initial battery capacity. We consider that nodes crash due to energy depletion only. We assume the events for predictions to be happening over a longer period of time, for example, events that may take hours, days or even months to develop. We consider that events are not spontaneous, spatially correlated, do not depend discretely on a single node and are predictable.

3.2 Requirements on the MPM

We identify the following three requirements on the MPM. First, the MPM should be *lightweight*, i.e., its creation, management and usage require minimal resources with respect to energy. Second, we desire the MPM to *long-term* predict the network status and hence the events *accurately*. Depending on the context of the problem, long-term may mean hours to days or even months that should be enough for the preventive mechanism to activate a self* mechanism to support autonomic actions. Third, we desire the framework to be *generalized* to adapt to prediction of varied event types.

3.3 Definitions

Here we give some basic definitions that are necessary to develop the MPM framework.

Definition 1. *A* time series *is a sequence of data points x_t considered as a sample of random variable $X(t)$, typically measured at successive times. The time series can be modeled to predict future values based on past data points.*

Definition 2. *A* stationary random process *exhibits similar statistics in time, characterized as constant probability distribution in time. However, it suffices to consider the first two moments of the random process defined as weak stationary or wide sense stationary (WSS) as follows:*

1. *The expected value of the process ($E[X(t)]$) does not depend on time. If $m_x(t)$ is the mean of $X(t)$ then*
   $$E[X(t)] = m_x(t) = m_x(t + \tau) \; \forall \tau \in \Re$$

2. *The autocovariance function for any lag τ is only a function of τ not time t*
$$E[X(t_1)X(t_2)] = R_x(t_1, t_2) = R_x(\tau, 0) \ \forall \tau \in \Re$$

Definition 3. *$X(t)$ is an Autoregressive Moving Average Process $ARMA(p, q)$ process of order (p, q) $p, q \in \aleph$, if $X(t)$ is WSS and $\forall$ t,*

$$X(t) = \phi_1 X_{t-1} + \cdots + \phi_p X_{t-p} + \theta_1 Z_{t-1} + \cdots + \theta_q Z_{t-q} \tag{1}$$

where Z_t is white noise with mean zero and variance σ^2, denoted as $WN(0, \sigma^2)$.

Definition 4. *To quantify the continues space of WSN and construct the map a grid is virtually placed over the WSN field and each grid cell represents the aggregated attribute of all the nodes located within the grid cell. We define the process of network space quantification as* Segmentation *and resultant quantification as* Grid Map.

Definition 5. *The quantification of WSN space and the conversion of a WSN to a map level abstraction is the key to detecting generic events. The abstraction of WSN as map, transforms the event into a region of a map as shown in Fig. 1. For example an event of fire will be a region of a map in which the value of temperature exceeds a given threshold. In our framework we define an event as a region of map whose values fall in the range of attribute values for which event is defined.*

4 Predictive Monitoring: The MPM Approach

We present here our MPM Framework that can be used to support self* actions. To keep the event prediction as generic as possible, we have proposed it as a four phases process. In each phase we have proposed a technique that is independent of the attribute to be monitored. It is important to highlight that the use of proposed techniques in the framework does not imply to limit the framework to only these, rather for a particular implementation specialized techniques can always be easily accommodated due to its modular structure. The four phases of the framework are summarized here. *The segmentation* phase specifies the properties of the grid maps (Def. 4) such as grid cell size. In *the data collection* phase, data is periodically but efficiently fetched from the network on the sink. *The prediction* phase is used for predicting future status of the network in the form of future grid maps. *The event detection* phase is used to detect events (Def. 5) in the predicted grid maps. These phases are individually detailed in the following sections.

4.1 The Segmentation Phase

In order to reach an acceptable spatial resolution with higher level abstraction of network as a map, we considered virtual grids and Voronoi diagram [16] techniques to segment (Def. 4) WSN space. Voronoi-based segmentation depends only on sensor node distribution and is static for a given node distribution. However, we require a segmentation strategy that allows variable spatial sampling to accommodate both the physical and network parameters. Such variability allows to investigate prediction accuracy and profiling efficiency tradeoffs. Grid allows such flexibility therefore, we base our segmentation on grid.

The virtual grid or simply *grid* divides the WSN area into fixed size squares or grid cells as shown in Fig. 1. Thus nodes that fall within a cell are grouped.

For the grid maps construction, two parameters must be specified. The first parameters is the grid cell size γ, which is a spatial sampling or resolution parameter. The second parameter is the aggregation value ξ that a grid cell represents. Both parameters are essential for event detection. γ defines the geographic area covered by the grid cell. The number of nodes being grouped in a grid cell is dependant on γ. It can also be seen as a zooming parameter. Hence it can be used to decide at which level the user intends to detect the event, i.e., very detailed (zoomed-in) level of node or an overview at the level of regions. Depending on the application the appropriate value of γ is affected by (1) physical parameters such as attribute's spatial distribution, (2) network parameters, such as communication range, (3) application requirements such as the (zoom) level at which to detect the event. In applications such as temperature and humidity, the grid size can be selected big enough that it represents the patches of the geographic areas, each differing considerably in the attribute values.

The grid cell value ξ is an aggregate of the attribute values of the set of nodes in a cell. The choice of the exact function depends on the application. For example to sense temperature or pressure, it is most appropriate to average the values of the nodes in the grid cell. If ξ_{ij} is the grid cell value in the $(i,j)^{th}$ grid cell g_{ij} and v_n represents attribute value of node n in g_{ij} then ξ_{ij} is an aggregation function such as average, min, max of v_n

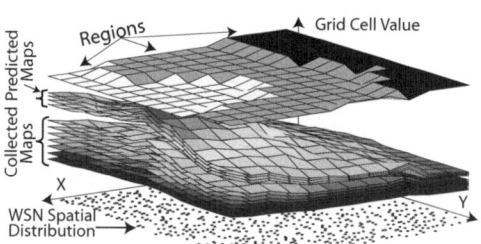

Fig. 1. Temporal stack of the grid maps

$$\xi_{ij} = f(v_n) \ \forall \ n \in g_{ij} \tag{2}$$

During map construction the nodes undergo coordinate transformation. If a node has (x, y) coordinates then the grid cell coordinates (i, j) can be calculated as

$$i = \lfloor x/\gamma \rfloor, j = \lfloor y/\gamma \rfloor \tag{3}$$

All the nodes in the area of a certain grid cell have same grid cell coordinates.

We do not impose assumptions on the selection of γ and f, highlighting the generality of our framework (requirement on our framework). An illustration for the selection of both parameters is given in the case study in Section. 5.

4.2   The Data Collection Phase

In this phase the profiles are obtained by collecting the attribute values from the network for which the event is to be predicted. The grid technique also aids in reducing the number of the reporting nodes by forming clusters of nodes. Once the grid size has been specified (and disseminated) in the network, the nodes calculate their corresponding grid cell using coordinate transformation. Consequently, in-network aggregation can be performed according to f by the nodes within the same grid cell. Only single node should then report ξ to the sink. This reduces the amount of data to be reported to the sink enormously. Simple criteria can be used for selection of reporting node such as, node with maximum energy. It is not efficient even for reporting nodes to send the ξ values periodically to the sink. To further reduce the overhead on the network as per our design requirements, the aggregated attribute are modeled as in [8] and only the model parameters are sent to the sink, which can be used to regenerate the actual data. The model is recalculated only when the prediction error exceeds a predefined threshold. With each model update the outliners of the last model are also reported. To achieve this the reporting nodes maintain a limited history of the aggregated attribute values to be reported to the sink and fits 3rd order autoregressive model ($AR3$) which is only a particular case of the $ARMA(p, q)$ model (Def. 3), when $p = 3$, $q = 0$. This data compression approach uses fixed simple model because nodes have too low computational resources to determine the parameters of a general ARMA model.

As per requirement on our framework we emphasize the generality of MPM with respect to data collection, since we abstract the attribute type as a generic time series (Def. 1). Using this technique any attribute type can be reported to sink with a comparable efficiency, making our framework independent of the attribute type to be reported. We fulfill our first requirement of framework to be lightweight using the segmentation and data collection techniques. These techniques collectively reduce the overhead on the network tremendously.

4.3   The Prediction Phase

The models received on the sink in data collection phase from each reporting node are used to regenerate the variation patterns or attribute history through reverse transformation. The regenerated history is essentially the grid map representation of the WSN. This forms a temporal stack of the grid maps as shown in Fig. 1. Each grid cell in the grid maps stack can be treated and modeled as a separate time series for prediction. Individual models of each grid cell can then be used to predict future values by fitting a prediction model, effectively predicting grid maps. The time series can be modeled in different ways [13]. In this paper, we use the widely used time domain modeling because of its general applicability.

It is important to point out that time series modeling performed in this section is different from that done on node level for data collection, which is only short-term prediction to compress the data using the fixed $AR3$ model. In this phase, we perform a full scale modeling of the collected data to predict the future states using the complete history and model each component separately.

Modeling Time Series. A time series $X(t)$ can be modeled as a process containing following components

$$X(t) = T_t + S_t + R_t \tag{4}$$

where T_t is a trend, S_t is a function of the seasonal component with known period, and R_t is the random noise component. To keep the notion of generality valid for the framework we use a well known generalized technique Box-Jenkins Model to model a time series containing any of these components.

Box-Jenkins (BJ) Model. Box-Jenkins model predicts a time series by fitting it an ARIMA process (Autoregressive Integrated Moving Average). The term integrated here means differencing the series to achieve stationarity (Def. 2). To fit an ARIMA process the model and the order of the model needs to be specified. The BJ model provides a guideline to select the appropriate model, i.e., either Autoregressive (AR, Eq. 5.1) or Moving Average (MA, Eq. 5.2)

$$X(t) = \phi_1 X_{t-1} + \cdots + \phi_p X_{t-p} \ \{5.1\}, \ X(t) = \theta_1 Z_{t-1} + \cdots + \theta_p Z_{t-q} \ \{5.2\} \tag{5}$$

or combination of both, i.e., ARMA process as given in Eq.1. It also gives the guideline for the model order selection. BJ modeling is a four steps procedure:

i.) Data Preparation: As Box-Jenkins model requires a time series to be stationary (Def. 2). If it contains trends and seasonal components then these should be appropriately removed. This can be achieved by either Least Square Polynomial Fitting (LSPF) or differencing as $X(t) = X(t) - X(t + u)$. For a simple linear trend, u is 1. For higher order trends or seasonal component of period s, u equals s. This operation is repeated until stationarity is achieved.

ii.) Model Identification: At this stage run-sequence plot or Autocorrelation Function (ACF) can be used to identify the stationarity of the time series and the order of the AR model. ACF for k lag is given by

$$\rho_k = \frac{\sum_{i=1}^{N-k} \left(X_i - \bar{X} \right) \left(X_{i+k} - \bar{X} \right)}{\sum_{N}^{i=1} \left(X_i - \bar{X} \right)^2} \tag{6}$$

where $\bar{X}$ is the mean value. Non-stationarity is often indicated by an ACF plot with very slow decay. Order of the AR and MA models are determined with the help of ACF and Partial Autocorrelation Function (PACF) [17]. To automate the model selection process either Akaike's Information Criterion (AIC) or Akaika's Final Prediction Error (FPE) [18] can be used. Various models can be computed and compared by calculating either AIC or FPE. The least value of AIC or FPE ensures the best fit model.

iii.) Parameter Estimation: In this step the values of the ARMA model coefficients that give the best estimate of the series are determined. Iterative techniques are used for model parameter estimation [18].

iv.) Prediction: Once the modeling is complete, it is simple to predict the series values using the estimated model. It comprises of calculating the future values at next time instances and reversing all the transformations applied to the series in phase 1 for data preparation.

To fulfill our second requirement of long-term prediction and to accommodate any variation patterns in the time series we use a generic time series modeling techniques that models each component (trend, seasons, random) to facilitate long-term predictions. We keep the generality of framework valid also in this phase by abstracting the attribute to be predicted as time series, which makes this phase also independent of attribute type.

4.4   The Event Detection Phase

From Def. 5 we know that events appears as regions in a map. We introduce here a generalized regioning technique that can detect the regions and their perimeters which leads to generic event detection.

The regions are formed due to the fact that the attribute values fall into certain class of values. For example we normally classify the temperature as freezing, low, normal, high or very high. These classes also contain event class (range of values belonging to event such as temperature above 90° for fire). This gives us a more acceptable abstraction than the exact values themselves. Therefore the thresholding of values into classes becomes logical representation for event detection. Thus to detect these events we define the *class maps* that thresholds the exact values of the cells in grid map with their class denominations. If we define class map values C as $c_1, c_2, \cdots$ for the range of the values of grid cell g_{ij} between $(\xi_2, \xi_1]$ and $(\xi_3, \xi_2] \cdots$ respectively, then a class map value is defined by

$$C = \begin{cases} c_1 & \text{if } \xi_2 < \xi_{ij} \leq \xi_1 \\ c_2 & \text{if } \xi_3 < \xi_{ij} \leq \xi_2 \\ \cdots \end{cases} \tag{7}$$

Our regioning algorithm (Fig. 2) takes the grid map as input and determines perimeter and regions belonging to different classes and hence events. We refer to the resultant output as the *regions map*. The grid cells in the grid map that belong to the same class are grouped to form the regions. The regioning technique essentially needs a class map to group all the same class cells and determine the boundary. The process of converting to class map and determining the regions boundary are both carried out concurrently. In order to merge the cells into regions, we define attribute classes as in Eq. 7. Neighboring cells are merged to form the same region if they belong to the same class. The definition of attribute classes and fusion of same class grid cells makes the regioning algorithm independent of the shape that a region takes or the number of regions (hence the number of events) in the map.

The algorithm starts by defining all the cell as *not assigned a region* by initializing the variable regionsMap[]=-1. The algorithm next searches a grid cell that has not been assigned a class yet and lists it as the border of the region, as

1:(mapX, mapY)=dimensions(map);	15: for each neighbor in neighborsList[] do
2:regionsMap[]= -1;	16: if (currentCell & neighbor are in the same class)
3:mapBorders[][];	& neighbor in regionsMap[]==-1 then
4:while there is cell not assigned region yet do	17: neighbor in regionsMap[]=regionId;
5: regionBorder[]= (Find cell with regionsMap=-1)	18: include cell in the newRegionBorder[];
6: dilateRegion(map,regionBorder[],regionsMap[],regionId)	19: changeInBorder=1;
7: mapBorders[regionId][]=regionBorder;	20: end if
8: regionId++;	21: end for
9:end while	22: if currentCell and (1 or more) neighbors are not
10:dilateRegion(map,regionBorder[],regionsMap[],regionId)	in the same class then
11:repeat	23: add cell to newRegionBorder[];
12: changeInBorder=0;newRegionBorder[]=0;	24: end if
13: for (currentCell=each cell in regionBorder[]) do	25: end for
14: neighborsList[]= eightNeighborsOf(currentCell);	26: regionBorder[]=newRegionBorder[];
	27:until changeInBorder

Fig. 2. Regioning algorithm

the region itself and region border at this moment consists of a single cell (line 5, cell with -1 in regionsMap is not assigned a region yet). Algorithm then starts expanding/*dilating* the region (line 6). To expand the region, the neighboring eight cells around the this region cell are checked if they already belong to the any class (line 14-16), if not then they are also classified according to Eq. 7. If they belong to the same region they are assigned the same region ID and the new qualifying cells are listed as the region border, otherwise the previous cells retain their status as region border (line 17-18). To further expand the region neighboring cells of each cell in the border cells are searched iteratively until no change occurs in the border of the region (line 11,27), which implies the completion of the construction of a single region with its boundary. The whole process repeats again by searching a new cell that has not been assigned a region yet. It keeps on repeating until all the cells in the map are classified into their corresponding regions (line 4,9).

We maintain the generality of the framework by devising a technique that does not depend on shape, size or the number of events occurring in the WSN.

5   Case Study: MPM Adaptation for Predicting Network Partition

To use our framework for network partition prediction, the problem needs to be formulated according to the abstractions (maps, classes etc.) in the framework.

5.1   Problem Formulation

Partition detection is a complex problem as a physical and a network parameter are being coupled, i.e., energy level of the nodes and communication range necessary to maintain connectivity. Given that sensor nodes are resource constrained, eventually a WSN has to consider the depletion of node batteries leading to the partitioning of the network. The energy dissipation however, is generally spatially correlated. Therefore, groups of nodes form hotspots that deplete to coverage holes. A hole can be defined as a part of the network, which due to the energy depletion is no longer covered. These holes can sometimes disconnect a part of network from accessing the sink defined as a partition.

If the network energy state can be modeled and predicted, then we can predict the occurrence of the holes and consequently the partitions. The holes and partitions appear as regions in an energy map. Our framework has all the tools to profile the energy dissipation patterns, predict the network future energy state and detect the regions formed due to partitioning. Therefore, partition prediction becomes a natural candidate problem to be solved using our framework.

We can now define the problem according to the abstraction of our framework. A grid cell gets disconnected from the network if it has energy below a minimum threshold so that it can not communicate anymore. These depleted grid cells form a region that represents a hole in an eMap. Partition however, is a group of non-depleted grid cells that can not access sink due to the holes. It is therefore sufficient to profile the energy status of the network during its lifetime by collecting the eMaps in order to predict network partitioning. As per definition the adaptation of the MPM framework to predict network partitioning consists of four phases that we discuss as follows.

5.2   The Segmentation Phase

The first step towards the abstraction of the WSN network as a grid map (eMap in this case) is the selection of resolution ,i.e., grid cell size at which this event (network partitioning or holes) is to be detected. From the formulation of the problem we know that we have two coupled parameters, i.e., energy and communication range. Therefore, an upper bound for γ is the communication range (R). To accommodate a worst case scenario of two nodes lying on opposite corners of two grid cells the $\gamma < R/2\sqrt{2}$, as shown in Fig. 3. The lower bound can be obtained from

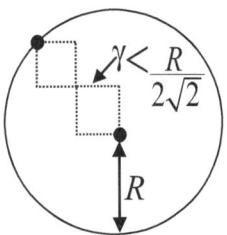

Fig. 3. Max grid size

the node density, it should be selected such that the network area is not over sampled, as we show in simulation Section. 6.2.

5.3   The Data Collection Phase

Once the grid cell size is specified, the nodes determine their corresponding grid cells using Eq. 3. As we discussed earlier energy dissipates in a spatially correlated manner, the hotspot energy dissipation model is considered [9]. Therefore, the nodes in a grid cell are expected to have similar energy dissipation pattern. A cell is connected to the network until at least a single node has enough energy to communicate. The node having the highest energy level is selected as reporting node and Eq. 2 becomes

$$\xi_{ij} = max(v_n) \ \forall \ n \in g_{ij} \tag{8}$$

All reporting nodes start aggregating the energy values. $AR3$ model is fitted to the data as per the scheme given in Section. 4.2. Model parameters are sent to the sink. The sink regenerates the time series (data of reporting sensor node) by applying reverse transformation. The regeneration of the data of the reporting nodes actually generates the grid maps according to the given parameters.

5.4   The Prediction Phase

The time series of each grid cell is modeled and predicted as described in Section. 4.3. Energy dissipation is a decaying process so the time series contains trends but no seasonal components. The trends are removed by fitting polynomials. ARMA models are fitted to random components, selecting the best fit model using AIC criteria. After completion of modeling the grid cell values are predicted and hence the future grid maps.

5.5   The Event (Holes/Partition) Detection Phase

The regioning algorithm developed in Section. 4.4 is used for both partition and hole detection. As per the given scheme we define two energy classes at 10% and below as the partition (or hole) class, and above 10% as non-partition class. This definition of energy classes gives the areas that are vulnerable to partitioning because of low energy. The regioning algorithm detects these regions along with the perimeter. The regions with energy 10% and below are holes in the network. To detect partitioning however, the algorithm does not need to find all the regions, therefore it executes only two iterations. In the first iteration with two energy classes it starts from the area connected to the sink and merges the non partitioned area around it. In the next iteration the classes are omitted and rest of the area is merged to find its perimeter that represents the partitioned region.

6   Evaluation – Viability of Our Approach

To evaluate how well our framework meets to design requirements, we evaluate it for the problem of partition prediction as formulated in the case study. To determine accuracy and efficiency of MPM we compare it with ideal case situation in which the data from all nodes is assumed. In the ideal case we predict the future energy states of every node separately and hence the future profiles of the network. The future profiles are then converted to maps. We denote these maps predicted using profiles of all the nodes as ideal grid maps Gi. We denote maps generated through our approach as optimized grid maps Go.

6.1   Evaluation Metrics

The transformation of a value spatial distribution into a map is a three stage process, i.e., a grid map, then a class map and finally a regions map. The regions map is physically same as a class map with additional information of region perimeters. Hence, we use two error criteria for the grid map and the class map. We use two more metrics to assess the accuracy of regions and efficiency in terms of number of fetched packets from the network. Our first metric is the **mean sum of absolute error** *(Eq. 9.1)* between the reference grid map Gr (the actual data generated on the nodes) and the test grid map $(Gi$ and $Go)$, defined as

$$Ge = \frac{\sum_i \sum_j abs(\xi r_{ij} - \xi t_{ij})}{m}\{9.1\}, Ce = \sum_i \sum_j count(Cr_{ij} - Ct_{ij})\{9.2\} \quad (9)$$

where (i, j) are grid cell coordinates, Ge is the mean sum of absolute error, ξr_{ij} is the grid cell value of reference map and ξt_{ij} is the grid cell value of test grid map and m is the number of occupied grid cells. Gr is the true data generated on the nodes, while Gi and Go are the gathered data from network, which undergo local modeling and hence will deviate from true data due to modeling. Ge determines the relative accuracy of our approach against the ideal case.

The second metric **misclassification cell count** (Eq. 9.2) counts the number of misclassified cells between the reference and the test class map. Ce is the total count of class cells that differ between the reference Cr and test class map Ct (Ct are ideal class map Ci and optimized class map Co). 'count' function returns '1' if the two cells do not belong to the same class else it returns '0'. Ce is the direct measure of correct classification of the grid cells into the classes and indirect measure of the accuracy of area and perimeter of the detected event area. Our third metric is the misclassified cells percentage for each region to assess the accuracy the framework on regions level that we call **regional percentile error**.

The fourth metric **message count** is the efficiency metric, where we count the messages required to profile the network.

6.2 Simulation Settings

As three phases of the framework are carried out on the sink, therefore we performed our simulations on Matlab. It is a very well known simulation tool and suits our work as it facilitates to model energy dissipation patterns of very huge number of nodes. The network that we used in our simulations is generated as a random non-uniform distribution of nodes. The node distribution, as shown in Fig. 4, was selected to cover many

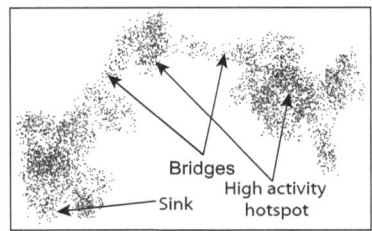

Fig. 4. Node distribution

possible scenarios in a real deployment. It contains some areas with high node density and some with low node density. It also contains two narrow bridges between two parts of the network that may lead to network partitioning. For energy dissipation modeling the common hotspot model [9] was used. The energy dissipates in a spatially correlated manner around the hotspot. The nodes nearest to the hotspot are more active and hence dissipate more energy. The parts of the network that act as the coverage-bridge between two parts of the network and around the sink show relatively high energy dissipation rates. Subsequently these areas are modeled as hotspots.

We used a network containing 5000 nodes that span an area of $50 \times 100 \ unit^2$, with each node having a communication range $R = 2$ units. For $R = 2$ the upper bound for grid cell size is 0.7 units. We found 0.3 as the lower bound because if we take a grid size smaller than 0.3 then we have more occupied grid cells than the number of nodes that over samples the network area. We therefore selected three grid sizes between upper and lower bounds 0.3, 0.5 and 0.7 units. Energy dissipation history of 164 profiles was collected for the ideal case from all the nodes and then using our approach 164 grid maps were collected from a subset of reporting nodes. To evaluate the statistics we divided the history of profiles into two parts. 139 profiles were used for modeling purposes and 25 used for validation. 164 profiles represent the network lifetime history. If we scale 164 lifetime profiles to 164 days then 139 days of network operation are used to predict the next 25 days network status. First, we considered the ideal situation and 139 profiles of each node were used to predict next 25 states. Each ideally predicted profile of the network was transformed to grid map. Then using MPM approach 139 collected grid maps were used to predict next 25 grid maps.

6.3   Simulation Results

Fig. 5(a) shows the graphs for mean sum of absolute error for 25 prediction steps for 3 different grid sizes. It shows that the optimized grid maps are almost as accurate as ideal grid maps. The mean error ranges from 0.6 to 0.9 even after 25 prediction steps, showing the level of accuracy of the prediction model. The lower bound of error 0.6 is actually the maximum error that was allowed in the local models on sensor nodes. The increasing trend is natural, as an increasing number of prediction steps makes the prediction model less accurate. In ideal case due to the cumulative error of considerably more number of nodes, the error is slightly more than our approach.

Fig. 5(b) shows the misclassified cells count. The results of the mean sum of absolute error imply that we can not expect much inaccuracy in misclassification graphs. The highest count is naturally in the case of grid size 0.3, which touches 67 at the peak. The total number of occupied grid cells at this resolution is 4146, so a worst case misclassification of 67 cells accounts to less than 2% of the total cells. We also see a slight increasing trend in the misclassification for each prediction step because of the increasing error between model approximation and the actual data. The peak in the graph gives interesting insight. We have defined two classes of energy and as soon as the grid cells cross the class threshold (10% of energy) they are classified into the partitioned class. From Fig. 5(a) it is obvious that there is a minimum difference of 0.6 between the reference data and test data, which creates a lag in the value of both data sets. This peak appears when some cells in the actual data (Cr) cross the threshold of 10% but the cells from the modeled data (Ct), due to the lag in value do not cross the threshold at the same time. Therefore, many cells from the reference class are classified

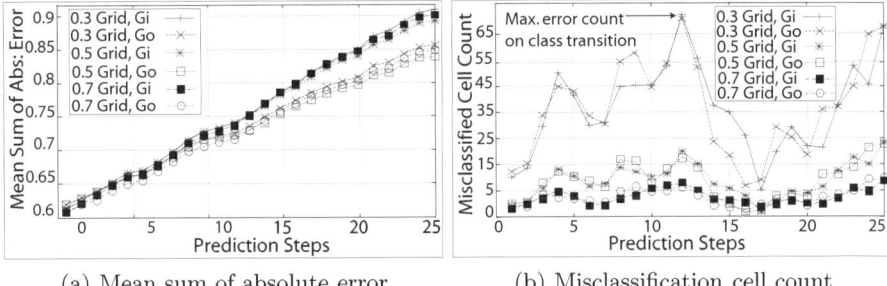

(a) Mean sum of absolute error   (b) Misclassification cell count

Fig. 5. Predictability and accuracy measures of MPM

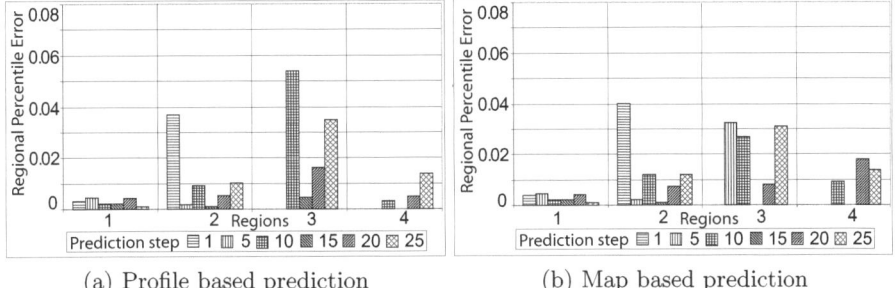

(a) Profile based prediction   (b) Map based prediction

Fig. 6. Misclassification percentile error per region

in the partitioned class but corresponding cells in the test class are still in the non-partitioned class, which increases the count of difference cells. As soon as the modeled data crosses the threshold this peak disappears but a slight trend in increase of error continues.

Fig. 6(a) gives account of the error in the detected regions predicted through profiles of the nodes and Fig. 6(b) is the regional error predicted through collected maps. To summarize the results we have selected prediction profiles/maps separated by five prediction steps. On first prediction step there are only two regions with less than 0.04% max percentile error. With each next prediction step the number of regions increases and errors distribute between the different regions. In the worst case scenario a region has a maximum percentile error of less than 0.06%. The results however show that each region is very accurately detected. Moreover our approach of maps that uses only a subset data is almost as good as the profiles that consists of the data from all the nodes. Misclassification per region on the average is less than 0.03% which shows the accuracy of our approach to detect the regions and their boundaries.

Now, we summarize the results w.r.t. efficiency metric, i.e., number of packets needed for profiling. To profile the whole network of 5000 nodes and to collect 164 profiles for the entire lifetime, requires nearly 1 million data points. This overhead is reduced dramatically by sending the models instead of raw samples. It is further reduced to 14214 packets with the utilization of the gridding technique.

This equates to each node sending less than 3 packets instead of 164, which is less then 1.8% of raw data to be collected. If the network is scaled to 100 nodes the packets to be sent scale down to 284. These results satisfy our requirement on the framework to be lightweight.

6.4   Discussion

The results obtained in the evaluation are in accordance to the design requirements for the framework. It is very lightweight as data compression and gridding cumulatively reduce the data to less than 1.8% of the raw data needed to profile all nodes. The prediction is very accurate, represented by max prediction error of approximately 0.06% in misclassification of the areas of the maps for 25 prediction steps. The 22 days (in scaled time as explained in Section. 6.2) earlier prediction of partition is also feasible for proactive self reconfiguration, enabling autonomicity of WSN.

Our framework ensures reliably accurate information for proactive action and well before the event takes place. The proactive action can be triggered using regioning algorithm results. If there exists an event region, it will be detected by the regioning algorithm along with the perimeter of the event regions. The regioning algorithm determines the perimeter of the holes and the partition, it implicitly provides the exact information about the area, location and the affected nodes that lie within that perimeter. As soon as the sink successfully detects an event, it can either trigger an early warning or initiate a proactive action like move some nodes to affected areas, redeploy new nodes etc. if applicable. The MPM framework gives vital information beforehand to act proactively, but this proactive action itself is beyond the scope of this work.

7   Conclusion and Future Directions

We have developed Map based Predictive Monitoring, a generalized framework for event prediction to support an autonomic self* system for WSN. To demonstrate the feasibility and validity of approach we predicted the network partitioning as a case study. We were able to detect multiple holes and resulting partitioned area of network; information necessary to initiate proactive self reconfiguration. Simulations support the practicality of our approach by showing its high accuracy and low monitoring overhead on the network. We plan to extend our approach for proactive reconfiguration of network entities to enhance functionality and dependability through the predicted events.

References

1. Yick, J., et al.: Wireless sensor network survey. Computer Networks 52(12), 2292–2330 (2008)
2. Yu, L., et al.: Real-time forest fire detection with wireless sensor networks. In: WCNM, vol. 2, pp. 1214–1217 (2005)

3. Shrivastava, N., et al.: Detecting cuts in sensor networks. In: IPSN, p. 28 (2005)
4. Rost, S., Balakrishnan, H.: Memento: A Health Monitoring System for Wireless Sensor Networks. In: IEEE SECON, pp. 575–584 (2006)
5. Shih, K.P., et al.: PALM: A Partition Avoidance Lazy Movement Protocol for Mobile Sensor Networks. In: Proceedings of the IEEE WCNC, pp. 2484–2489 (2007)
6. Wang, X., et al.: Contour map matching for event detection in sensor networks. In: SIGMOD, pp. 145–156 (2006)
7. Achir, M., Ouvry, L.: Power consumption prediction in wireless sensor networks. In: 16th ITCS (2004)
8. Tulone, D., Madden, S.: PAQ: Time series forecasting for approximate query answering in sensor networks. In: Römer, K., Karl, H., Mattern, F. (eds.) EWSN 2006. LNCS, vol. 3868, pp. 21–37. Springer, Heidelberg (2006)
9. Zhao, J., et al.: Residual energy scan for monitoring sensor networks. In: WCNC, pp. 356–362 (2002)
10. Banerjee, T., et al.: Fault tolerant multiple event detection in a wireless sensor network. Journal of Parallel and Distributed Computing 68(9), 1222–1234 (2008)
11. Landsiedel, O., et al.: Accurate prediction of power consumption in sensor networks. In: EmNets, pp. 37–44 (2005)
12. Mini, A.F., et al.: A probabilistic approach to predict the energy consumption in wireless sensor networks. In: IV Workshop de Comunicao sem Fio e Computao Mvel, So Paulo, pp. 23–25 (2002)
13. Wang, X., et al.: Robust forecasting for energy efficiency of wireless multimedia sensor networks. Sensors 7(11), 2779–2807 (2007)
14. Khelil, A., et al.: MWM: A map-based world model for event-driven wireless sensor networks. Autonomics, 1–10 (2008)
15. He, T., et al.: Range-free localization and its impact on large scale sensor networks. Transaction on Embedded Computing Systems 4(4), 877–906 (2005)
16. Aurenhammer, F.: Voronoi diagrams - a survey of a fundamental geometric data structure. ACM Computing Surveys 23(3), 345–405 (1991)
17. Montgomery, D.C., et al.: Introduction to Time Series Analysis and Forecasting. John Wiley and Sons, New Jersey (2008)
18. Ljung, L.: System Identification: Theory for the User, 2nd edn. Prentice-Hall, New Jersey (1998)

Integrating Autonomic Grid Components and Process-Driven Business Applications

Thomas Weigold[1], Marco Aldinucci[2], Marco Danelutto[3], and Vladimir Getov[4]

[1] IBM Zurich Research Lab., Zurich, Switzerland
twe@zurich.ibm.com
[2] Computer Science Dept., University of Torino, Italy
aldinuc@di.unito.it
[3] Computer Science Dept., University of Pisa, Italy
marcod@di.unipi.it
[4] School of Electronics and Computer Science, University of Westminster,
London, U.K.
V.S.Getov@westminster.ac.uk

Abstract. Today's business applications are increasingly process driven, meaning that the main application logic is executed by a dedicate process engine. In addition, component-oriented software development has been attracting attention for building complex distributed applications. In this paper we present the experiences gained from building a process-driven biometric identification application which makes use of Grid infrastructures via the Grid Component Model (GCM). GCM, besides guaranteeing access to Grid resources, supports autonomic management of notable parallel composite components. This feature is exploited within our biometric identification application to ensure real time identification of fingerprints. Therefore, we briefly introduce the GCM framework and the process engine used, and we describe the implementation of the application using autonomic GCM components. Finally, we summarize the results, experiences, and lessons learned focusing on the integration of autonomic GCM components and the process-driven approach.

Keywords: Autonomic computing, components, parallel applications, distributed applications, process-driven applications.

1 Introduction

Today's businesses are increasingly process driven. Ideally, all actions within an enterprise are explicitly defined as processes with the goal to improve control, flexibility, and effectiveness of delivering customer value. Additionally, business processes are oftentimes supported or even fully implemented by software applications [1]. In many cases, the business processes are turned into software such that they are hidden in the application's source code. However, there is a trend towards separating the main business logic from the functional code such that the resulting applications become more transparent and more flexible. The approach is to embed a so-called process engine into the application,

A.V. Vasilakos et al. (Eds.): AUTONOMICS 2009, LNICST 23, pp. 96–113, 2010.

which then executes process definitions representing the main control logic of the application. Functional code is then triggered from the process engine in accordance with the process definition. Such applications are called process-driven or workflow-driven applications [2,3,4]. The main advantages of this approach are the fact that the application logic can be modified without re-compiling the application, even at runtime, the business logic is more evident, and monitoring features of the process engine can be explored.

Besides the trend towards process-driven applications, enterprises seek ways to benefit from resources available from computing Grids/Clouds, in particular in all those cases were parallel computing is required to guarantee fair performances. The development of grid applications is not an easy task, however. Grid architectures present peculiar features such as dynamicity, heterogeneity and non exclusive access to resources, that require substantial effort to be suitably handled; furthermore, this effort is in addition to the normal effort required to develop efficient parallel/distributed applications. Within the *plethora* of programming environments targeting Grids, GCM (the Grid Component Model developed within CoreGRID [5] and whose reference implementation has been provided by GridCOMP) [6]) supports Grid programmers in designing parallel/distributed grid applications. In particular, GCM provides pre-defined composite components modelling standard parallel/distributed computation patterns that users can instantiate just providing the components implementing the sequential computations involved in the parallel pattern. In addition, these pre-defined composite components implement proper autonomic managers that completely take care of non functional aspects related to application execution according to what is specified in user supplied contracts.

In this work, we discuss a process-driven application, which makes use of GCM autonomic components to solve the problem of large-scale biometric identification[7], that has been developed as part of the activities of the GridCOMP project [6]. In particular, we discuss how process-driven application development exploits the autonomic features provided by the underlying Grid software as well as the results, experiences, and lessons learned during application development focusing on the integration of autonomic GCM components and the process-driven approach.

The paper is organized as follows: Sec. 2 introduces GCM and Behavioural Skeletons (BS), i.e. the autonomic composite components modelling notable parallel/distributed patterns within GCM. Sec. 3 introduces the process engine used to implement the biometric identification application discussed in Sec. 4 on top of GCM/BS. Eventually, Sec. 5 discusses the overall results achieved and Sec. 6 drafts the conclusions of the paper.

2   The GCM Framework

The *Grid Component Model* (GCM) is a component model explicitly designed to support component-based autonomic applications in distributed contexts [5]. The main features of this component model can be summarised as follows:

- **Hierarchical:** GCM components can be composed in a hierarchical way in *composite* components. Composite components are first class components and they are not distinguishable from non composite components at the user level. Hierarchical composition greatly improves the expressive power of the component model and is inherited by GCM from the Fractal component model [8].
- **Structured:** In addition to standard intra-component interaction mechanisms (use/provide ports [9]) GCM allows components to interact through *collective* ports modelling common structured parallel computation communication patterns. These patterns include broadcast, multicast, scatter and gather communications operating on collections of components. Also, GCM provides *data* and *stream* ports, modelling access to shared data encapsulated into components and data flow streams. All these additional port types, not present in other well known component models, increase the possibilities offered to the component system user for developing efficient parallel component applications.
- **Autonomic:** GCM specifically supports implementing autonomic components in two distinct ways: by supporting the implementation of user defined component controllers and by providing behavioural skeletons. Component controllers can be programmed in the component membrane (the membrane concept, as the place where component control activities take place, is inherited from Fractal [8]) and controllers can be components themselves. This provides a substantial support to the development of reusable autonomic controllers. Behavioural skeletons, thoroughly discussed in Sec. 2.1, are composite GCM components modelling notable parallel/distributed computation patterns and supporting autonomic managers, i.e. components taking care of non functional concerns affecting parallel computation.

Due to the presence of controllers and autonomic managers, GCM components implement two distinct kinds of interfaces: functional and non-functional ones. The functional interfaces host those ports concerned with the implementation of the functional features of the component. The non-functional interfaces host the ports related to controllers and autonomic managers. These ports are the ones actually supporting the component management activity in the implementation of the non-functional features, i.e. those features contributing to the efficiency of the component in obtaining the expected (functional) results but not directly involved in result computation.

GCM has been designed within the Programming Model Institute [10] in CoreGRID [11] and a reference implementation of the component model has been developed within the GridCOMP project [6]. Within the same GridCOMP project, a Grid Integrated Development Environment (GIDE) has been developed to support development and maintenance of GCM programs.

2.1 Behavioural Skeletons

Behavioural skeletons represent a specialisation of the algorithmic skeleton concept for component management [12]. Algorithmic skeletons have been

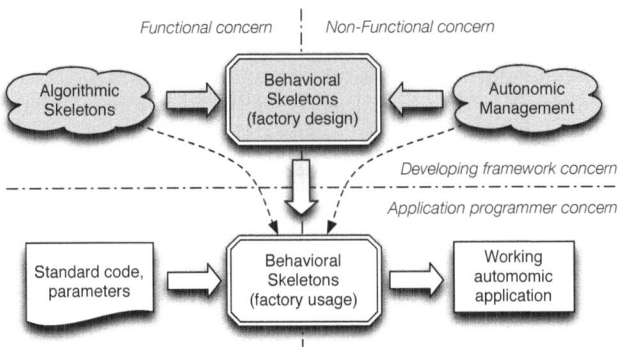

Fig. 1. Behavioural skeleton rationale

traditionally used as a vehicle to provide efficient implementation templates of parallel paradigms. Behavioural skeletons, as algorithmic skeletons, represent patterns of parallel computations (which are expressed in GCM as graphs of components), but in addition they exploit the inherent skeleton semantics to design sound self-management schemes of parallel components.

Behavioural skeletons are composed of an algorithmic skeleton together with an autonomic manager (see Fig. 1). They provide the programmer with a component that can be turned into a running application by providing the code parameters needed to instantiate the algorithmic skeleton parameters (e.g. the code of the different stages in a pipeline or the code of the worker in a task farm) *plus* some kind of Service Level Agreement (SLA, e.g. the expected parallelism degree or the expected throughput of the application). The code parameters are used to build the actual code run on the target parallel/distributed architecture, while the SLA is used by the autonomic manager that will take care of ensuring this SLA (best effort) while the application is being computed.

The choice of the skeleton to be used as well as the code parameters provided to instantiate the behavioural skeleton are functional concerns: they only depend on what has to be computed (i.e. on the application at hand) and on the *qualitative* parallelism exploitation pattern the programmer wants to exploit. The autonomic management itself is a non-functional concern. The self-management and self-tuning activities taking place in the manager to ensure user supplied SLA both depend on the application structure (the one defined by the algorithmic skeleton) and on the target architecture at hand. The implementation of both the algorithmic skeleton and the autonomic manager is in the charge of the "system" programmer, i.e. the one providing the behavioural skeleton framework to the application user.

In the programming model provided by behavioural skeletons, the application programmers are in charge of picking up a behavioural skeleton (or a composition of behavioural skeletons) among those available and of providing the corresponding parameters and SLA. The system, and in particular the autonomic managers of the behavioural skeletons instantiated by the application programmer, are in charge of performing all those activities needed to ensure the user supplied SLA. These activities, in turn, may include varying some implementation parameters (e.g. the parallelism degree, the kind of communication protocol used among different parallel entities or scheduling/mapping of the parallel activities to the target processing elements) as well as changing the behavioural skeleton (composition) chosen by the application programmer (e.g. using "under the hoods" an equivalent, but more efficient (with respect to the target architecture and user supplied SLA) behavioural skeleton (composition)).

Autonomic management of non-functional concerns is based on the concurrent execution (with respect to the application "business logic") of a basic control loop such as that shown in Fig. 2. In the *monitor* phase, the application behaviour is observed, then in the *analyse* and *plan* phases the observed behaviour is examined to discover possible malfunctioning and corrective actions are planned. The corrective actions are usually taken from a library of known actions and the chosen action is determined by the result of the analysis phase. Finally, the actions planned are applied to the application during the *execute* phase [13,14,15,16,17,18].

Currently, two kind of behavioural skeletons are implemented in GCM: a *task farm* BS and a *data parallel* BS (see Fig. 3). The former models embarrassingly parallel computations processing independent items x_i of an input stream to obtain items $f(x_i)$ of the corresponding output stream. The latter models data parallel computations by computing for each item of the input stream x_i an item $f(x_i, D)$ of the corresponding output stream, where D represents a read only data structure and the result of $f(x_i, D)$ can be computed as a map of some function $f'(x_i)$ on all the items of D followed by a reduce of the different $f'(x_i, D_j)$ with an associative and commutative operator g.

Both BS implement an AM taking care of the performance of the parallel computation at hand. In particular, the AM may ensure contracts stating the expected service time (or throughput, i.e. the time between the delivery of two consecutive items on the output stream) of the BS (both task farm and data parallel BS) or the expected partition size of data structure D (data parallel BS only). Currently, the contracts must be supplied to the BS AMs through the BS non functional ports as a(n ASCII string hosting a) set of JBoss rules

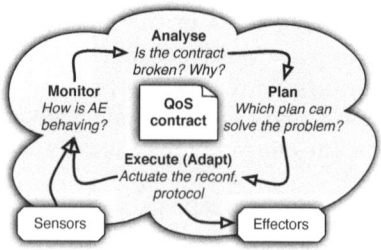

Fig. 2. The classical control loop implemented within Autonomic Managers in GCM Behavioural Skeletons

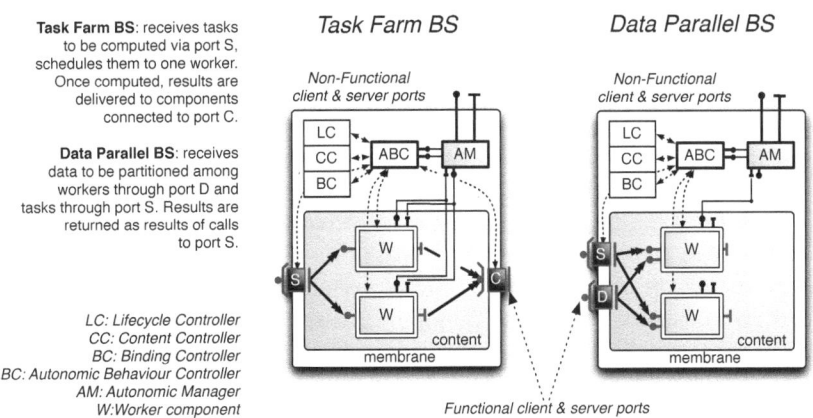

Fig. 3. Behavioural skeletons currently implemented in GCM

defined in terms of the operations provided by the ABC controller bean. In fact, the AM control loop is implemented by running an instance of the JBoss business rule engine at regular intervals of time. At each time interval, all the *pre-condition-action* rules supplied to the AM are evaluated and those that turn out to be fireable (e.g. whose with the pre-condition holding true) are executed ordered by priority (or *salience* according to JBoss jargon). The pre-conditions are evaluated using values provided by the monitoring system implemented in the ABC controller beans, actually. The period used to run the JBoss engine is determined in such a way it is neither too fast (reacting when it was not the case to react to small changes in the system, thus increasing overhead to the autonomic management) nor too slow (poorly reacting to actual changes in the system, thus decreasing efficiency of autonomic management).

Current AMs manage the contracts varying the parallelism degree of the BS, i.e. the number of worker instances actually used to implement the BS. The variation of the number of worker instances happens adding/removing a fixed amount of workers. This fixed amount is a BS user configurable constant (Δ_w). Rules supplied to the AM in the BS also consist in specific rules avoiding to perform (probably) useless adaptations (e.g. avoiding to adapt BS parallelism degree immediately after another adaptation took place) as well as rules default actions basically only taking care of updating monitored values when no other, more significant actions turn out to be fireable.

3   The ePVM Process Engine

The embeddable Process Virtual Machine (ePVM) is a research prototype process engine [4] basically built upon two core concepts. Firstly, a process model which is rooted in the theoretical framework of communicating extended finite state machines (CEFSM). Secondly, whereas many efforts have been made to create the ultimate process language, ePVM provides in contrast a low-level run-time environment based on a JavaScript interpreter where higher-level domain specific process languages can be mapped to.

The idea of ePVM can be considered to follow a bottom-up or micro-kernel type of approach for building process-driven applications, Business Process Management Systems (BPM), or workflow systems. This means that ePVM is a basic framework for building such systems rather than a complete off-the-shelf application that can run stand-alone. It consists of a library including a lightweight, generic, and easily programmable process execution engine. Lightweight hereby means that the engine is small in size and imposes minimum requirements on its environment, namely the host application it is embedded in. ePVM has its own process model resembling networks of communicating state machines running in parallel, which makes it an inherently asynchronous, event-driven run-time system. Every state machine is implemented by one JavaScript function, has an associated thread executing it, has a state object which is passed every time the function is invoked, and can communicate with other processes as well as entities external to the process engine via some messaging system. An arbitrary number of external entities, so-called host processes, can be attached to the engine to become visible for ePVM processes. The ePVM programming model based on the theory of CEFSM combines the simplicity of JavaScript with an easy and powerful way of defining complex concurrent business processes. More details can be found in [4].

4   Process-Driven Distributed Biometric Identification

In recent years biometric methods for verification and identification of people have become very popular. Applications span from governmental projects like border control or criminal identification to civil purposes such as e-commerce, network access, or transport. Frequently, biometric verification is used to authenticate people meaning that a 1:1 match operation of a claimed identity to the one stored in a reference system is carried out. In an identification system, however, the complexity is much higher. Here, a person's identity is to be determined solely on biometric information, which requires matching the live scan of his biometrics against all enrolled (known) identities. Such a 1:N match operation can be quite time-consuming making it unsuitable for real-time applications. In order to tackle this challenge, a distributed biometric identification system (BIS), which can work on a large user population of up to millions of individuals, has been developed. It is based on fingerprint biometrics and allows real-time identification within a few seconds period by taking advantage of the Grid, in particular via GCM components.

4.1   Application Architecture

The BIS can be considered a process-driven application, as it is centrally driven by the ePVM process engine. Fig. 4 outlines its high-level architectural design.

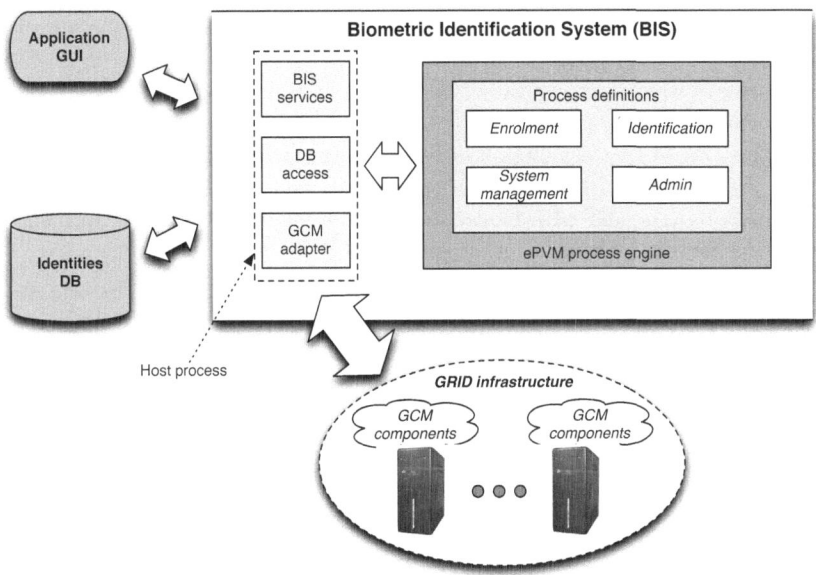

Fig. 4. BIS high-level architecture

A number of ePVM process definitions describing the main control flow for operations such as starting up the system or identifying a person are loaded into the process engine. These processes co-operate with external entities such as the GUI, the database (DB) of known identities, and the distributed GCM component system via a number of host processes to implement the overall functionality of the BIS.

4.2   Process-Engine/GCM Interfacing

The actual distributed fingerprint matching functionality is implemented via a set of GCM components deployed within a Grid/Cloud infrastructure as indicated in Fig. 4. Processes running within the process engine must be able to create, deploy, configure and interact with these components. For this purpose, a dedicated host process named *GCM adapter* (c.f. Fig. 4) has been developed, which receives messages from ePVM process instances, turns these messages into method invocations on GCM framework methods or GCM components, and generates appropriate reply messages returned to ePVM. The GCM adapter represents the main interface between ePVM and GCM. As ePVM process definitions are implemented in JavaScript and the GCM framework is available as a Java library, the GCM adapter essentially converts between JavaScript messages and Java method invocations.

An alternative option would have been to export the GCM components as Web services, as supported by the GCM implementation, and invoke them from within the GCM adapter. However, this would have increased the number of required type conversions going from Java Script over SOAP to Java and vice versa. Also, the GCM framework only supports exporting GCM components as Web services. Other framework services, for example, functionality for deployment and component creation, cannot be turned into Web services automatically. Finally, the ePVM process engine does not necessarily require working on Web Services level like, for instance, process engines based on the Business Process Execution Language (BPEL). Consequently, we decided not to use Web services as interfaces between the process engine and GCM.

The functionality provided by the GCM adapter includes:

- Activate a given GCM deployment descriptor to start the nodes available in the Grid.
- Modify architecture description language (ADL) files describing the GCM components used.
- Create GCM components within the Grid.
- Invoke methods on GCM components, for example, to configure the quality of service (QoS) contract, distribute the DB of known identities, or submit the biometrics of a person for identification.

The GCM adapter is triggered by ePVM process instances to implement the overall application logic. As an example, the activity flow chart shown in Fig. 5 illustrates the control logic implemented within an ePVM processes as it is executed during BIS initialization. For each of the activities a message is being sent to a host adapter which implements the functionality. Some of the activities execute in parallel, for instance, activity 1.1 to 1.3, some are sequential.

4.3   Using Autonomic GCM Components

The problem of biometric identification can be considered a search problem where the compare function is a biometric matching algorithm, here fingerprint

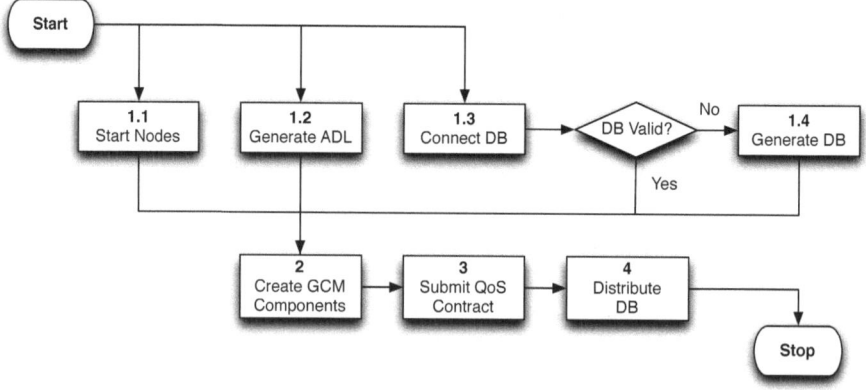

Fig. 5. BIS initialization process flow

matching. To distribute the problem within a Grid infrastructure, the DB of known identities needs to be distributed such that each computing node in the Grid receives a partition of the overall DB and can match a given identity against this partition. The time spent in matching the given identity against the local portion of the database is clearly proportional to the size of this local DB portion. Therefore, considering that the distribution of the DB among the grid nodes is performed once and for all, and considering negligible the time spent to broadcast the fingerprint that has to be matched with those in the distributed database, the ability to perform fingerprint matching in real time roughly depends on the ability to distribute local portions of the database small enough to allow real time matching of the broadcasted fingerprint. More precisely, the time spent in matching a single fingerprint against the local database also depends on the computing power and on the load of the machine used to perform the matching. The machine power and the local database sizes are somehow static properties. The load of the machine is instead a dynamic property. Thus, in order to keep the matching time perceived by the application user within a given range (i.e. satisfying a given service level agreement (SLA) or *performance contract*), our BIS application should i) properly dimension the number of distributed resources used to host database portions and ii) dynamically adapt to the varying load of the grid resources involved in such a way a user supplied performance contract (such as *match fingerprint in less than 30 secs*) is ensured. Both features are supported within the GCM Behavioural skeletons presented in Sec. 2.1: if the user instantiates a Behavioural skeleton to implement the BIS search process, and if he/she provides a contract stating the expected latency of the fingerprint matching process, the AM of the behavioural skeleton will start with a predefined number of workers (i.e. a predefined parallelism degree) and then adapt this number to achieve the matching latency adding (removing) workers from the BS composite component. In case of overload of some of the resources used in the matching, the AM of the behavioural skeleton will also manage to increase the number of resources recruited to the parallel matching, in such a way the contract can be ensured again. In this case, the recruitment of a new processing resource induces a physical redistribution of the database among the resources. This redistribution is completely implemented/managed by the behavioural skeleton AM.

In order to implement our BIS application, we used a data parallel (DP) behavioural skeleton. Referring to Fig. 3 (right), the DP skeleton is a composite component which includes an autonomic behaviour controller (ABC) and an autonomic manager (AM). The AM periodically evaluates certain monitored properties of the skeleton to ensure that a given QoS contract is satisfied. If this is not the case, it triggers appropriate reconfiguration operations provided by the ABC. To apply the DP skeleton for our application scenario, it must be parameterized with a worker component and a QoS contract. The worker component, here named *IDMatcher*, implements the actual fingerprint matching functionality and the skeleton allocates one instance of this worker component per node. The QoS contract consists of a set of rules interpreted by the JBoss Drools rule engine.

For our BIS prototype we chose to implement a QoS contract requiring to keep the partition size of the workers constant, independently of the size of the database presented to the BS through port D. The contract is provided before starting the computation through the non functional server port attached to the BS AM. The AM, in this case, adds or removes workers from the BS in case the partition size exceeds or is less than the value supplied by the user within the contract provided through the non functional BS ports.

Before identification requests can be processed, the identity DB is distributed across the worker components using port D. As a consequence, the DB is partitioned on the inner W components. The identity DB holds information such as name, address, and fingerprints of all enrolled (known) people.

Once the skeleton has been initialized, identification requests can be submitted via the second port provided by the BS, port S, the so-called broadcast port. Fingerprints of a person to be identified are broadcasted via this port to all worker components and each worker matches them against its partition of the DB. Results are returned synchronously via method return values.

If the AM triggers reconfiguration via the ABC, for example, to increase the number of worker components, the AM collects all DB partitions from the workers, modifies the number of workers, and finally redistributes the DB to the workers. This way the DB is redistributed during each reconfiguration operation.

The submission of the contract through non functional interfaces, the DB through BS port D, and the fingerprints to be matched through port S are all interactions with the GCM BS triggered by ePVM processes via the GCM adapter.

4.4   Deployment and Component Creation

When the BIS application is started, activation of the GCM deployment descriptor is triggered by the process engine as indicated in Fig. 5, activity 1.1. The GCM framework then defines virtual nodes, creates a mapping to real nodes, and starts JVMs on all of them. The DP skeleton uses virtual nodes listed in the descriptor for allocating worker components. Afterwards, when the initialization process reaches activity 2, the GCM component system is created according to the ADL files of the BS.

4.5   Application Monitoring

Monitoring is one of the core features of every process engine and it is an important argument for using one when building an application. The ePVM engine supports monitoring processes by registering monitor objects for one or more process definitions. Furthermore, it can be specified which events shall be monitored. Available are a number of standard events such as a process instance being created, a message being processed, or a process becoming idle. Furthermore, custom events can be defined such that more fine-grained monitoring can be implemented, for example, multiple events can be trigged while a single message is processed.

In the BIS application, a monitor object is used to track the progress of ePVM process instances, for example, while the system initialization process is executed (c.f. Fig. 5). The monitor object is triggered by the process engine whenever activities start or finish and it updates the GUI to reflect the state of the system. Furthermore, it is desired to monitor the GCM component system with the goal to visualize AM actions and the number of workers used in the DP skeleton. A system administator could observe this and, if required, trigger reconfiguration or add resources manually. For monitoring the skeleton, functionality provided by the GCM framework can be used. However, monitoring in GCM is based on a pull model where information about components and their states can be retrieved on request. On the contrary, the ePVM approach can be considered a push model where a monitor is registered and receives events. To integrate GCM monitoring with the event-driven paradigm applied in ePVM some adaptation is necessary. A first solution is to create a dedicated ePVM process which regularly retrieves information about the component system via the GCM API and creates events for the monitor object. A second solution is to instrument the component implementation to actively send events to an ePVM process. The first approach is more generic with respect to distribution, as the GCM framework handles remote method invocations required to query for component states automatically. The second approach is more efficient, as communication only takes place if an event to be monitored occurs. However, a component might not be able to easily communicate with the process engine if it is running on a remote machine, since the process engine itself is not a GCM component. In the BIS we used the first approach to implement monitoring the number of workers, as the workers are typically distributed. For monitoring AM actions, we use the second approach exploiting the fact that in our deployments the AM is always co-located with the process engine such that no remote communication is necessary.

In general, the requirement to monitor actions within the DP skeleton to some extend is contradictory to the idea of autonomic components. On one hand the goal of using the DP skeleton is to take advantage of its built-in functionality without taking care of the implementation details. On the other hand, we still want to be able to monitor certain internal details such as reconfiguration operations and the number of workers. From the perspective of the process-driven applications paradigm all important actions which shall be monitored should be centrally controlled by the process-engine. However, in real-world applications a trade-off between central process control and autonomy must be made.

4.6   Automatic Futures vs. Message Passing

When integrating process-engines and distributed computing frameworks, it is very important to be aware of their communication and synchronization paradigms. The GCM framework is based on Java RMI and implements the concept of automatic futures [19]. This means that method invocations always return immediately, whereas results which are not yet available are represented by so-called future objects. Program execution is then blocked automatically if a future object is being accessed as long as the value represented is not yet available. The goal is to ease parallel programming by hiding synchronization details within a meta object protocol implemented in GCM. The ePVM process engine, however, uses message passing for communication and synchronization between concurrent control flows. If these two paradigms are interweaved, as it is the case in the BIS application, process flows can easily become distorted. For example, if a process definition assigns two activities to be carried out sequentially (c.f. activity 2 and 3 in Fig. 5), it must be ensured that no more future objects resulting from the first activity exist before the second is triggered.

This issue becomes obvious when an identification process is triggered within the BIS. In this case, an ePVM process sends a message to the GCM adapter including fingerprints of a person to be identified. The GCM adapter forwards this information to the component system by invoking the broadcast interface of the DP skeleton (port S, Fig. 3). This interface is a so-called collective interface, which turns one method invocation into N method invocations on all the bound IDMatcher components to broadcast the identification request. The return value is a list of result objects, one from each IDMatcher component. When the interface is invoked, it immediately returns a list of future objects, which at the beginning are all unavailable and then by-and-by become available as the IDMatcher components return their results. It is important that the GCM adapter waits for the futures to become available and generates messages to be returned to the ePVM process instance accordingly. It must not report the identification as completed before all futures are available. Effectively, the GCM adapter retracts automatic synchronization in order to make the actual progress visible to the process engine, which must to be informed whenever an IDMatcher component has searched its part of the DB. Obviously, converting from one paradigm into the other must be handled with care as the semantics of the process definitions can be broken due to delayed synchronization within GCM.

4.7   Integrated Development

On one hand, the advanced features offered by both technologies, the process engine and the GCM framework, significantly reduce the development effort required for the BIS. On the other hand, it requires handling a large number of different development artefacts including plain Java code, JavaScript process definitions, XML deployment descriptors and ADL definitions, and JBoss Drools rule files. As the process engine does not mandate the use of high-level modelling tools, developers can use the Java/JavaScript toolset of their choice. For GCM development, the Grid IDE (GIDE) [20] is available, which consists of a set of plugins to the famous Eclipse development environment. It also includes support for graphical GCM component composition and ADL code generation. Consequently, all artefacts can be developed within Eclipse with appropriate plugins installed. This reduces the complexity to a manageable level, such that once the knowledge about both technologies is available, integration work can be carried out smoothly.

5   Results, Experiences, and Lessons Learned

The primary result of this work is the fully functional prototype of the BIS application, which acts as a use case demo for the process engine as well as for the GCM framework. Additional results have been gained by critically evaluating the application and experimenting with it. Firstly, it has been successfully deployed on various hardware platforms ranging from one multicore PC to heterogeneous sets of clusters as provided by the Grid5000 project [21]. Switching hardware platforms did not require changing a single line of functional code, only the infrastructure part of the XML deployment descriptor required modification. The strict separation of concerns and the autonomic functionality implemented within the GCM framework have turned out to be the main factors leading to this flexibility. The former ensures that resources are never directly referenced in the source code while the latter provides autonomic adaptation to the performance properties of the hardware in use.

Secondly, functionality and autonomic behaviour of the application has been verified using Grid5000. The BIS has been started using 50 workers (one per node), a DB holding 50000 identities (approx. 400 MB), and a QoS contract mandating a partition size of 1000 identities/worker. At runtime, the contract has been updated to 800 ($\pm$ 10%) identities/worker. Thereupon, the AM has successfully detected 7 contract violations and each time reconfigured the DP skeleton by adding one additional worker until a partition size of 877 identities/worker was reached at 57 workers/nodes. During this experiment, every reconfiguration operation took about 9 seconds in which the complete DB has been redistributed (from the node hosting the whole database to the nodes hosting the workers of the data parallel BS) by the ABC. When identification requests where issued during reconfiguration, they where queued automatically by the skeleton and processed as soon as reconfiguration was completed. For the given DB size, each identification request required around 10 seconds to be processed. This means that each reconfiguration operation roughly decreases the throughput of the BIS by one identification for any given timeframe. Therefore, if the BIS is used in a very dynamic environment requiring frequent reconfiguration, the number of occurrences of reconfigurations may be sensibly reduced by adopting more aggressive parallelism degree variation policies, in such a way the overall overhead is reduced. Such more aggressive policies at the moment consist in varying the constant Δ_w that defines the number of workers to be added/removed when reconfiguring the parallelism degree of a BS. In the BS/GCM framework we are currently investigating the possibility to use a kind of exponential backoff increase/decrease protocol. All those cases, of course, rely on the possiblity to effectively monitor the increase/decrease achieved in the BS performance as a consequence of the parallelism degree adaptation.

Finally, evaluating the application's source code, including the deployment descriptor required to run on 50 nodes of Grid5000, unveiled the source code breakdown illustrated in Fig. 6. The functional code mainly includes the host processes (c.f. Fig. 4) providing DB access, the GUI functionality, and the interfacing to the GCM components. Its absolute size is about 2500 lines of code,

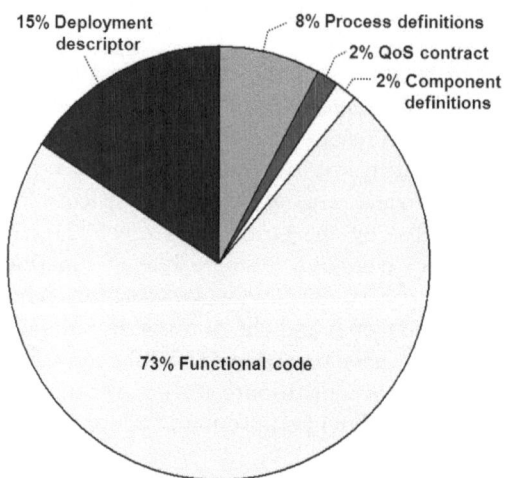

Fig. 6. BIS source code breakdown

which is very small considering the the overall functionality provided by the application. This is due to the fact that the GCM framework provides all the functionality for distribution and autonomicity. Implementing this functionality from scratch not using GCM would have been significantly more effort. In particular, adding autonomic control to an application is virtually effortless if a matching behavioural skeleton is available. Only the QoS contract must be provided and a few non-functional interfaces used by the controller must be implemented within the worker component. In case of the BIS application, only about 200 lines of code where necessary for that. Furthermore, it is to be noted that more than a quarter of the source code (27%) consists of code interpreted at runtime. This code, including the deployment descriptor, the process definitions, the QoS contract, and the GCM component definitions, contains the main control logic and infrastructure definition of the application. As a result, the application can be adapted significantly without recompilation - a very important property required for operation in today's dynamic business environments. Hard-coding this part of the application would clearly decrese the applications flexibilty as achieved through the combination of GCM and ePVM.

During application development, we have made a number of experiences with regards to the integration of process technology and the GCM framework. The interfacing between the two technologies went rather smoothly, since the ePVM engine is available as a Java library and it does not dictate the use of Web services. Also, the DP skeleton fits well to the given biometric identification problem. However, application monitoring turned out to be challenging. One must be aware that the idea behind components is hiding complexity and this can be a problem if component internals need to be monitored. The GCM framework supports querying the state of a component system, however, it does not support monitoring activities within components, for example, reconfiguration within a BS. Solving this problem by instrumenting component implementations (c.f. Sec. 4.5) requires comprehensive knowledge of the GCM framework. Furthermore, the monitoring support of GCM follows a pull model while process engines are mostly event driven. Joining the two paradigms in a sensible way requires an extra effort and can have a performance impact. For example, regularly traversing component hierarchies to detect newly created components is not very efficient.

Another lesson we have learned is that the two different synchronization paradigms applied in GCM and ePVM can interfere if not handled with care. The concept of automatic futures implemented in the GCM framework follows the wait-by-necessity idea. This means that unavailable results are replaced by future objects such that synchronization is delayed as long as possible. Therefore, it must be carefully checked if results of activities within a process flow include one or more future objects before the next activity of a sequence is triggered, otherwise the process semantics can easily become distorted. In other words, if a GCM component returns an object it does not necessarily mean that all the related operations have completed.

Finally, we realized that working with the advanced features of both frameworks, ePVM and GCM, requires working with a large number of different development artefacts and acquiring related skills. The GIDE eases this to some extend and provides a jump start into GCM. Nevertheless, combining process technology with GCM allows producing extremely flexible and complex distributed applications with minimum effort.

6   Conclusions

Process-driven application development is increasingly gaining attention in the business environment. At the same time, software development frameworks for the Grid/Cloud are raising interest in the course of the Cloud computing wave. In this paper we have considered combining the two approaches to produce a process-driven distributed biometric identification system. In discussing the application we have made the following contributions:

– We provided a brief overview of the GCM framework, its support for autonomic components and behavioural skeletons, and the ePVM process engine.
– We described the architectural design and implementation of the process-driven biometric identification system utilizing the DP autonomic behavioural skeleton available in GCM.
– We presented the results, experiences, and lessons learned while integrating both technologies, the process engine and the GCM framework.

We believe that this use case application demonstrates that combining process technology and autonomic Grid/Cloud components represents a powerful approach for developing flexible distributed applications with minimum effort. Obviously, the application could have been developed without using GCM and ePVM. However, the development effort would have been much higher and the resulting application would have been less flexible due to the hard-coded application logic and autonomic strategy.

References

1. zur Muehlen, M.: Process-driven management information systems - combining data warehouses and workflow technology. In: Proc. of the 4th Intl. Conference on Electronic Commerce Research (ICECR-4), Dallas, TX, USA, pp. 550–556 (2001)
2. Bukovics, B.: Pro WF: Windows Workflow in.NET 3.0. Apress (2007)
3. Faura, M.V., Baeyens, T.: The Process Virtual Machine (2007),
   http://www.onjava.com/pub/a/onjava/2007/05/07/
   the-process-virtual-machine.html
4. Weigold, T., Kramp, T., Buhler, P.: ePVM - an embeddable Process Virtual Machine. In: Proc. of the 31st Intl. Computer Software and Applications Conference (COMPSAC), Beijing, China, pp. 557–564 (2007)
5. CoreGRID NoE deliverable series, Institute on Programming Model: Deliverable D.PM.04 – Basic Features of the Grid Component Model (assessed) (2007),
   http://www.coregrid.net/mambo/images/stories/Deliverables/d.pm.04.pdf
6. GridCOMP Project: Grid Programming with Components, An Advanced Component Platform for an Effective Invisible Grid (2008), http://gridcomp.ercim.org
7. Weigold, T., Buhler, P., Thiyagalingam, J., Basukoski, A., Getov, V.: Advanced grid programming with components: A biometric identification case study. In: Proc. of the 32nd Intl. Computer Software and Applications Conference (COMPSAC), Turku, Finland, pp. 401–408. IEEE, Los Alamitos (2008)
8. ObjectWeb Consortium: The Fractal Component Model, Technical Specification (2003)
9. Armstrong, R., Gannon, D., Geist, A., Keahey, K., Kohn, S., McInnes, L., Parker, S., Smolinski, B.: Toward a common component architecture for high performance scientific computing. In: Proc. of the 8th Intl. Symposium on High Performance Distributed Computing, HPDC 1999 (1999)
10. CoreGRID NoE: Home page of the Institute on Programming model (2009 - last accessed), http://www.coregrid.net/mambo/content/blogcategory/13/292/
11. CoreGRID NoE: Home page (2009 - last accessed), http://www.coregrid.net
12. Cole, M.: Bringing skeletons out of the closet: A pragmatic manifesto for skeletal parallel programming. Parallel Computing 30, 389–406 (2004)
13. Kephart, J.O., Chess, D.M.: The vision of autonomic computing. IEEE Computer 36, 41–50 (2003)
14. Danelutto, M.: QoS in parallel programming through application managers. In: Proc. of Intl. Euromicro PDP: Parallel Distributed and network-based Processing, Lugano, Switzerland, pp. 282–289. IEEE, Los Alamitos (2005)
15. Aldinucci, M., Danelutto, M.: Algorithmic skeletons meeting grids. Parallel Computing 32, 449–462 (2006)
16. Aldinucci, M., Campa, S., Danelutto, M., Dazzi, P., Kilpatrick, P., Laforenza, D., Tonellotto, N.: Behavioural skeletons for component autonomic management on grids. In: CoreGRID Workshop on Grid Programming Model, Grid and P2P Systems Architecture, Grid Systems, Tools and Environments, Heraklion, Crete, Greece (2007)

17. Aldinucci, M., Danelutto, M., Kilpatrick, P.: Towards hierarchical management of autonomic components: a case study. In: El Baz, D., Tom Gross, F.S. (eds.) Proc. of Intl. Euromicro PDP 2009: Parallel Distributed and network-based Processing, Weimar, Germany, pp. 3–10. IEEE, Los Alamitos (2009)

18. Aldinucci, M., Danelutto, M., Kilpatrick, P.: Autonomic management of non-functional concerns in distributed and parallel application programming. In: Proc. of Intl. Parallel & Distributed Processing Symposium (IPDPS), Rome, Italy, pp. 1–12. IEEE, Los Alamitos (2009)

19. Caromel, D., Henrio, L.: A Theory of Distributed Object. Springer, Heidelberg (2005)

20. Basukoski, A., Getov, V., Thiyagalingam, J., Isaiadis, S.: Component-based development environment for grid systems: Design and implementation. In: Danelutto, M., Frangopoulou, P., Getov, V. (eds.) Making Grids Work. CoreGRID, pp. 119–128. Springer, Heidelberg (2008)

21. The Grid5000 Project: An infrastructure distributed in 9 sites around France, for research in large-scale parallel and distributed systems (2008),
http://www.grid5000.fr

Using a Teleo-Reactive Programming Style to Develop Self-healing Applications

James Hawthorne and Richard Anthony

Dept. Computer Science,
The University of Greenwich, London, UK
{J.Hawthorne,R.J.Anthony}@gre.ac.uk

Abstract. A well designed traditional software system is capable of recognising and either avoiding or recovering from a number of expected events. However, during the design phase it is not possible to envision and thus equip the software to handle all events or perturbations that can occur; this limits the extent of adaptability that can be achieved. Alternatively a goal-oriented system has the potential to steer around generic classes of problems without the need to specifically identify these.

This paper presents a teleo-reactive approach for the development of robust adaptive and autonomic software where the focus is on high level goals rather than the low level actions and behaviour of software systems. With this approach we maintain focus on the business objectives of the system rather than the underlying mechanisms.

An extensible software framework is presented, with an example application which shows how unexpected events can be dealt with in a natural way.

Keywords: Robust software, Goal-based systems, Software frameworks, Error recovery, Context awareness, Self-healing.

1 Introduction

The process involved in achieving a goal for a human or other living creature is very different to the way in which a goal is achieved in a computer programming language. The level of robustness in a computer program is determined by software developers, in the sense that errors can be prevented or caught by a system of try-catch exception blocks or if-else statements. However, the programmer must explicitly implement these techniques and it is very easy to miss some errors or catch and deal with one error only to fail to deal with that same error if it reoccurs whilst dealing with it at a different level in the code; also there is the decision as to which error to deal with at the point they are detected, and which need to be thrown up to some higher level handler.

In short, there are many ways for many types of errors to cause failure and it is almost impossible for a programmer to deal with them all (some of which were not even known at design time) using these built-in techniques. The problem is that the types of standard techniques and methods available to programmers offer a quite un-natural way of producing robust systems.

A.V. Vasilakos et al. (Eds.): AUTONOMICS 2009, LNICST 23, pp. 114–129, 2010.
© Institute for Computer Sciences, Social-Informatics and Telecommunications Engineering 2010

Consider how humans achieve a goal; for example, loading the dish washer. We have several items to load into the dish washer and we know that if the dish washer door is closed we must first open it. So opening the dish washer could be considered a sub-goal towards the main goal of loading the dish washer.

Next, we consider if there are items already in the machine so we must take the clean ones out first. We must then put the heavier items in the bottom sturdier drawer and the lighter items in the top drawer but your partner unintentionally puts a large plate in the top drawer which means you now have to take out that plate and put it in the bottom drawer before continuing to load the rest in. You are then about to place a cup in the dish washer but your partner does this task before you get a chance to pick up the cup. This means that you are no longer concerned with that cup, so continue with the other items. Humans continue to monitor the situation like this so that we can prioritize tasks in this way. We continue until the goal of loading the dish washer is complete and all the items are loaded.

As another example, you can hear someone calling your name behind you, but you do not exactly know their position. If you continue to look round you will eventually see the person and can respond. You know the general direction of this person but there is no discrete instruction, like rotate 167 degrees as there might be in a computer program. We just continue to rotate until the goal "Can see person" is met.

Within this human process there are many low level decisions made and many possible variations on the actual sequence of turning and looking, yet this seemingly straightforward task (with all its inbuilt contextual adaptation) is very difficult to express using traditional procedural or object oriented code - even if well structured.

Teleo-reactive (T-R) programming [1] was initially developed for the robotics domain but the techniques involved produce systems which react and resolve problems in a more natural way, similar to how humans head towards a goal (with continuous context aware micro-adaptation). With T-R programs, a system heads towards a goal without knowing how to get there precisely. The basic methods involved in these programs produce more robust code which can gracefully recover from errors or unexpected events.

From the users point of view this gives the impression of self-healing. Though faulty code is not strictly replaced with 'fixed' code, the program will continue to run after an unexpected event or error occurs. T-R programs effectively 'fall back' to an earlier stage and work back to this point. The term 'Self-healing' is used in this work to mean a recovery and continuation from an unexpected event, rather than a physical fix or replacement of faulty logic. A more in-depth description of how this works is provided in section 3.

T-R programs have been shown to be highly effective in building robotic agents, but in higher level systems it has not been extensively tested. A Java based software framework variation of a T-R system and example implementation have been built to allow exploration of the feasibility and benefit of using T-R programs in higher level systems. The example implementation described in this paper shows how we can write software without the need to know precisely how individual goals will complete and that we do not need to know exactly what errors will occur to deal with them.

In fact, we find that to some extent, low level error handling is replaced by higher level goal tracking logic. This is important because goal tracking seems to be a lot more intuitive for humans to express accurately than exhaustive error identification and handling. Thus we propose this should be the foremost aspect during software development.

The rest of the paper is organized as follows. Section 2 discusses similar work, including work which achieves similar results to our own with different methods and techniques. We also highlight some of the varied domains in which T-R programs have been applied. Section 3 gives an overview of the components of our framework and how to apply them. We also demonstrate the difficulty of achieving similar results with traditional approaches. Section 4 shows the levels of design involved in the construction of programs using our framework. We show some of the design features and the simplicity of our framework. We then extend the framework to produce a simple example in section 5, show some of the ways we can extend the framework in section 6 and conclude in section 7.

2   Related Work

In our framework and in T-R programs in general, the focus is on *goals* as the most influential part of a program. Maintaining focus on these high-level aims is essential in delivering a valid product. Goal-oriented requirements analysis and reasoning is the main subject of [2] who use the Tropos methodology [3,4] to make goal analysis more complete, developing a formal model for this aim. The authors have developed a goal reasoning tool, which allows algorithms for forward and backward reasoning to be run on goal models. The backward reasoning in Tropos [3,4] is used to analyse the goal models to find the minimum cost goal that could guarantee the achievement of top level goals. This could be important for our framework as we try to guarantee goal achievements.

[5] places a high degree of importance on high level goals with the focus on functional and non-functional goals, such as performance and quality of service. We have mainly been targeting functional goals although our framework could be used to address non-functional issues as well. The authors of [5] also focus on how to generate these goals in the first place, saying that many goals can be obtained simply by asking HOW and WHY questions to obtain parent and sub-goals. Simple Use Case diagrams described in [6] are a good way to obtain and focus on initial goals.

The current work on self-healing systems has largely been directed towards enabling learning and dynamic updates to find better fitting solutions to problems. This is a logical method because the main goal of autonomics is to reduce the reliance on human assistance. However, many of these models and architectures offer very heavyweight solutions to self-healing or only offer a finite selection of alternative models.

As an example [7] uses a large number of connected components to support adaptation. The required changes are simulated to verify the proposed changes. If successful, the changes are merged using an architectural 'diff' tool. [8] also uses a complex model of interconnected components to monitor, analyse, perform adaptation etc. The model is 'externalized' from the running system providing a way to monitor and understand the system in a high level way. This is quite similar to many meta-modelling approaches where the level of abstraction allows changes to be made without the need to know the precise details of the program code.

In contrast to these self-healing systems, [9] uses case and rule based reasoning to make a correct decision about how to proceed in the case of problem events. Storing and retrieving a solution from the case-based-reasoning (CBR) system if it has previously been encountered, or contacting the domain expert if not. They use the example of diagnosing a faulty printer. However, many printer problems require human intervention such as adding more paper to the tray or replacing an ink cartridge and in many situations the system asks the user some questions to help it find the fault. In several cases a technician is called to make the fix. Although this technique could be useful for these 'non-automated' problems, self-healing is usually applied to software oriented errors and the recovery from them.

2.1   Teleo-Reactive Programs

T-R programs developed by Nilson [1] are designed for autonomous control of mobile agents. T-R programs continually accept feedback from the environment, performing actions based on this current state. A T-R program is structured with a hierarchical list of condition and action pairs with each action fulfilling or partly fulfilling the condition of immediately higher precedence. An action will end execution if it ceases to be the highest true condition, either because the action has fulfilled the next condition or some other circumstance has caused this case. Figure 1 is a graphical representation of this production rule structure as shown in Nilsson's work.

This structure directs a design so that the top level condition (the goal condition) of a program is worked towards and will eventually be satisfied. The robot

$$
\begin{aligned}
K_1 &\rightarrow a_1 \\
K_2 &\rightarrow a_2 \\
&\cdots \\
K_i &\rightarrow a_i \\
&\cdots \\
K_m &\rightarrow a_m
\end{aligned}
$$

Fig. 1. Condition-Action production rule list. K_i are conditions and a_i are actions. Condition-Action rules are evaluated from the top down, with the higher rule taking precedence over a lower one.

example shown in [1] and block stacking example applet shown at [10] further illustrates this technique.

The author of [11] has produced a reasoning framework, supporting verification of T-R programs. This is especially important in the case of safety and mission-critical software, but it does increase the complexity in developing T-R programs, unnecessarily in many cases as the system is quite robust in the first instance.

Basing their approach on T-R programs, [12] have designed GRUE, an architecture for controlling game characters (agents). With GRUE the agents are able to react to competing goals with conflicting requirements. GRUE is able to generate new goals in response to the changing current situation and allows multiple actions to be run in parallel in pursuit of several simultaneous goals, thus producing more convincing and believable game agents. The work supports the idea that the T-R approach not only allows programs to be written more naturally, but also produces more natural behaviour from agents.

3   Method

The example application presented in section 5 was created from a base framework. This framework is a generic architecture designed to make it simple to design applications using the methods of T-R programs and gain self-healing benefits. The example extends the actions and conditions in this base framework as required for the specific program needs. Each action must evaluate its condition and obtain a positive result before it is permitted to execute. The actions of the program should build on one another. That is, an action, once executed, should complete or go some way to completing part of the condition of the next action. This action may be executed more than once before the next condition is satisfied. Once this next condition is evaluated to 'true' its associated action can be executed. This action builds towards completing the next condition, and so on until the program is complete.

One important difference between our framework and regular T-R programs is that a condition which fails for whatever reason will return 'false' when queried. Designing code this way, produces self-healing behaviour; for example, a condition which fails for whatever reason will produce a negative evaluation and the program will be forced to 'drop' to a lower level. If the lower condition fails, the program will drop back further, until a true condition can be evaluated. Evaluating these conditions and executing the corresponding actions in order will 'build' the program back up to the original failing condition. On this next attempt, the condition should be 'true'. If not, it will continue with the previous action until it is. In the same way, if a higher priority condition becomes 'true', then there is no longer a need to perform the previous steps and the program will continue from the highest priority 'true' condition-action pair.

This method is a more natural way to produce code that is aligned closely with the way humans solve problems. We continually monitor the 'state of play' to determine what our next move should be. We will not simply fail if we meet a problem. If we need to complete a goal, we often perform a simpler action first, which makes the more complex action easier to address when we meet it again. The TR approach is similar but it is important to understand how this method works in order to effectively use the framework.

Following are short descriptions of the framework elements and classes.

3.1 Goals

In the framework there is no specific type (class) 'goal', but the highest priority condition is effectively the goal, i.e. all the lower prioritized actions work directly or indirectly towards it which marks the end of the program. Another way to view goals is that each T-R program is itself a goal and each T-R program consists of actions and conditions as shown in section 4.2.

We should make clear that the goals should be described in a high level manner, for example a condition 'Has connected client' in the server side of a network application can be treated as a simple boolean condition, but the underlying code to perform this query may be complex and technical. The check might involve sending small packets back and forth to verify that a connected client is still 'alive' or we might just be checking a list of clients which were previously connected and presumed to be still connected. In the later case, more errors are possible later as there is no 'pre-emptive' checking as in the first case. In which case a well designed T-R program will recover. The T-R approach supports functional division of logic and modular development, for example there could be two levels of programmers for the application, non-technical ones to write the T-R program in terms of high level goals and technical ones to provide the low level implementation.

3.2 Conditions

Conditions are linked tightly to actions. An action will only be executed when the associated condition is true. The action directly preceding should work to 'complete' this condition. As an example, a condition could check that a link to some external database is established before its action to gather some records is executed. The previous action would work towards establishing that link and therefore the action to gather records will not be executed until the connection is established.

A condition will return false if an error occurs whilst it is processing. This prevents the program from failing when an error is encountered. In this case the T-R program 'drops' to a lower condition and it is likely that the previous action will be re-performed. If we knew exactly what the error was then we could deal with it directly but to catch and deal with all errors we must perform some broader scoped operations. In short, we do not need to know what the error is to be able to recover from it, but the recovery process might need to go back more levels than is strictly necessary. An analogy would be to replace your car engine because there is an unusual noise coming from it. It would fix the unknown problem that we might not be able to fix ourselves, but it is a big change just to fix a possibly small problem (fan belt just need tightening perhaps). A T-R program of finer granularity would likely be able to perform simpler actions to recover from an unexpected event.

3.3   Actions

Actions in a T-R program are performed only when the corresponding condition is true. Actions are arranged within a T-R program in order of 'closeness' to a goal condition. This way a T-R program builds towards this goal.

Actions will gracefully exit if an error occurs. In such cases, the program will be evaluated again from the top condition down. It may be the case that the error is a rare event and the action will likely succeed when met again. It may also be the case that the action took a relatively long time to complete and the condition which was true before execution is no longer true (a resource is no longer available perhaps). In this case the previous action will likely restore the resource and the action can 'try again'.

It is important to state that actions could and should have simple names but the function of which can be quite involved and complex and could even be constructed with other TR programs. For example, an action called 'Draw Circle' in a graphics application might make several low and high-level calls to a graphics object and driver to perform the action and we should take this approach to naming and constructing actions when writing a TR program.

3.4   Contrast

Here we consider methods (available in most programming languages) for dealing with and avoiding errors. The simplest alternative method to avoiding errors would be a simple conditional statement. Consider a situation where we want to write some output to a stream, but we realize that if the stream has not yet been initialized it could cause problems:

```
if (printStream != null)
   printStream.write("Hello World!!");
```

The if statement avoids using the print stream if it has not yet been initialized. This is fine if that were the only possible error in the system, but this is very unlikely to be the case. As we recognize some other possible errors, we add these to our checks and end up with a large ball of spaghetti code, an untraceable structure consisting of nested if-else statements. Alternatively we could use an exception handler to catch any error within the try block. Something like this:

```
try{
   printStream.write("Hello World!!");
}
catch(Exception e){ \\ignore error }
```

The problem of not initializing the stream before we use it and any other problems would be handled and in this case they would be ignored. This might catch problems, but it does not implicitly fix or recover from the problem. We can add extra code within the catch block to reinitialize the print stream, but again the problem is only solved for one instance of one type of problem and there is no guarantee that this, now fixed code, will ever be accessed again. The next steps after the handler might cause the stream to become uninitialized, but because this code is no longer within the try-catch block, the error is no longer caught. It would be difficult and time consuming to write exception handlers at every point in the code where an error is possible (and clearly the fact that so much commercial software is released with 'bugs' is evidence that even very large software houses are unable to develop code to explicitly handle all possible faults).

Another problem with traditional approaches is that it is easy to assume that the problem, handled once is no longer a concern. This requires the handler to be either duplicated or called from all places the problem could occur (one of the authors previously worked at a software house who released products typically containing hundreds of 'known' bugs - on the basis that most of these had very minor impact and were 'unlikely' to occur, being triggered during very specific sequences of events, and were thus very costly to track down and fully eradicate). In contrast, the T-R approach deals with an error whenever and wherever it occurs at any level of seriousness and for as many times as it arises.

Figure 2(a) highlights the problem in procedural languages of dealing with an event which the programmer had never thought of as a possibility (if the event is not envisioned how can it be dealt with?). What usually happens is the program 'crashes' and the programmer(s) spend some time finding the cause of the problem. The 'bug' is usually fixed but the fix may have caused another error (perhaps more serious than the first) at some other point in the code and it is likely that there are still many more errors to be revealed.

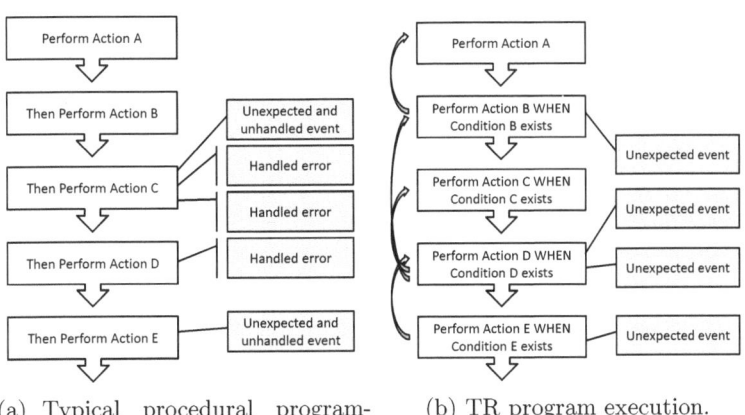

(a) Typical procedural programming language execution.

(b) TR program execution.

Fig. 2. Comparison of typical event and error handling processes in procedural and TR code execution

In contrast figure 2(b) shows that a TR style program executes each stage only when its condition is met. The program will not execute a stage if its condition is not fulfilled and will instead return to a previous step, perhaps executing previously performed code until the condition has been satisfied. The same steps are performed in the case of an unexpected event. Possible unexpected events are indicated in the diagram by the labels on the right and possible roll-backs are indicated by the arrows to the left of figure 2(b).

4   Design

There are effectively two stages of design when using the T-R method and framework. The first is the T-R program part and the second is the more familiar program structure design.

4.1   Teleo-Reactive Program Design

This part of design is the most important part, because it describes the flow of execution of the program. The design of the structure should be more obvious once this part is complete, although it is likely that this part of design will be modified through iterations of the system design phase.

There are some guidelines to follow when designing this part of the program which should make this phase a bit easier. Essentially we need to decide what actions and conditions there should be and in what order. The first condition should be the TRUE condition and the last condition should represent the goal.

The TRUE condition will always return 'true' and thus its action will always execute if there are no higher priority 'true' conditions. This means that if no other action is possible, the action associated will execute until one of the higher priority actions are possible. The TRUE condition provides a place where execution can begin, a definite starting position.

The final goal condition should be the state which all other actions are working towards. Here is an example where a delivery company uses GPS to track its delivery van and display the status of the van on a web site for customers to view (the software is running on a computer in the van depot):

$$Has\_van\_returned\_to\_depot \longrightarrow Nil$$
$$Has\_DB\_link \land Has\_van\_status \longrightarrow Write\_van\_status\_to\_DB$$
$$Has\_van\_status \longrightarrow Get\_DB\_link$$
$$Has\_van\_link \land Timer\_expired \longrightarrow Get\_van\_status$$
$$T \longrightarrow Establish\_van\_link$$

Now imagine we are at the stage where we about to write the van status to the database, but the 'Has van status' condition is no longer true for whatever reason. The program may 'drop back' to establishing the link to the van and getting its state again. From this point it can work back to the point of writing to the database, thus the program self-heals.

The framework allows each T-R program to be run as a thread so in this case the T-R program would have as many instances as there are vans to track, each running in the background, leaving the main program free for more important tasks. We can be confident that each instance will work unsupervised because we know errors will be handled. This is why confidence in this part of design is so important. This is important for autonomics applications, where programs need to handle themselves resulting in little maintenance for the programmer after deployment, and only low levels of external supervisory input are needed.

4.2   Program Structure Design

The framework has been designed to allow the concepts discussed in this paper to be realized and even integrated into current systems as easily as possible. Programmers are required only to extend the types *Action* and *Conditional* as shown in figure 3. It is important to recognize that the onus currently resides with the programmer to produce actions and conditions which can complete at least some of the time, given the expected state and context. The framework robustly handles events and errors, however, an action or condition which never completes will always cause a 'deadlock' situation and the T-R program will never progress past this point. In section 6 we discuss how future iterations of the framework might reduce or remove the chances of this problem occurring.

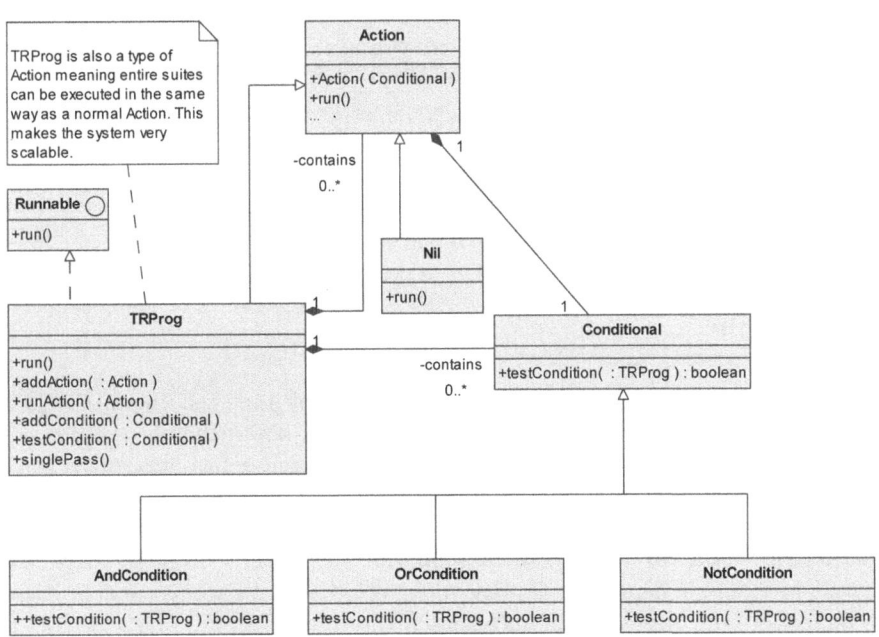

Fig. 3. Main framework presented as a UML class diagram

The design allows for easy modifications and extensions to be made to the framework without affecting current implementations. For example future versions of the framework could prohibit an Action which causes a deadlock from ever being added to the T-R program (See section 6).

The TRProg type is used as the main controlling class of the framework. This type contains an ordered list of Actions and an unordered list of Conditionals. Actions are executed in the order they appear in the list if its associated Conditional type returns TRUE. If not, the TRProg will 'try' the next Action. Conditionals are automatically added when the Action is added to TRProg unless the Conditional instance is already listed.

As can be seen in figure 3, TRProg extends the Runnable interface, meaning a TRProg can be run as a background thread. This can be useful to enable a T-R program to asynchronously run in the background performing some vitally robust task while your main thread of execution continues unaffected.

The figure also shows that the TRProg is itself a type of Action, meaning that an Action in the list could be an entire TRProg type, containing a different set of Actions. The TRProg may also call itself in a recursive fashion.

Actions that have been extended are required to implement the Run() method. If its condition evaluates to true, the code in this method is executed. The provided Nil Action is an empty action and is usually associated with the top level goal condition, i.e. if the goal condition is reached, perform no action.

When an action is constructed it must accept exactly one Conditional argument. However, this single conditional can contain many sub-conditionals when layers of And, Or and Not types are nested.

Conditionals contain the testCondition() method which must be implemented in extensions. This testCondition() method returns a boolean literal indicating whether the condition is true or false. A conditional also returns false when, during evaluation within the TRProg class, an error is encountered.

5   Example

The example presented here is a simple simulation of a file sending application presented from the point of view of the server side. The example extends the framework to simulate a simple protocol which we assume has little or no error handling abilities. The program contains several condition-action pairs leading to its main goal. For the sake of simplicity the file is always the same and is fixed length, with the aim being to send the file to the currently connected client. The example demonstrates how unexpected events such as random client disconnection and noisy communications are recovered from. This example is not designed as an accurate representation of networking components, it only serves to highlight some of the features and benefits of the approach. Figure 4(b) shows the example in mid-execution with one of the four clients connected.

$$IsFileComplete \longrightarrow Nil$$
$$IsFivePacketsSent \longrightarrow ChangePacketSize$$
$$IsConnected \longrightarrow SendNextPacket$$
$$IsAccepting \wedge IsWaiting \longrightarrow Connect$$
$$T \longrightarrow Accept$$

(a) Alternative T-R state view

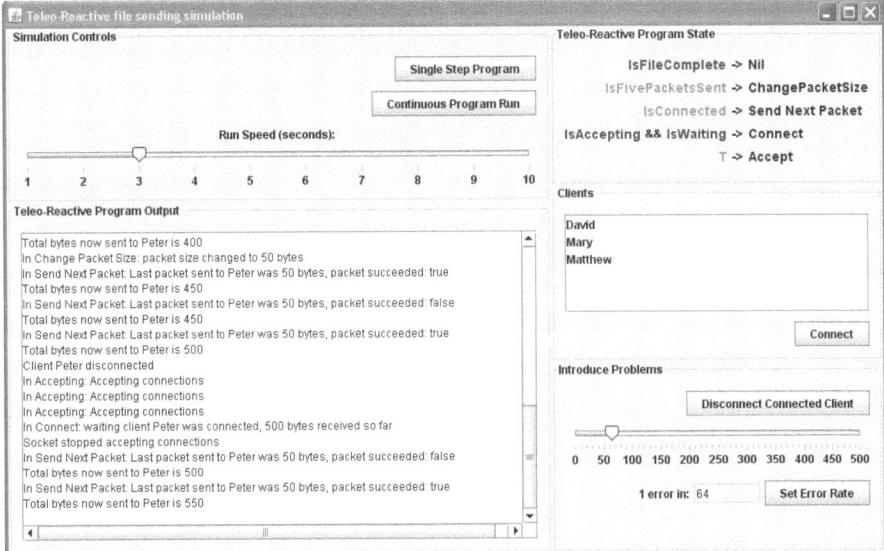

(b) 'True' conditions will be shown in green. 'False' conditions will be shown in red

Fig. 4. Screenshot from the example program

5.1 T-R Elements

The goal of the example is the same as the top level condition, to finish sending the file to the connected client. With the file sent the program does nothing until the top level condition turns false again. Figure 4(a) shows an alternative view of the 'Teleo-Reactive Program State' section from the example.

The conditions in the example are:

- **T** - As this is always true its action will always execute if no higher precedence condition is true.
- **IsAccepting** - The example simulates a socket which accepts incoming connection attempts. This condition returns the accepting state of the socket.
- **IsWaiting** - This condition reports whether there is a client in the waiting state.
- **IsConnected** - The client is connected after it is moved from waiting to the connected state.
- **IsFivePacketsSent** - Returns true when a multiple of 5 packets has been sent to the connected client.

- **IsFileComplete** - True once the whole file has been transferred to the connected client.

The actions in the example are:

- **Accept** - Simply turns the accepting state of the socket to on.
- **Connect** - Any waiting client is connected and the Accepting state is switched off.
- **Send Next Packet** - A packet is sent to the connected client at the current packet size.
- **Change Packet Size** - The packet size is adjusted using a simple algorithm based on statistics from the previous packet sending attempt.
- **Nil** - Performs no action.

5.2   Controls

The simulation GUI controls enable the user to set a run speed by changing the run speed slider and clicking the 'Continuous Program Run' button. This sets the simulation to perform a full pass through its action list every X seconds. The 'Single Step Program' button performs a single pass at each press and turns off automatic running.

The program state area gives visual feedback on the current progress in the T-R program. In the GUI a 'true' condition is coloured green, whilst a 'false' condition is coloured red. The action associated with the highest precedence 'true' condition is executed and will continue to execute at each pass while it remains so.

There are four possible clients that can be connected to the file sending service. Clicking on one of these four names followed by 'Connect' performs the connection attempt.

The 'Introduce Problems' controls are used to inject 'unexpected' conditions. This includes a random disconnection of a connected client and the introduction of degraded communication. The level of degradation is set with the slider and 'Set Error Rate' button.

5.3   Working through the Example

In this test, we make a connection and begin sending the file to this client in blocks. We demonstrate how the program copes with the unexpected events of disconnection and changing quality of service. This simple example is intended to illustrate the benefits and robustness of the approach.

When the example application is started the only 'true' condition is the first 'T' condition. On the first pass, the accept action will be executed, making the 'IsAccepting' condition 'true'. This will be the only executed action on subsequent passes until a higher precedence condition evaluates to 'true'. If we now connect one of the clients, the 'IsWaiting' condition becomes 'true'. The conditions for the action 'Connect' are now completely satisfied and takes precedence over any lower actions. Its execution on the next pass will connect the waiting client, stop the socket from accepting connections, and cause 'IsConnected' to return 'true'. The service will continue to send packets to the client unless a higher precedence condition becomes 'true'.

There are a variety of ways in which the connection to the client can be lost. Clicking 'Disconnect Connected Client' causes the only 'true' condition to be 'T' again. The program will then work back to the point where the problem occurred and continue sending packets until the entire file is sent. The file send is resumed at the point where the last successful packet was sent. If another client connects before the fist client, they must wait until this client disconnects.

Every multiple of five packet sending attempts the service has the opportunity to adjust the packet size to suit the current communication quality and keep the service running at optimal levels. By changing the error rate in the simulation we can show how this works. The packet size will increase or decrease as appropriate. Once the 'IsFileComplete' condition is satisfied no other action will be performed and the goal is complete.

It was simple to add self-optimising behaviour to the program, which is demonstrated by the changing packet size program function. We had not intended to implement self-optimization within the first version of the example, but it turned out to be very simple to add this function. Using our framework, constructing the application was remarkably easy, and with future iterations of the framework (adding further robustness and inherent validation checking), this will be an easier task.

A more elaborate example scenario could be chosen with the opportunity to introduce a greater variety of problems and some of further features of the framework such as ease of reuse and recursion. However, the example is simple enough to demonstrate the main advantages which a more complex scenario might have blurred.

6   Future Work

A possible way to extend the framework is in use of policies to specify and possibly dynamically replace high level goals. A more advanced method of use of policies in T-R agent control is in [13] where situation graphs determine good policies for groups of cloned T-R agents. It is claimed that the use of situation graphs enables policies to be evaluated taking into account objective states and not just perceptions, yielding a high degree of discrimination.

We are interested in further applications of the framework in the development of autonomic and self managing systems [14,15] where the system needs to continually adjust its behaviour to suit its operating circumstances. It is far easier to describe these systems by their high level goals than by their actual behaviour at any given moment, and thus we suggest that T-R programs have great potential for this domain.

T-R programs also provide a lot of opportunity for automatic learning techniques and much of the work based on T-R programs has been addressing this issue. For example [16,17]. Automatic learning of new goals and dynamic adaptation would reduce human reliance and this is another direction for future work on the framework we would be interested in pursuing.

In our view and from an autonomic viewpoint beneficial adaptations should, where possible, be transparent to the framework user. Automatic validation techniques could be incorporated to inform the programmer of the problem at the point where an action is added to the program, perhaps using simulation of T-R programs. This is a logical progression for the framework and would provide further guarantees about the level of robustness and reduce or eliminate the possibility of a dead/live-lock situation.

7   Conclusion

T-R programs were originally invented for the robotics domain and for agent control. This work shows how T-R programs can be effectively applied at a higher level and with many benefits over more traditional approaches. This method of programming focuses on goals as a driving factor and produces code more closely associated with a thought process. This in turn produces more natural behaviour and increases the potential for adaptability.

The T-R framework makes it simple to produce programs that use this technique as we demonstrated with our example scenario. T-R programs can form the main structure of a program with additional T-R programs called hierarchically. Or a T-R program could be used on a subsection where the program is started as a background asynchronous thread.

This initial work demonstrates the viability of the approach to programming high level software systems and we have illustrated how this approach contribute's to robust program design and self-healing capabilities. At this early stage in development the framework has been shown to be very effective in recovering from errors and using natural goal progression to improve system design. We believe that the complex models and architectures in some other self-healing approaches are not necessary and in many cases these methods are limited in the number and types of errors which can be handled.

References

1. Nilsson, N.J.: Teleo-reactive programs for agent control. Journal of Artificial Intelligence Research 1, 139–158 (1994)
2. Giorgini, P., Mylopoulos, J., Sebastiani, R.: Goal-oriented requirements analysis and reasoning in the tropos methodology. Engineering Applications of Artificial Intelligence 18(2), 159–171 (2005)
3. Bresciani, P., Perini, A., Giorgini, P., Giunchiglia, F., Mylopoulos, J.: Tropos: An agent-oriented software development methodology. Autonomous Agents and Multi-Agent Systems 8(3), 203–236 (2004)
4. Castro, J., Kolp, M., Mylopoulos, J.: Towards requirements-driven information systems engineering: the tropos project. Inf. Syst. 27(6), 365–389 (2002)

5. Van Lamsweerde, A.: Goal-oriented requirements engineering: A guided tour. In: RE 2001: Proceedings of the Fifth IEEE International Symposium on Requirements Engineering, pp. 249–262. IEEE Computer Society, Washington (2001)

6. Fowler, M.: UML Distilled: A Brief Guide to the Standard Object Modeling Language. Addison-Wesley Longman Publishing Co., Inc., Boston (2004)

7. Dashofy, E.M., van der Hoek, A., Taylor, R.N.: Towards architecture-based self-healing systems. In: WOSS 2002: Proceedings of the first workshop on Self-healing systems, pp. 21–26. ACM, New York (2002)

8. Garlan, D., Schmerl, B.: Model-based adaptation for self-healing systems. In: WOSS 2002: Proceedings of the first workshop on Self-healing systems, pp. 27–32. ACM, New York (2002)

9. Hassan, S., Mcsherry, D., Bustard, D.: Autonomic self healing and recovery informed by environment knowledge. Artif. Intell. Rev. 26, 89–101 (2006)

10. Nilsson, N.J.: Teleo-reactive programs web site
http://robotics.stanford.edu/users/nilsson/trweb/tr.html (last accessed 2009)

11. Hayes, I.J.: Towards reasoning about teleo-reactive programs for robust real-time systems. In: SERENE 2008: Proceedings of the 2008 RISE/EFTS Joint International Workshop on Software Engineering for Resilient Systems, pp. 87–94. ACM, New York (2008)

12. Gordon, E., Logan, B.: Game over: You have been beaten by a GRUE. In: Fu, D., Henke, S., Orkin, J. (eds.) Challenges in Game Artificial Intelligence, Technical Report. Papers from the 2004 AAAI Workshop, pp. 16–21. AAAI Press, Menlo Park (2004)

13. Broda, K., Hogger, C.: Determining and verifying good policies for cloned teleo-reactive agents. Int. Journal of Computer Systems Science and Engineering 20(4), 249–258 (2005)

14. Kephart, J.: Research challenges of autonomic computing. In: International Conference on Software Engineering (ICSE), pp. 15–22. ACM, New York (2005)

15. Sterritt, R., Parashar, M., Tianfield, H., Unland, R.: A concise introduction to autonomic computing. Advanced Engineering Informatics 19(3), 181–187 (2005)

16. Kochenderfer, M.J.: Evolving hierarchical and recursive teleo-reactive programs through genetic programming. In: Ryan, C., Soule, T., Keijzer, M., Tsang, E.P.K., Poli, R., Costa, E. (eds.) EuroGP 2003. LNCS, vol. 2610, pp. 83–92. Springer, Heidelberg (2003)

17. Choi, D., Langley, P.: Learning teleoreactive logic programs from problem solving. In: Kramer, S., Pfahringer, B. (eds.) ILP 2005. LNCS (LNAI), vol. 3625, pp. 51–68. Springer, Heidelberg (2005)

Sensor Selection for IT Infrastructure Monitoring

Gergely János Paljak, Imre Kocsis, Zoltán Égel, Dániel Tóth, and András Pataricza

Department of Measurement and Information Systems
Budapest University of Technology and Economics
Magyar tudósok krt. 2, H-1117 Budapest, Hungary
paljakg@sauron.inf.mit.bme.hu,
{ikocsis,egel,dtoth,pataric}@mit.bme.hu

Abstract. Supervisory control is the main means to assure a high level performance and availability of large IT infrastructures. Applied control theory is used in
physical and virtualization based clustering, autonomic-, self-healing and cloud
computing, but similar problems arise in any distributed environment.

The selection of a compact, but sufficiently characteristic set of control variables is one of the core problems both for design and run-time complexity.
Most results in the literature are based on a single algorithm for variable selection, but our measurements indicate that no single algorithm can generate faithful estimates for all the different operational domains.

We propose to use a combination of different model extraction techniques
on benchmark-like data logs. The main advantages of this multi-paradigm approach are twofold: it provides good parameter estimators for predictive control
in a simple way; and supports the identification of the actual operational domain facilitating context-aware adaptive control, diagnostics and repair.

Keywords: Autonomic computing, control theory, signal processing, artificial
intelligence, benchmarking, performance and performability control.

1 Introduction

Modern system management aims at guaranteeing a high service level in terms of all
operational aspects, primarily of performance and availability by applying a feedback
control loop scheme. Feedback control in autonomic computing continuously monitors
the service level and upon an unwanted deviance triggers optimization/health maintenance actions according to a predefined control policy.

Trustworthy autonomic performability management necessitates establishing a formal relationship between certain *monitored* and *influenced* attributes of a system even
for rough granular control; for fine granular approaches utilizing classic control theory
it is even more so (for some examples, see [1]). However, when first principles based
modeling is infeasible – what is rather the rule, than the exception for IT systems in
general – a prerequisite of system identification is establishing the set of underlying
attributes for the model. While this naturally occuring task of autonomic performability
control design seems to be quite neglected, it is by no means a trivial one. This paper
proposes an AI inspired approach to address this requirement.

A.V. Vasilakos et al. (Eds.): AUTONOMICS 2009, LNICST 23, pp. 130–143, 2010.

A monitored configuration consists of the application and its runtime platform instrumented with additional sensor and actuator agents *Sensors* report on the run and health state of the application and its run time platform The monitoring node processes these sensor values and initiates active diagnostic probing, repair actions, like dynamic allocation or reallocation of resources or even a reconfiguration of the application deployment in the system to be executed by actuator agents.

The monitoring scheme covers on the one hand functional and extra-functional discrete state change events (like beginning or termination of a job or detection of an error manifestation, respectively) and on the other hand platform, application, component and system service level quantitative performance and dependability measures.

While this overall control algorithm problem appears in a general form in all large scale systems and the principle of our approach remains valid for this more general context, we confine our subsequent discussion to datacenter-like infrastructures (and cloud environments).

Large IT infrastructures and even the monitoring functions are large-scale distributed systems. The objects of the control in server farms and clouds, the applications, their deployment with the monitoring and control agents and the local control functions in the application nodes are all distributed. The control functionality has typically a hierarchical structure composed of domain and system controller nodes processing the raw sensor data and preprocessed data from the subordinate monitor nodes.

A *monitoring and supervisory control node*

- collects the raw information directly incoming from the sensors and the potential preprocessed data from subordinate monitoring nodes,
- *correlates the events,* estimates the metrics and,
- identifies (and possibly predicts) the *situation,*
- compares them with those in an anticipated use case (e.g. prediction of a potential overload of a particular resource or diagnoses fault in a component) based on estimates,
- decides on the reactions to be executed by the actuators according to a predefined *control policy* usually formulated in a rule-based manner.

    The candidate actions triggered by the actuator agent deployed into the monitored infrastructure consist of *tuning the resources* available for the individual application tasks (priority reassignment in multitasking, modification of the resource arbitration in virtualization) or *structural reconfiguration* (dropping non-critical tasks, task replication and/or migration).

The prevailing industrial approach is still dominated by the former age of manual control for configuring system supervision. It deploys and activates a very wide set of sensor agents onto the platform under control, as the operator may select the relevant ones and simply ignore all the others. On the processing side, some application and infrastructure specific, quite ad-hoc thresholds and simple empirical rules are provided, aiming primarily at defining "normal" operational intervals on a per metric basis and raising an "out of range" type of alarm in the case of a deviance.

However, the automated identification (prediction), diagnosis and reaction on problems needs precisely formulated rules of metric aggregation and correlation.

Over-instrumentation, i.e. monitoring too many metrics of a system poses significant problems, as a large number of threshold estimation, quantification, aggregation, situation identification and diagnostic rules exclude reliable manual design and maintenance, especially in evolving applications. On the other hand monitoring too many metrics also causes unnecessary performance overhead on the monitored systems, and data collection nodes especially in case of historic data collection.

Under-instrumentation, i.e. the improper reduction of the set of monitored metrics, on the other hand can significantly compromise the capabilities of supervision, manifesting in large reaction times to workload changes, significantly reduced availability due to late error detection and diagnosis.

Heuristic manual control based monitoring does not scale well for large, heterogeneous IT systems from many aspects; as emphasized by industry initiatives as IBM Autonomic Computing [2] or the evolving "cloud computing". Such systems increasingly employ almost fully automated structural reconfiguration and other adaptive techniques borrowed from control theory to guarantee the performance and dependability of services.

Consequently, a theoretically well-founded approach is needed for selecting a minimal or sufficiently small set of metrics and associated points of measurement out of the technically measurable ones, which characterizes the system "adequately". Selecting such a metric set is certainly only the precursor to setting up e.g. diagnostic rules. More precisely, given a *control objective metric* (e.g. throughput of a particular service as an influenced attribute), we seek a corresponding, near-minimal subset of metrics and an appropriate approximation function delivering enough information to assure the fulfillment of the control objective.

We illustrate the core problem and our approach by a simple example of curve approximation:

Given a series of observations y co-recorded with all the parameters potentially forming its cause, we have to select the principal factors in an estimator of this objective function sufficiently closely matching it.

After selecting the set of independent variables for the estimator, we have to select a single function or a family of functions for a best fit estimator (for the sake of simplicity we assume that a single independent variable is sufficient to create a faithful estimator of the observation series in Figure 1.). Here the "best fit" is measured by means of an approximation error metrics quantifying the deviance of the estimator from the observation.

Given the set of independent variables, one option is to use a single function for curve fitting (curve 1 and 2 in Figure 1.) An additional degree of freedom is given when a family of functions is offered for fitting instead of a single one. Here the accuracy of the approximation can be further improved by piecewise fitting, i.e. by selecting a particular function for a given interval resulting in the best match (the partially linear/non-linear/linear curve 3 in Figure 1.).

The online approximation delivers as byproduct the identification of the best matching estimator function in addition to the expected value of the objective variable, as well.

As the splitting of the domain of the independent variable corresponds to the different operation domains, the information on the best fitting function identifies the actual state at a rough granular level at the resolution of the operational domain. An adaptive control policy may fine granular evaluation of the causal variables for diagnostics after the appearance of degradation, as indicated by the best fit of the corresponding non-linear approximation function of a phenomenological variable.

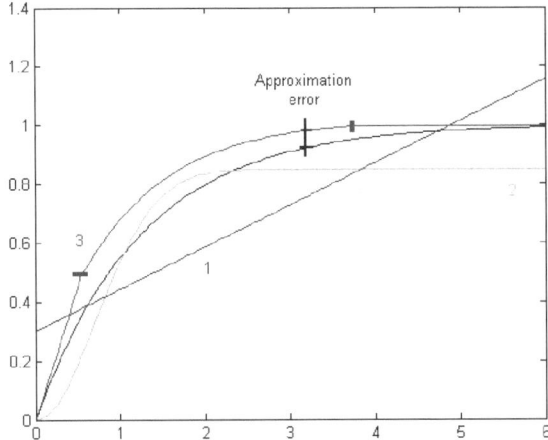

Fig. 1. Curve fitting examples with respect to the objective function in black

This paper proposes to combine linear estimators with the powerful minimum-Redundancy-Maximum-Relevance (mRMR) nonlinear feature selection scheme for the selection of such a small set of metrics that still adequately characterizes an objective metric.

Measurements on a testbed implementing an industrial OLTP performance benchmark equipped with a fully instrumented, commercial enterprise system monitoring product were used for the experimental validation of the approach.

The remainder of this paper is organized as follows. We first briefly describe prior research underlying our experiment and introduce existing approaches and then discuss monitoring instrumentation issues. Section 4 describes the test-bed, subject of our investigations that are described in Section 5. Section 6 highlights the most important results which we then evaluate and conclude.

The experiment was partially funded by the AMBER EU FP7 project, "Assessing, Measuring and Benchmarking Resilience" [3].

2 Related Work

Evaluating large amounts of measured metrics by statistical methods and methods from artificial intelligence can be effectively utilized to improve enterprise systems' dependability by allowing fault detection [4, 5] and the forecasting of the system's behavior [6, 7]. Unfortunately, the dimensional problem of such approaches has not been sufficiently addressed by the community. Either simple, linear methods are used either in forward selection or backward elimination fashion or a wrapper approach [8] is utilized that requires the presence of a learning algorithm and thus the speed of the process is significantly reduced while results are dependent on their further usage. Munawar et al. [9] are suggesting a Mutual Information based method and show that non-linear correlations exist between metrics and those can be effectively utilized to enhance fault diagnosis.

Our approach applies all the measures offered by the large set of sensors in industrial tools (e.g. IBM Tivoli Monitoring) in a benchmark-like experiment. The set of the measures to be monitored in the operational environment will be reduced by intelligent log analysis to those few ones which sufficiently characterize the system by themselves for fast reaction or early error detection. We use the systematic, well-tunable mRMR algorithm for variable selection. It is also based on mutual information and has been shown to scale well for large problem spaces [10, 11].

There are numerous approaches to utilize the data obtained this way. [12] shows an entropy based – like mutual information in case of mRMR – fault detection method that is dependent on the window size and thus may not always be sufficient for early fault-detection. [13] presents collected data from a web server under overload and builds time series ARMA (autoregressive moving average) models to detect aging, and estimate resource exhaustion times. [14] presents a way of on-line discovery of quantitative models, based on linear least-squares regression and shows its application for a database system. However, no established investigation is known that would validate these approaches across different operational domains and evaluate their performance.

3   Instrumentation Support of Metric Selection

We apply the mRMR feature selection scheme for multi-tiered online transaction processing systems. All major components of the system (operating system instances, middleware, server software, network interfaces and components) are instrumented with sensors in the initial data log acquisition phase.

Commercial off-the-shelf monitoring products offer a large selection of candidate *sensor agents* out-of-the-box for each major component type. IT infrastructure components and services have typically the option to be associated with a wide set of *metrics* and emitted *events* delivered in a raw form by the *instrumentation* of the controlled node.

Local metrics measured by the sensor agents and derived metrics used in the control nodes can be grouped jointly, independently of their source into two main classes:

- *Phenomenological metrics* deliver the measured or derived results in an implementation independent form in the terms of some standard (logic) units (for instance the average transaction time in a database).

  Such metrics are typically used to characterize the extra-functional characteristics of the services delivered by the individual software components and applications, computing nodes and the entire system.

  As the objective of the feedback control is keeping the overall service level characteristics within the range allowed by the specification (frequently expressed in the form of an SLA), these are the primary control variables at the topmost level of control.

- *Causal metrics* are able to "look inside" the component internals (for instance buffer pool attributes for the version x of type y of a database). The main advantage of their use is the high level of observability and controllability provided at a price of high maintenance and version control costs originating in the strong implementation platform dependence.

Their typical use is on the one hand the reduction of error latency in critical applications, as monitoring and checking the internal state may detect an error in a component prior of the degradation of the services delivered by it; on the other hand they are used in fine granular diagnostics.

The components of the target system are treated as providing either a "resource service" or a "request-response" service for other components. These two service types have some associated metrics that are meaningful in all cases, regardless of the specific service provided or its implementation. The former category is typically associated with quantitative metrics that are utilization aggregates, originating from the behavior of multiple clients. The latter case can be characterized by workload (faultload) and output performability metrics.

As a rule of thumb, all these "implementation independent" metrics (arguably of a phenomenological nature) should be recorded for each component. This guideline is to ensure that there is a uniform set of metrics that applies for all components, comprising at least a black-box characterization of all the individual system components.

Additionally, in most cases the COTS instrumentation of the components offers insight into the internals of the component implementation supporting a much earlier problem detection and actual fault diagnosis, like their manifestation in the services. However, our approach should not solely rely on these, as the behavioral coverage they provide is quite hard to reliably assess.

As part of the necessary system instrumentation, the examined objective metrics forming the core factors of the service level agreement offered to the end user are also to be chosen and their measurement implemented. For OLTP systems, service response time and throughput are the most natural choices for selecting sensors for managing service performability, which is our current focus.

4 Experimental Setup

Our experimental testbed is a small, three-tier virtualized server architecture having two additional nodes: one for workload generation, the other for monitoring and processing the captured data (Figure 2.). The infrastructure contains 6 virtual servers, each of them running the CentOS 5 GNU/Linux distribution with Apache, Tomcat, MySQL and Sequoia (a database clustering middleware) installed, respectively. All servers are deployed on a single VMware ESX host.

The environment runs an implementation of the TPC-W standard benchmark [14]. The workload used in the experiment is the TPC-W "Shopping Mix".

Two objective metrics were chosen: response time and throughput, using Web Interaction Response Time (WIRT) and Web Interactions Per Second (WIPS) metrics of the TPC-W specification for exact definition.

IBM Tivoli Monitoring 6.1 (ITM) is used for monitoring purposes. ITM is a centralized, agent-based monitoring solution: central monitoring server(s) collect measurement data and event notifications provided by monitoring agents running on the supervised hosts. On a single host multiple agents may be deployed, as every platform and software component covered by the product is supported by a separate agent. Numerous platforms, software components and devices that are not supported by the product

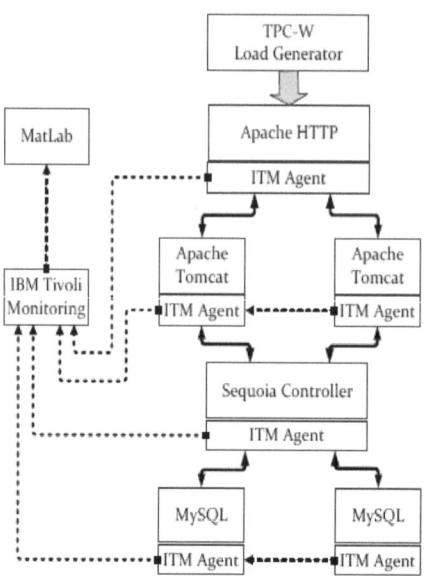

Fig. 2. Architecture of the experimental setup

out-of-the box (or by product extensions) have agents freely available on the Tivoli Open Process Automation Library (OPAL) site. Most of these utilize the Tivoli Universal Agent, a special agent type with the purpose of enabling the development of custom sensors against documented interfaces. Altogether over 1000 metrics were measured by the agents, with a sampling interval of 30 seconds.

A Java importer has been implemented for in-MATLAB execution that queries data samples from the central ITM server and transforms them to MATLAB-format time series for further processing.

5   Experimental Methodology

As our goal is early fault-detection and pro-active prevention we opted for the regression of the selected objective metrics i.e. throughput and response time. First of all, we have to reduce the number of sensors/metrics considered in order to avoid over-instrumentation and to simplify the regression problem.

Dimension reduction is the generic term for methods that aim at reducing the number of considered variables in the mathematical model of a given problem. The dimension of the task at hand is the number of variables to be measured for some further action. The problem is well-known in the statistical and machine-learning communities who were the pioneers facing the problem of high-dimensional datasets. Here the impact of a given variable can frequently not be determined on sole human expertise; all and any of them can be "important" for the understanding of the examined process/system.

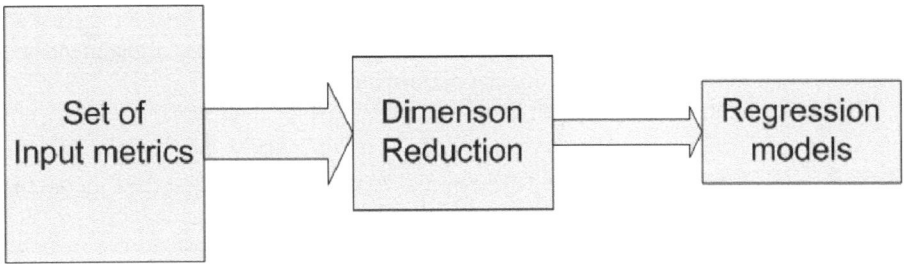

Fig. 3. Experimental methodology

Dimension reduction methods are traditionally divided into two groups: feature selection- and feature extraction approaches. Feature selection aims at finding a subset of the measured variables while feature extraction is applying a projection of the multidimensional problem space into a space of fewer dimensions thus resulting in aggregate measures that did not exist in the measured environment [16, 17].

So the dimension of the problem – i.e. the number of attributes processed – shall be reduced. A selection of few attributes is required: finding those that mostly influence system-level metrics (e.g. throughput, response time) and thus enable the construction of a control algorithm with relatively unambiguous rules. As a basic approach we implemented a greedy forward selection method that uses linear regression as an evaluative measure in the incremental process. We also selected the relatively new mRMR algorithm [18], a feature-selection method to identify candidates that are likely to have influence on high-level performance metrics for its high accuracy and fast speed [19], presenting a promising approach to grab a descriptive set of metrics considering various aspects of the system.

mRMR is based on the concept of mutual information, that for two probabilistic variables x, y, can be calculated as:

$$I(x;y) = \iint p(x,y)\log\frac{p(x,y)}{p(x)p(y)}\,dxdy \tag{1}$$

In our case we want to select a set of variables, S, so that the mutual information between each element of S and the objective metric c is maximal (maximum relevance):

$$\max R(S,c), \quad R = \frac{1}{|S|}\sum_{x_i \in S} I(x_i;c) \tag{2}$$

and the redundancy is minimal inside S, which means the mutual information between the elements is minimal:

$$\min r(S), \quad r = \frac{1}{|S|^2}\sum_{x_i,x_j \in S} I(x_i;x_j) \tag{3}$$

We intend to find those attributes that have the highest mutual information against an objective metric, and keep the mutual information low among the set of the identified attributes in order to find signs of distinct performance issues.

In practice an iterative algorithm optimizes the following condition:

$$\max_{x_j \in X - S_{m-1}} \left[I(x_j; c) - \frac{1}{m-1} \sum_{x_i \in S_{m-1}} I(x_j; x_i) \right] \tag{4}$$

where $m = |S|$, the size of S in the current iteration and S_{m-1} being the set of metrics selected prior to the current iteration.

So the algorithm in the first step selects the variable $x \in X$ with the greatest mutual information with respect to the objective metric. In the second step it selects the variable $y \in X - x$ with the smallest mutual information with respect to x while maximizing the S_2 subset's mutual information according to the objective metric c. It carries on iteratively until the pre-determined number of variables is reached.

Please note that calculating the mutual information for a set of time series is a very computation intensive task. The mRMR algorithm is incremental, gradually selecting the target variables by choosing the next best fitting one for extending the variable set. This way, it is only optimal in a local sense for each iteration step but does not ensure global optimality.

As for the regression part we decided to utilize two different methods: linear regression that aims at approximating the objective metric as the weighted sum of the selected variables and two-layer feed-forward neural network that works similarly but has non-linear capabilities. Traditionally the linear regression equation is as follows:

$$Y(t) = \sum_{i=1}^{K} w_i X_i(t) + \varepsilon(t)$$

Assuming that we selected K variables X the method computes the weights w to calculate the objective value Y in the given time t with error $\varepsilon(t)$. In case of prediction, the right hand side is shifted back in time and thus the result is estimated based on the available values of the past i.e. using the values (t-k), k = 1..N.

6   Experimental Results

In order to gain an insight into the setups internal relations, we stressed the system with different load scenarios, including normal and extreme loads and some abrupt changes as well and then evaluated the acquired time-series with the methods introduced above.

First we examined the available features and those selected. Calculating the correlation matrix we find a lot of high coefficients, clearly confirming the base assumptions in [20] of lower dimensions. On the other hand it is also suggests that due to that and the large number of measured metrics we are unlikely to find matchings in the individual scenarios between the features selected by mRMR and those by the greedy algorithm. However, that is not the case. By selecting 50 features we find that 17% of them are present in both cases and in general the simpler the case (practically: the lower the load) the more matches are present. Finally, the approaches tend to select the same metrics (although with different ranks) across different load scenarios (around 40% of the selected metrics) thus highlighting those that should be considered under most circumstances.

To evaluate the methodology we selected 6 different load scenarios, performed the feature selection and executed the approximation with constantly growing number of features. A typical curve is depicted in Figure 4. while the Mean Square Error results are shown in Table 1. where 'R' stands for Linear Regression, 'N' for Neural Network, 'F' for the Forward Selection and M for the mRMR feature selection respectively.

Table 1.

	MSE - RF	MSE - RM	MSE - NF	MSE – NM
LOW	0.0233	0.0326	1e-30	1e-4
MID	0.0510	0.0887	1.86e-4	1e-29
HIGH	0.2361	0.3139	1e-25	1e-26
VHIGH	0.9309	1.0020	0.7746	0.8111
DROP	0.2806	0.4990	0.0227	0.0516
STEP	0.1908	0.2300	0.0961	0.1818

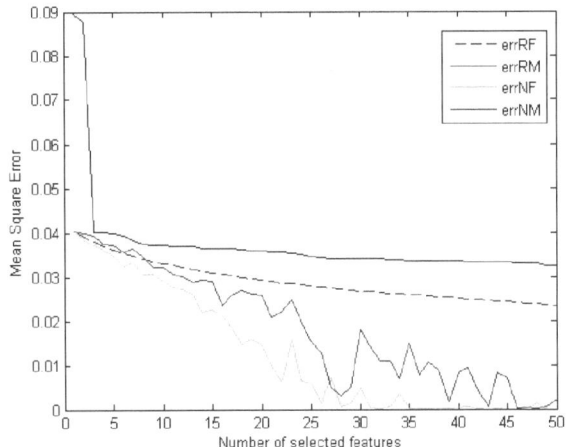

Fig. 4. MSE as the function of selected features

The overall results seem to show that despite the successes of using mRMR in bio-informatics applications, it is inappropriate in our case where a simple greedy algo-rithm can outperform it.

If we take a closer look at the targeted throughput (see Figure 5.), our objective met-ric, we can discover the intervals where the system begins to saturate. Note the abrupt falls in performance at time instants 33, 74, 110 and 170. Those are the typical times indicating that non-linear phenomenon like resource pooling, swapping and caching do occur and the mutual information based non-linear capabilities of mRMR come in handy.

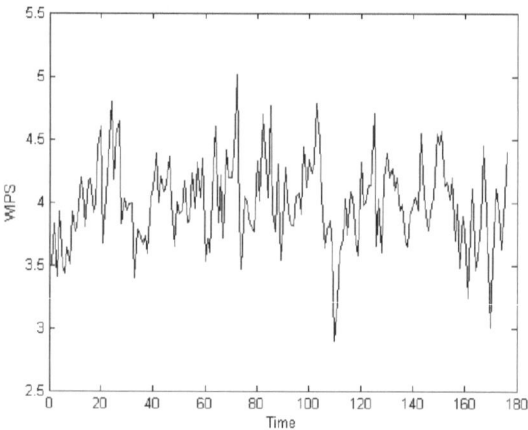

Fig. 5. The 'Throughput' objective metric in the MID scenario

Considering this we can assume the following about the system:

- NORMAL operational states can be adequately approximated by means of linear methods, providing good approximations in a simple and computationally inexpensive way.
- DEGRADATION system states however can be more effectively treated by non-linear means that can provide their earlier detection and thus a more time for pro-active actuation.
- SATURATION (over-loaded) system states can also be approximated by linear methods as the systems performance will be mainly influenced by its physical parameters and limits rather than the internal relations of its metrics.

7   Conclusion and Future Work

The most important conclusion of our work is that no single approach is sufficient for system management and early diagnostics, but a combination of the approaches best fitting to the individual qualitatively different operational domains is needed for this purpose. Related efforts [12, 13, 14] all seem to lack this consideration and while [13, 14] show convincing results their impact is limited to the normal and saturated operation states, disregarding degradation, and thus seem insufficient from the pro-active actuation's point of view. [12] on the other hand may lack the benefit of early diagnosis in some cases where linear methods could raise alarm in a more prescient way.

All of the solutions above exploit implicitly some mutual interdependency between the operation domain and the best fitting function. However, our measurements indicate that the behavior of a system is so much different in the individual operational domains that a homogenous approach using a single kind of function fails to faithfully approximate it.

The solution should be a heterogeneous monitoring and control system that utilizes linear and non-linear methods in parallel and switches them according to the current system behavior (Figure 6.).

An additional benefit of this approach is that the growing error between the estimate delivered by the model in use and the observed value indicates simultaneously the necessity of a switchover from one approximation function to another one and correspondingly detects a transition in the operation domain.

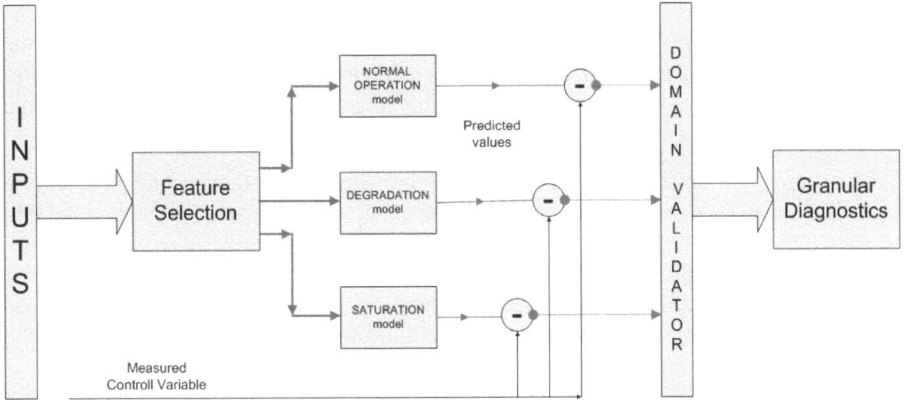

Fig. 6. Blockscheme

While our measurements validated the soundness of the multi approximate based approach further questions are raised for checking the practical usefulness:

- As in each case in benchmark-based methods, the representativeness of the benchmark scenarios and measurement setup drastically influence the results, thus determine whether they can be used in a generalized form under all circumstances.
- This approach is expected to scale well with respect to the number of attributes to be processed. Still, a monolithic approach for large systems seems to be disadvantageous. The two main arguments supporting hierarchical modeling and feature selection are the following.

  On the one hand, it is reasonable to believe that the number of distinct "operating points" – the sets of significant variables as the function of system state, current load and time – will become so big that it becomes unpractical for system management design. To counter this, systems can be subdivided into a hierarchy of subsystems so that a higher level (and the feature selection of that) sees only specific, service access related attributes of the subsystems.

  On the other hand, in sizeable heterogeneous, distributed systems the compensation and repair mechanisms usually operate on multiple levels of granularity; thus, a hierarchical approach with intra-subsystem feature selection is also of paramount importance, not only for localized early warning, but also to support decisions of compensation or repair.

Refinement and experimental validation of the hierarchical approach sketched above shall be performed.

- Our measurements indicated that the linear approximation fits well to the normal operation mode and saturation, mRMR is flexible enough to support a good match to the behavior in the degradation phase. Further experiments are needed to identify the best fitting functions for abrupt changes in the system, like those caused by critical resource faults.
- The number of sufficient features should be determined in a methodical way, e.g. using the Lipschitz-index [21] or some other approach. Here the robustness of the control and its impacts has to be assessed both in the terms of selecting a low number of input variables and the sensitivity to errors in the parameter estimation (this second question is a traditional topic in control theory).

In this paper, we have shown in a methodology experiment that the mRMR feature selection scheme combined with linear approximation can be employed for selecting the few, "most significant" quantitative aspects of a system for the purpose of supervisory instrumentation.

We also have to address the question that how can be the results systematically used for configuring simple rule-based supervision and even more importantly, helping the design of autonomic control schemes.

References

[1] Diao, Y., Hellerstein, J.L., Kaiser, G., Parekh, S., Phung, D.: Self-managing systems: A control theory foundation. Engineering of Computer-Based Systems, 441–448 (April 2005)

[2] IBM Autonomic Computing Initiative,
    http://www.research.ibm.com/autonomic/

[3] Assessing, Measuring and Benchmarking Resilience, FP7 ICT CA 216295,
    http://amber.dei.uc.pt/

[4] Cohen, M., Goldszmidt, T., Kelly, J., Symons, J., Chase, J.S.: Correlating instrumentation data to system states: A building block for automated diagnosis and control. In: Proc. 6th USENIX OSDI, San Francisco, CA (December 2004)

[5] Zhang, S., Cohen, I., Goldszmidt, M., Symons, J., Fox, A.: Ensembles of models for automated diagnosis of system performance problems. Technical Report HPL-2005-3, Hewlett-Packard (January 2005)

[6] Powers, R., Goldszmidt, M., Cohen, I.: Short term performance forecasting in enterprise systems. In: ACM SIGKDD Intl. Conf. on Knowledge Discovery and Data Mining (August 2005)

[7] Hoffmann, G.A., Trivedi, K.S., Malek, M.: A best practice guide to resource forecasting for computing systems. IEEE Transactions on Reliability 56, 615–628 (2007)

[8] Kohavi, R., John, G.: The wrapper approach. In: Liu, H., Motoda, H. (eds.) Feature Extraction, Construction and Selection: A Data Mining Perspective. Springer, Heidelberg (1998)

[9] Jiang, M., Munawar, M.A., Reidemeister, T., Ward, P.A.S.: Information-theoretic modeling for tracking the health of complex software systems. In: Proceedings of the 2008 conference of the center for advanced studies on collaborative research: meeting of minds. ACM, New York (2008)

[10] Ding, C., Peng, H.: Minimum redundancy feature selection from microarray gene expression data. In: Proceedings of the Computational Systems Bioinformatics Conference, pp. 523–529 (2003)

[11] Zhou, J., Peng, H.: Automatic recognition and annotation of gene expression patterns of fly embryos. Bioinformatics 23(5), 589–596 (2007)

[12] Jiang, M., Munawar, M.A., Reidemeister, T., Ward, P.A.S.: Automatic Fault Detection and Diagnosis in Complex Software Systems by Information-Theoretic Monitoring. Will appear in IEEE International Conference on Dependable Systems and Networks (2009)

[13] Grottke, M., Lie, L., Vaidyanathan, K., Trivedi, K.: Analysis of software aging in a web server. IEEE Trans. Reliability 55(3), 411–420 (2006)

[14] Keller, A., Diao, Y., Eskesen, F., Froehlich, S., Hellerstein, J.I., Surendra, M., Spainhower, L.F.: Generic On-Line Discovery of Quantitative Models. IEEE Transactions on Network and Service Management 1(1), 39–48 (2004)

[15] TPC-W official page, http://www.tpc.org/tpcw/default.asp

[16] Fodor, I.K.: A survey of dimension reduction techniques". Technical Report UCRL-ID-148494, Lawrence Livermore National Laboratory, Center for Applied Scientific Computing (2002)

[17] Molina, L., Belanche, L., Nebot, A.: Feature selection algorithms: a survey and experimental evaluation. In: International conference on data mining, Maebashi City, Japan (2002)

[18] Peng, H., Long, F., Ding, C.: Feature selection based on mutual information criteria of max-dependency, max-relevance, and min-redundancy. IEEE Transactions on Pattern Analysis and Machine Intelligence 27, 1226–1238 (2005)

[19] Li, S., Zhu, Y., Feng, J., Ai, P., Chen, X.: Comparative Study of Three Feature Selection Methods for Regional Land Cover Classification Using MODIS Data. In: Proceedings of the 2008 Congress on Image and Signal Processing, vol. 4 (2008)

[20] Chen, H., Jiang, G., Yoshihira, K.: Monitoring High-Dimensional Data for Failure Detection and Localization in Large-Scale Computing Systems. IEEE Trans. Knowl. Data Eng. (TKDE) 20 (2008)

[21] He, X., Asada, H.: A new method for identifying orders of input–output models for nonlinear dynamic systems. In: Proc. Autom. Contr. Conf., pp. 2520–2523 (1993)

Context-Aware Self-optimization in Multiparty Converged Mobile Environments

Josephine Antoniou[1], Christophoros Christophorou[1], Augusto Neto[2,3],
Susana Sargento[2], Filipe Cabral Pinto[4], Nuno Filipe Carapeto[4], Telma Mota[4],
Jose Simoes[5], and Andreas Pitsillides[1]

[1] University of Cyprus,
75 Kallipoleos Str., 1678 Nicosia Cyprus
{josephin,christophoros.andreas.pitsillides}@cs.ucy.ac.cy
[2] Instituto de Telecomunicações, Univerity of Aveiro,
Campus Universitario Santiago, Aveiro, Portugal
susana@ua.pt, augusto@av.it.pt
[3] Universidade Federal de Goiás, Instituto de Informatica
Bloco IMF I, sala 239, Campus II – Samambaia, Goiânia, Brazil
augusto@inf.ufg.br
[4] PT Inovação, Rua Engenheiro José Ferreira Pinto Basto, 3830 Aveiro, Portugal
{filipe-c-pinto,nuno-f-carapeto,telma}@ptinovacao.pt
[5] Fraunhofer FOKUS, Kaiserin-Augusta-Allee, 31, 10589 Berlin, Germany
jose.simoes@fokus.fraunhofer.de

Abstract. The increase of networking complexity requires the design of new performance optimization schemes for delivering different types of sessions to users under different conditions. In this scope, special attention is given to multi-homed environments, where mobile devices cross areas with overlapping access technologies (Wi-Fi, 3G, WiMax). In such scenario, efficient multiparty delivery depends upon the grouping operation (creation of a set of users to receive a given session), which must be done based on several parameters. We propose sub-grouping of content-based service groups, so that the same service session can be delivered using different codings of the same content, to adapt to the current network, users, session and environment context. The context-aware information is used to improve the sub-grouping process. This paper aims to describe these sub-grouping techniques, in particular how they improve network performance and user experience in the future Internet, in the scope of cognitive autonomic networks.

Keywords: Context-awareness, self-optimization, multiparty, sub-grouping, convergence, session & network management.

1 Introduction

Increasing demands in group-based multimedia sessions and market forces are fuelling the design of the future Internet, which is expected to fundamentally change the networking landscape in the upcoming years. In order to preserve profitability while

A.V. Vasilakos et al. (Eds.): AUTONOMICS 2009, LNICST 23, pp. 144–159, 2010.

increasing revenues, network/service providers must optimize costs and provide new sessions operating in a mixture of access technologies, which is not trivial and demands complex control. Attention must be given to group-based multimedia sessions, since the strong requirements on Quality of Service (QoS) must be fulfilled simultaneously for all group users and be kept during the entire session lifetime.

These requirements increased interest of the research community in cognitive and autonomic/self-managed networks. Main functional components of such networks include self-configuration and context-awareness, deploying auto-learning implemented by means of network-aware middleware distributed across network components. Benefits of cognitive autonomic networks include, but are not limited to, network performance optimization, automatic and seamless reconfiguration, fast and efficient resilience, etc. Applications and devices are able to exploit such adaptations while being agnostic of underlying reconfigurations, in accordance with the seamless service provision paradigm. The proposed architecture plans to support automatic and seamless session reconfiguration, through self-management, recognizing context and acting on it to sub-group users. Context refers to information collected dynamically over time, describing the user, its environment and the network's current state.

In terms of group-based sessions, efficiency of session setup requires a correct definition of user groups. Nowadays, most mobile devices are produced with multi-homed capabilities, and it is common to cross areas where there exists overlapping of different network access technologies, such as Wi-Fi, 3G and WiMax. The efficiency of the grouping operation (creation of a set of users to receive a given session) may depend on parameters, such as access technology, since for instance, 3G networks have lower bandwidth capabilities than Wi-Fi and WiMax networks. Thus, sub-grouping could be performed and the same service session could be delivered with different throughput (e.g., using different codings of the same content) to adapt to the current network capabilities. In addition to network traffic, other types of context should also be used to improve sub-grouping, such as noise, terminal location and speed, user's priority and network preferences, user's terminal capabilities, quality of received signal etc. Moreover, history context can also be used for the improving sub-grouping. For example, previously received context can be compared with current context for patterns to be located. Using some intelligence, forecasts of undesirable events are possible and sub-groups may be created so that such events are avoided.

The FP7 Context Casting (C-CAST) project [1] proposes such innovative sub-grouping process to enable context-awareness and consequently self-optimization in multiparty, converged mobile environments. This paper describes these sub-grouping techniques, in particular how they improve network performance in the future Internet, in the scope of cognitive autonomic networks. Related work is presented in Section 1.1. Section 2 overviews context-aware multicasting (C-CAST), introducing sub-grouping. Sections 3 and 4 diverge into self-management through sub-grouping as designed at the session and network layers. Section 5 presents how context is used by the content, and finally Section 6 offers conclusions and directions for future work.

1.1 Related Work

Personalized sessions can be influenced by varying context, allowing users access sessions based on their location, preferences, profile and capabilities [2]. In next generation networks, multiple access networks coexist, thus, access selection using context-based algorithms is necessary to enable the optimization of both terminal and network [3]. Although many proposals base the decision process on radio signal properties (e.g. [4]), this is only one of the many criteria in such selection schemes. Some proposals suggest context-aware decisions [3]. Moreover, the majority of related work focuses entirely on network selection algorithms, not concerning other important mechanisms crucial to support the decisions e.g. QoS management, to enable the complete network re-configuration triggered by context. This lack of high-level perspective is addressed in more recent proposals, [5]. We consider the support of context-aware selection, in a multicast environment, where the group membership is a main issue, but being flexible enough to support any parameter envisioned.

The integration of class-based QoS and IP multicast is promising, since the former allows a scalable QoS approach while the latter saves bandwidth [6]. However, this is not trivial [7], e.g. while QoS achieves scalability by pushing unavoidable complexity to edges routers, IP multicast operates on a per-flow basis throughout the network. Also, dynamic addition of new group members may affect existing traffic [8].

On the session layer, most of the solutions proposed use the Session Initiation Protocol (SIP) as the main signalling protocol, e.g. in MPLS-based next generation networks [9], or as an enabler for session mobility in converged networks [10], integrating QoS management and mobility management as the basis for overall session management. Enabling session management mechanisms with context-aware information, the approach of [11] exploits strategies involving the use of contextual information, strong process migration, context-sensitive binding, and location agnostic communication protocols for "follow-me" sessions. Although interesting, these do not cover QoS and efficient multiparty delivery systems.

Finally, much effort has been put recently on the autonomic network concepts [12], where autonomic processes can perceive network conditions, plan, decide, and act on these conditions. They can learn from the impact of former adaptations and accordingly make future decisions, while considering end-to-end goals. Autonomic networks are promising for wireless networks, which are highly dynamic and complex to manage. Our approach is towards the autonomic concept by enabling the dynamic optimization of the use of the network taking into account also the history and instantaneous context of the users, network, sessions and environment.

2   Context-Aware Multiparty Service Provision

This section presents the overall context-aware multiparty architecture, able to support context from the user, network, sessions and surrounding environment in future multiparty mobile communications. We also show how the context can influence grouping and sub-grouping of users at different levels from the environment and user-levels, to the session and network-layers.

2.1   System Architecture

Figure 1 depicts the context-aware multiparty reference architecture. It aims at providing an end-to-end context-aware communication framework specifically for intelligent multicast/broadcast services. The three main parts comprise: the context and group management service enablers, with reasoning and grouping users based on context; content adaptation and delivery based on context; and context information collection through sensors, context distribution and context aware multiparty transport.

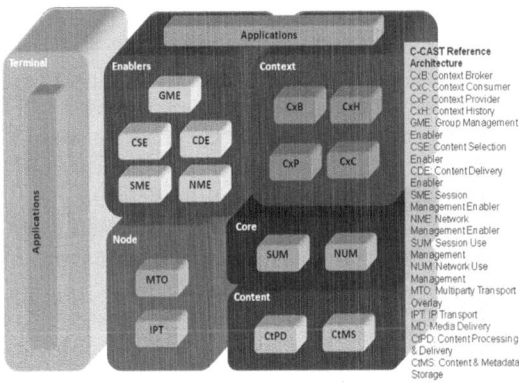

Fig. 1. Context-Aware Multicasting Reference Architecture

In this paper we focus on the multiparty delivery layers and how they support session delivery through efficient network mechanisms, both supporting context information and grouping/sub-grouping of users based on the different sources of context information. In the following paragraphs we briefly describe the main entities of the multiparty delivery process.

Context Providers (CxP) – They obtain contexts from sensors and networks, map them to information in an interpretable manner and deliver this information to the several components.

IP Transport (IPT) – It controls the integrated QoS and IP multicast enforcement in the nodes along a communication path for the efficient delivery of multiparty sessions to groups of users, with QoS-guaranteed over the time, able to build delivery trees in heterogeneous environments. The performance limitations of existing proposals motivated the design of IPT with support to distributed per-class resource control. For scalability, network edges coordinate resource allocations, and interior routers remain simple by reacting only upon both signalling and network events (e.g., link failures, re-routing or mobility), where these events can be local or triggered by context changes.

Multiparty Transport Overlay (MTO) – It provides a generic, scalable, and efficient transport service for group communications by applying the overlay paradigm at the transport layer. It hides the heterogeneity of underlying networks in terms of IP multicast capabilities or IPv4/v6 support, enabling the dynamic creation of an overlay tree at the transport level, between the source and the group members.

Network Management (NM) – It performs context-aware decision of the best connection of multimode terminals in heterogeneous networks. It makes use of user, environment, network and session context, to drive intelligent network selection, in terms of communication path, terminal interface and access technologies (Network Use Management – NUM). Moreover, it is expected to achieve more efficient resource utilization, as well as more uniform distribution of data load, while fulfilling the QoS required by sessions and experienced by the users (Network Management Enabler – NME). For instance, the Wi-Fi incoming interface of a terminal should be changed to the WiMax during congestion periods, or after terminal mobility to an area without Wi-Fi coverage area. Associated with MTO, Network Management (NM) allows generalized transport by assigning MTO trees and controlling packet transport along them. Thus, end-to-end multiparty content transport over network segments with different transport technologies (i.e., unicast and multicast) is deployed with context-driven self-organization and seamless resilience support.

Session Management (SM) – It manages user-to-content and content-to-user relationships, through session control. It is intelligently designed to enable the use of context information for session control, using the SIP, in terms of network-specific, environment-specific and user-specific contexts, without in fact knowing the actual network, environment or service details (through Session Management Enabler – SME - and Session Use Management – SUM).

Mobility Controller – It applies the decisions issued by the NUM related to the network interface to be used in the terminal, for vertical handover decisions.

2.2 Context-Based Sub-grouping

In this context-aware architecture, grouping and sub-grouping of users in the same session does not only depend on the desired content (performed at higher layers, such as the service and application ones), but also, due to the variety of context information, it is also performed at Session and Network levels, supported by the session-related and network-related context. As an example of session-related context, due to session context on availability of codecs support in the terminals, users may need to be sub-grouped in different sessions with different codecs but with the same content. As an example of network-context, due to the context information, the best access technologies for the users in the same group may be different, since user context and environment is also taken into account. The different networks may have different guarantees, which may require the content to be delivered with different quality and codecs, hence requiring users sub-grouping.

The Session Management is the element that accepts context information through requests or triggers and answers by creating/modifying/terminating the sub-group sessions. Once it determines the matching between available content formats and user capabilities in terms of supported content formats, resulting in candidate files per user in the service group, it is responsible of first inviting the users to a session and to invite the content provider to deliver the content to these users.

When some context changes, and this change is relevant to the way the service is consumed, the session management should be capable to adapt the user's session accordingly (move a user to a different sub-group, or eventually create or delete sub-groups). The trigger can have different sources (network conditions change, device change, handovers, etc).

The Session Management should be able to notify the Network Management to modify sub-groups/sessions based on the above notifications. This might make the Session Management terminate the SIP and Media session both with the client and the content provider. It should also be able to modify sessions based on notifications coming from other components, through changes in context. According to the trigger, when only sub-groups are affected, it is important to modify the multiparty session as well. Therefore, it will communicate all changes to the Network Management to adapt the trees (and the correspondent overlays) accordingly, and possibly change the network and technology to the new sub-groups.

On the network side, each of the created sub-groups represents a multicast delivery tree or, in case of a single user, a unicast connection, that requires network resources. The network Context Provider improves the network operator view about network resource usage, to optimize network resource management and QoS. It is then possible for the network to provide different QoS levels to the members of a group that consume identical content, but experience various network contexts. This is the process of network level sub-grouping.

The Network Management is triggered through network level context, or by the Session Management to adapt the network and users in the network. It performs an optimization process whose output is the decision of the networks/technologies to be used by the several users. If the users in the same group change the attached network, and the levels of QoS cannot be fulfilled in the networks of a group, sub-grouping is performed at network level, and different QoS, and possibly different codecs, will be assigned to the different sub-groups. For some cases, where not all the users or links have the required multicast capabilities, overlay nodes are enabled to abstract the grouping decision of these considerations. Moreover the sub-grouping flexibility allows, for instance, in the same session, two users to receive the same audio, but different video streams with different codecs or rates, just depending on the available resources or preferences. In this case both would be in the same audio sub-group, but in different video sub-groups.

Sections 3 and 4 will deeply describe the context-aware sub-grouping processes, both from the session and network sides.

3 Enabling Context-Awareness at the Session Level

3.1 Session Management Overview

Session Management is the entity in the core of a multicast-enabled, converged mobile architecture that provides the necessary signalling to deliver a specific content to its consumers and can handle different types of events regarding session control, specifically: session establishment, session renegotiation (upon a given trigger or change), session termination and session mobility. Session management may thus participate in dynamic changes such as switching between different content, and is closely interlinked with media delivery; the two functionalities achieve cooperatively a system delivery of the appropriate content for a given user (or group of users) and act as an intermediary platform between the content provider and the content consumer.

This work extends the session management functionality to consider context triggers in the creation, modification and teardown of sessions. Context-aware Session Management is a key functionality of the converged system. Specifically, the session management deals with context that is relevant in order to select the right content for its consumers. Therefore, the first step for enabling context-awareness in the multicast-enabled, converged network is to recognize and use content-related context such as user capabilities and preferences regarding content support (content formats/codings). Such context may be categorized as Device Context (e.g. supported coding options) and User Context (e.g. user preferences). Thus, the Session Management entity will eventually be able to:

- Recognize a context trigger for session setup
- Recognize a context trigger for session renegotiation
- Recognize a context trigger for session termination
- Recognize a context trigger for session mobility
- Make additional requests for context when necessary to the appropriate functional entity within the system architecture.

3.2 Context-Aware Session Management

The context-awareness in the Session Management entity comes from the recognition of context as triggers or the capability to receive any context requested, e.g. for initiating a context-aware session. Therefore, the Session Management entity has the functionalities described in the subsequent paragraphs (based on interfaces with other system functional entities), necessary to enable this awareness of context information:

Primarily, the Session Management needs to interact with the entity responsible for identifying the service groups, i.e. the groups of users that will receive the same content. Through this interface the Session Management must receive identification for the service group, together with separate identifications for the individual users that comprise the particular group.

Consequently, the Session Management must be able to interact with the entity responsible for Content Processing and Delivery. The Session Management sends the group identification received in the step described above, to the Content Processing and Delivery entity, and the identification is used by this entity to collect some general content information for the group e.g. whether the content that will be received by the particular group is video or audio, as well as some more specific information, e.g. the coding(s) and bitrate(s) in which the particular content is available. Thus, the Session Management entity acquires descriptive information on the content that will be transmitted to the group.

Once the content description has been acquired, the Session Management needs to check which of the available content codings the users are capable (device context) and willing (user context) to support. Device and User context information is collected at a broker system entity, i.e. an entity in the converged architecture that accumulates all context information received from various context producing entities (for instance the user terminals in this case). The Context Broker entity contains the required device and user context, therefore the Session Management entity requests for each user in the service group its capabilities and preferences, including coding options supported/preferred by the user device (e.g. resolution, coding options supported).

Once this particular context information is obtained per user, the Session Management entity can match the content codings/formats in which the content is available, i.e. received from the Content Processing and Delivery (CtPD) entity, to the content codings/formats that each of the group users is capable or willing to support. This matching will results in a list of particular content codings per each user, which may be viewed as an initial refinement of the original service group to sub-groups of users according to the supported content. This is the first step of the sub-grouping process. The sub-grouping will continue in the network where further refinements will additionally consider network and environment context (e.g. current QoS capabilities based on user location and current network load).

The sub-grouping is initiated in the Session Management entity but is further refined and concluded in the Network Management entity, since network and environment context are more appropriately collected at that system level. Consequently, the Session Management entity must support interface functionality to exchange information with the Network Management entity. Over this interface the Session Management entity will send the list of users and their supported content codings (once the matching of content availabilities and content capabilities is performed), and will receive the finalized context-based sub-grouping of the original service group by which only one content coding will be selected for each user in the original service group. This will enable a context-aware session for each multicast sub-group to be setup. Furthermore, in the case that context changes at any level, the session is modified accordingly, since the system allows the context information to be propagated through to the content. However, we focus on the session setup procedure and provide in the next session details on the signal flows.

3.3 Initiating a Context-Aware Session

In this section, we describe the session initiation in such a converged architecture, focusing on the Session Management functionalities as described above. In the subsequent discussion, the Session Management functional entity is separated in two functional modules: The Session Management Enabler (SME) and the Session Use Management (SUM). In terms of functionality the SME is the Session Management functional module that accepts context information through requests or triggers and responds by creating/modifying/terminating the sub-group sessions. In other words, it handles the interfaces between the SME and the entity providing the identification of the original service groups, the entity responsible for matching a group context to the appropriate content, the SME and Network Management, the entity that knows the network, the SME and the Context Broker entity, where all context information subsides, as well as the SME and CtPD, the content processing and delivery entity. Finally, the SME needs to interact with the SUM, which is the sub-module responsible for handling the SIP-specific tasks of the Session Manager, such as inviting the users and the Media Delivery Function to sessions. Once the Core Entity of the SME determines the matching between available content formats and user capabilities in terms of supporting content formats, resulting in candidate files per user in the service group, SUM takes over, which is responsible of first inviting the users to a session and to invite the CtPD to deliver the content to these users.

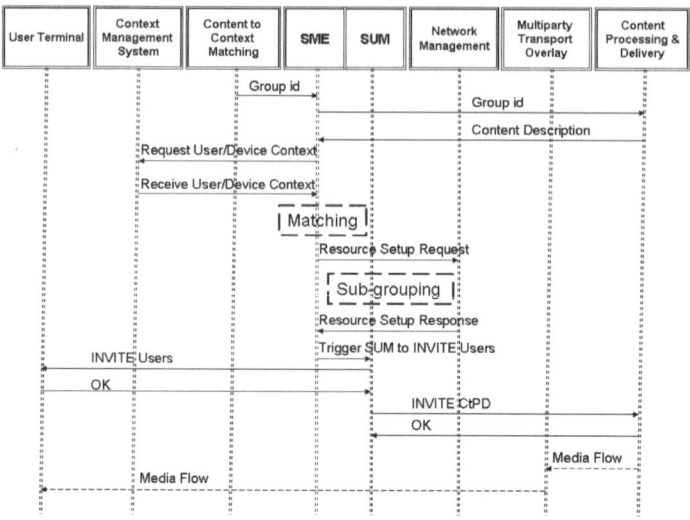

Fig. 2. Session Management related messages in session initiation

Figure 2 illustrates the messages exchanged during session initiation that are relevant to Session Management (both the SME and SUM). SME receives group identification and immediately requests group content information from the CtPD. For each group user the terminal capabilities and user preferences are collected from the Context Management System and particularly from the Context Broker entity. The matching between this information results in a list of users and corresponding available content codings, which is sent to the Network Management to proceed with more refined sub-grouping. The returned user list has one coding per user, with all users that support the same content coding to form a subgroup. Then, the SME interacts with the SUM to open a new session for each subgroup by inviting the users that belong to that sub-group to join the session, and by inviting the Media Delivery function (part of the CtPD) to deliver the particular content coding for that sub-group.

4   Enabling Context Awareness at the Network Level

4.1 Network Management Overview

The Network Management component is the element that provides intelligent network selection according to the user, network and environment context, allowing the terminals to always be best connected and be able to receive multiparty content sessions with satisfactory QoS. It uses different context information to keep multimode terminals always "best" connected, making use of context, to drive intelligent network selection, in terms of communication path, terminal interface and access technologies. The Network Management assumes complex and heterogeneous scenarios, where dynamic network events (link failures, handovers and traffic conditions) take place randomly. Such network dynamics and complexity require a new concept of network architecture to efficiently support for the users' sub-groups.

4.2   Context-Aware Network Management

This section describes the Network Management element in detail, which is divided in Network Management Enabler (NME) and Network Use Management (NUM), with different functionalities and roles in terms of context usage and sub-grouping.

The NME is the module responsible for the group management at network level. It splits the service groups into several network-based sub-groups according to user, network, operator policies, and a large set of context information. It receives requests from the Session Management, for resource setup, modification and teardown for session groups, i.e. service groups refined according to content availabilities matched with user capabilities. Once the NME receives a request, it triggers the network selection process for an intelligent access technology selection (using network, user and environment context dynamically obtained). Following, it is requested to NUM the selection of the best available communication paths, be them multicast or unicast which can provide sufficient resources throughout the network and support some of the network interfaces. With the possible end-to-end paths, the best network interfaces, the user and operator preferences and a set of context NME disposes the users in different sub-groups. This process is extendable to cope with new sets of context information that can refine the sub-groups. Each subgroup will reflect a specific unicast/multicast with a unique (to that content) quality of service. Furthermore these mechanisms were defined to cope and deal with updates and modification due to network problems or context changes. Once the sub-groups are created and provisioned by the underlaying modules, this information is returned to the Session Management entity with the respective addresses and ports and selected coding for each sub-group.

The NUM is the module responsible to manage network resource allocation functions. It encompasses the network selection, aiming at providing intelligent Radio Access Technology (RAT) selection based on context-aware information to groups of users, as well as selection of communication path, which is done based on network information. The network selection function deploys intelligent context-aware RAT selection for the users of a group, both at the session setup and also during the session. By taking into account both the history context information and the instant networking and environment context of the users (noise, interference, signal strength, signal strength alteration rate, speed, location, etc), as well as the overall network conditions (QoS capabilities, multicast capabilities, available capacity, current load of RATs, etc), the RAT selection algorithm estimates all possible transmission arrangements that can be used to distribute the multiparty session content to the users. Afterwards, NUM selects among them, based on pre-defined rules, the most efficient one (i.e. the one that enhances the overall network capacity and performance while at the same time fulfill the QoS requirements in all respects). These rules/constrains are dynamically defined by the NUM, using some intelligence, based on history and instantaneous context information of both, network and users. These rules aim to prevent NUM from selecting a transmission arrangement that might result in undesirable events like, congestion in the Network, overloading of certain RATs, users' QoS degradation etc. After deciding the transmission arrangement that will be used (i.e. per-user's RAT and content coding), NUM triggers the NME to control the network resources, also informing each terminal about the selected interface from which multiparty content will be provided (i.e. Wi-Fi, WiMAX, UMTS, LTE, DVB, etc.). The network selection process will also provide support during the multiparty sessions transport, i.e. users of on-going sessions can receive multiparty content from a different network interface, due to changing network conditions or even handovers (with the support of the mobility controller).

NUM is also responsible for selecting the best communication path within the network. It receives requests from the NME to decide on the best path for the multiparty connection. To do that, NUM maps the QoS requirement of the multiparty session into an available class of service, also taking into account all the network status when selecting the path (NetworkQoSCxP). In this sense, it deploys admission control operations along the network communication path. Afterwards, it returns them to the NME which finally takes the decision of enforcing the reservation. Consequently, NUM receives the enforcement order and commits the resource reservation by triggering each router to enforce both QoS and multicast. QoS enforcement consists in indicating amount of bandwidth and class of service for resource reservation, and Multicast enforcement comprises populating the Multicast Routing Information Base (MRIB) with the information about nodes of the selected communication path. At the end, NME is informed about the success of the operation.

4.3 Grouping as a Part of Network Management

Whenever SME triggers NME with a new request, it contains the users that were selected to receive the same content and the codings in which it is available. SME matches these to the user terminal capabilities and sends to NME. This starts the subgrouping process in NME that is composed by three different steps. Firstly, network selection is performed to determine, based on the instantaneous context of the network (i.e. RATs' current load, RATs' available capacity, RATs' QoS cababilities etc.) and of the users (RATs within reach, signal quality received, speed, etc.) and by considering the constrains set by NUM, the interfaces and the content codings that are more appropriate. This event takes into consideration all the access network context and might trigger an interface and IP address update.

Secondly, NME will decide on the multicast groups based on the user's and operator's context. Plus it chooses the optimal way to group the users based on optimization policies that can maximize the quality experienced by the user or the price he pays. Lastly NME wills send this result and updated interfaces to NUM and check for available paths between the source and users that respect the QoS constraints and optimize the resource allocation. This process is partially cyclic and may require some sub-iterations or adaptations of a decision previously taken. Still it allows the separation and simplification of processes towards the autonomy of the system. Afterwards, each of the multicast subgroups will trigger the resource reservation and multicast routing. In the end NME pushes the SME and SUM to invite all the users to the multicast groups and the content provider to start streaming. The process is depicted in Figure 3.

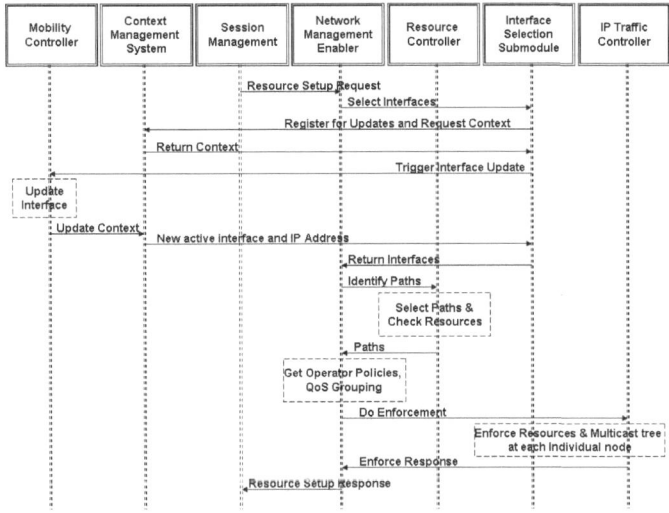

Fig. 3. Network Management related messages in session initiation

4.3.1 Network Role in Context-Aware Group Modification

The group modification capabilities are one of the strengths of this architecture since it is able to adapt to various environment changes. Network conditions can change rapidly and is virtually impossible to predict those changes. Thus the architecture must be prepared to adapt quickly. It is able to fast detect changes, and evaluate their impact on the architecture, by usually enforcing a modification of the group or sub-group of users receiving the same content. Actually three major triggering points were considered, updates to the group constitution, modifications on the network elements which may influence the experienced QoS and the mobility of users.

By assuming these different triggering possibilities, two levels of group modification must be considered. Group Session Modification, which represents the ability of users to join or leave the group session. This is initiated at SME and propagates through to the Network Management and other modules similarly to session initiation. The removal of users can reduce size or even terminate an existing sub-group. The addition of users has to take into consideration that the stream is already progressing and the content will only be viewed from that point forward. This is only considered when the new user viewing experience is not compromised by joining an already streaming context.

The second level is related with the Sub-Group Modification which only affects the actual QoS of a set of users and allows them to be switched between sub-groups. The actual session group is preserved and only the concerned sub-groups are updated. Whenever this happens, some users get "promoted" (or "demoted") to a group with different quality of service. This can be triggered by the implemented IPT resilience mechanisms, where the network conditions are significantly altered: a link or a router may go offline or back online, the QoS conditions may be altered, the access network

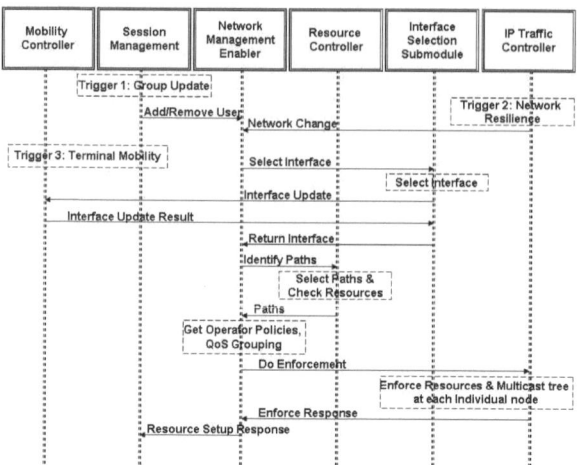

Fig. 4. Network Management in session modification

may become overloaded, or the terminal may be forced to move into a different network with different conditions. Eventually, session mobility between terminals with different characteristics and updates of operator policies will also trigger sub-group modifications. The process is depicted in Figure 4.

5 Propagating Context to Content

In the previous sections we presented how context-awareness was enabled at the session and network levels, describing the proceedings on how sub-groups are formed, the usage of context information in this process and further describing mechanisms activated in order to assure QoS across the whole transport and network layers. In this section we describe the mechanisms involved in delivering the correct media types to the right users and explain the importance of this process in the overall value chain of the content to context management and distribution. In order to better understand the relevant flows, Section 5.1 presents a use case that motivates this work, where one user is capable of receiving different media types (audio, video, text, etc.) and consequently is inserted into different sub-groups according to a myriad of context information. Section 5.2 explains the importance of this behavior and the impact on the Quality of Experience (QoE) for the end user.

5.1 Motivation: Same User, Different Media Types, Different Sub-groups

Although the sub-grouping is mainly done in the Network Management functional entity, the Session Manager is the entity responsible for managing the content-to-user and user-to-content relationships. Therefore, based on the sub-grouping information provided from the Network Manager, the Session Manager must initiate SIP sessions towards the end users as well as the CtPD entity (Media Delivery Function). Despite the complexity of this process, by using SIP, the session establishment, renegotiation, termination and mobility becomes quite straightforward. After receiving the notification from Network Management, Session Management needs to parse the message in order to initiate its back-to-back user agent (B2BUA) function.

Although all users will be receiving the same content, the media types may vary according to their preferences, needs or current conditions, allowing each user to receive its own personalized stream. Moreover, note that a subset of the users might belong to the same sub-group concerning one media type and different sub-groups in others. Assume the example depicted in Figure 5: i.e. that while doing the parsing, it is possible to identify two videos, three text and two audio sub-groups.

				Audio	Video	Text
Audio	Video	Text	john@open-domain.org	MP3	H.263	English
ACC	H.263	English	maria@open-domain.org	MP3	H.264	---
MP3	H.264		lena@open-domain.org	ACC	H.264	English

a)   b)

Fig. 5. Example of different media per user for the same content

After identifying the required sub-groups for a specific content (group with similar context) the Session Management entity initiates the session establishment process with the CtPD. On the other hand, Figure 6 shows a sequence flow from the messages between the Session Management entity and the CtPD concerning all sub-groups session initiation and corresponding session establishment. Simultaneously, Session Management triggers the session initiation towards the end user.

When this message reaches the user terminal, according to the network capabilities, the behavior may vary. If the terminal/network is multicast capable, the device will send a *JOIN* message towards the multicast addresses included in the initial *INVITE* message. If the device only supports unicast connections, it will answer the request

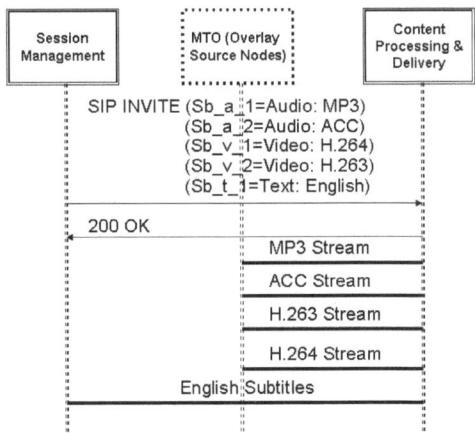

Fig. 6. Sequence Flow between Session Management and CtPD

with a *200 OK* message containing the ports it will be expecting media from. Consequently the Session Management entity will forward this information towards the Network Management entity, which will have the responsibility of updating the overlay leaf node with this information.

After the session is established towards the end user, Session Management activates the CtPD delivery process, which consists in creating unicast sessions with the overlay source nodes on the overlay trees. This step activates the media transmission across the entire multicast overlay trees. Consequently, the media flows are established, finalizing the process of context to content propagation.

5.2 The Importance of Sub-grouping

The importance of personalization in multiparty multimedia services strongly motivates content providers, operators and other players to adapt the user experience of these services. Often, adaptation raises scalability and performance concerns, which need to be addressed efficiently. Creating different sub-groups for different media types, allows users to receive the multimedia content that is most suitable to them, not only according to their personal, device and environmental contexts but also considering the current network context information. This compromise allows both users and operators to achieve an optimal point in what concerns personalization vs. scalability. Furthermore, this will boost the user perceived QoE as the sub-grouping mechanisms allow personalization, contextualization and adaptation of content/services, facilitating further interactivity and mobility tasks. By simultaneously allowing, the operators to optimize and save network resources, services become cheaper and consequently more attractive to the end user.

6 Conclusions and Future Work

In this paper we present an innovative way to achieve self-optimization through context-awareness in multiparty converged mobile system. This is achieved through the dynamic re-definition of service groups (sub-grouping) in a converged architecture. A general overview of the sub-grouping process was given with reference to the C-CAST system architecture, in order to place the proposed innovation within a system scope. Subsequently, the focus shifted to particular aspects of the sub-grouping process and the use of the reference architecture in this process; in particular, we have shown how the process begins at the session level, is refined at the network level, and is finally propagated at the service level to the content. As a result, the content received by each defined sub-group has been adapted to the users' preferences, situations and contexts on the one hand, and on their network capabilities on the other hand, improving both network resource usage and user experience.

This architecture is currently being implemented. More specifically, the elements required to perform and enforce sub-grouping both on the session and network level are being implemented: context providers, session and network management, multiparty transport overlay and IP Transport. The interfaces between the elements are also defined and currently in implementation.

Future work aims, within the scope of the C-CAST project, to finish implementation of the sub-grouping mechanism in order to collect relevant performance measures. Furthermore, specific sub-modules participating in this self-optimization process will be individually enhanced and evaluated.

References

1. ICT-2007-216462 C-CAST (Context Casting) project, http://www.ict-ccast.eu
2. Sigrid, B., Poi, M., Tore, U.: A simple architecture for delivering context information to mobile users. In: Workshop on Infrastructure for Smart Devices - How to Make Ubiquity an Actuality, Bristol (2000)
3. Jesus, V., Sargento, S., Aguiar, R.L.: Any-constraint personalized network selection. In: Personal, Indoor and Mobile Radio Communications Symposium, pp. 1–6. IEEE Press, Cannes (2008)
4. Pahlavan, K., Krishnamurthy, P., Hatami, A., Ylianttila, M., Makela, J.P., Pichna, R., Vallstron, J.: Handoff in hybrid mobile data networks. Personal Communications 7, 34–47 (2000)
5. Chen, Y., Yang, Y.: A new 4G architecture providing multimode terminals always best connected services. Wireless Communications 14, 36–41 (2007)
6. Yang, B., Mohapatra, P.: DiffServ-aware multicasting Source. Transactions of the High Speed Networks Journal 13, 37–57 (2004)
7. Bless, R., Wehrle, K.: IP Multicast in Differentiated Services (DS) Networks. IETF RFC 3754 (April 2004)
8. Bless, R., Nichols, K., Wehrle, K.: A Lower Effort Per-Domain Behavior (PDB) for Differentiated Services. IETF RFC 3662 (December 2003)
9. Kwon, Y., Park, H.J., Choi, S.G., Lee, H.S.: P2MP Session Management Scheme using SIP in MPLS-based Next Generation Network. Opt. Internet and Next Gen. Net., Jeju, 3–185 (2006)
10. Oberle, K., Wahl, S., Sitek, A.: Enhanced Methods for SIP based Session Mobility in a Converged Network. In: 16th IST Mobile and Wireless Communications Summit, Budapest (2007)
11. Handorean, R., Sen, R., Hackmann, G., Roman, G.C.: Context Aware Session Management for Services in Ad Hoc Networks. In: Intl. Conf. on Services Computing, Orlando, pp. 113–120 (2005)
12. Thomas, R., Friend, D., Dasilva, L., Mackenzie, A.: Cognitive networks: adaptation and learning to achieve end-to-end performance objectives. IEEE Communications Magazine 44, 51–57 (2006)

Context Discovery in Mobile Environments: A Particle Swarm Optimization Approach

Christos Anagnostopoulos and Stathes Hadjiefthymiades

Department of Informatics and Telecommunications, University of Athens, Greece
{bleu,shadj}@di.uoa.gr

Abstract. We introduce a novel application of Particle Swarm Optimization in the mobile computing domain. We focus on context aware applications and investigate the context discovery problem in dynamic environments. Specifically, we investigate those scenarios where nodes with context aware applications are trying to (physically) locate up-to-date context, captured by other nodes. We establish the concept of context quality (an ageing framework deprecates contextual information thus leading to low quality). Nodes with low quality context cannot capture such information by themselves but are in need for "fresh" context in order to feed their application. We assess the performance of the proposed algorithm through simulations. Our findings are quite promising for the mobile computing domain and context awareness in specific. We assess two different strategies for the PSO-based context discovery framework.

Keywords: Context-awareness, context-discovery, distributed systems, swarm intelligence, particle swarm optimization.

1 Introduction

Mobile and distributed computing has become increasingly popular during the last years. Many mobile applications exhibit self-organization in dynamic environments adopted from multi-agent, or *swarm*, research. The basic paradigm behind swarm systems is that tasks can be more efficiently dispatched through the use of multiple, simple autonomous agents instead of a single, sophisticated one. Such systems are much more adaptive, scalable and robust than those based on a single, highly capable, agent.

A swarm system can generally be defined as a decentralized group (swarm) of autonomous agents (particles) that are simple with limited processing capabilities. Particles must cooperate intelligently to achieve common tasks. We investigate a mechanism that exploits the collaborative behavior of the agents in order to deal with the Context Discovery Problem (CDP). Specifically, in CDP an agent (e.g., mobile node) needs to discover, locate and track the source that generates the required contextual information – *context* (e.g., environmental parameters like temperature, humidity, situations like fire outbreak) for the executing context-aware, mobile application (e.g., the control of a group of robots).

Swarm Intelligence (SI) introduces a powerful new paradigm for building fully distributed systems in which overall system functionality is attained by the interaction of

A.V. Vasilakos et al. (Eds.): AUTONOMICS 2009, LNICST 23, pp. 160–175, 2010.
© Institute for Computer Sciences, Social-Informatics and Telecommunications Engineering 2010

individual agents with each other and with their environment. Such agents coordinate using decentralized control and self-organization. Swarm systems are intrinsically highly parallel and exhibit high levels of robustness and reliability:

1. A SI-driven distributed system does not have hierarchical command and control structure and thus no single failure point or vulnerability. Agents are often very simple and the overall swarm is intrinsically fault-tolerant since it consists of a number of identical units operating (sensing context) and cooperating (sharing context) in parallel. In contrast, a conventional complex distributed system requires considerable design effort to achieve fault tolerance.

2. The key central concept in a swarm system is the simplicity of the agents -an agent can be a mobile phone carrying sensors. Simply increasing the number of agents assigned to a task (e.g., sensing context) does not necessarily improve the system's performance (i.e., efficiency and reliability). Agents collaborate by exchanging useful information in order to obtain the required context.

3. In a totally distributed environment agents collaborate for discovering context with certain validity (e.g., related to time and/or space constraints). Context periodically turns obsolete and has to be regularly determined and discovered. Moreover, the resources of simple agents are limited in terms of (1) memory; agents remember the history of their operation up to a certain extent, (2) sensing capabilities; for agents moving around, the sensing radius can be small enough relatively to the coverage area once possible neighboring agents can provide analogous local information, and (3) communication resources; communication among agents is intended solely to convey information on the swarm.

The above-mentioned points lead to the question: "Is the SI paradigm suitable for application in the CDP?" The aim of this paper is to address that question.

Many research efforts have examined multi-agent systems inspired by biology, e.g., flocking models [1, 2], emphasizing in fault tolerance [3], cooperative hunting [4] and ant colony optimization [5] for solving problems in distributed environments. Below we report some typical applications: 'covering' (explore enemy terrain), 'patrolling' (guarding a museum against theft), 'self-assembling' (reconfigurable robots), 'localization' (improvement of positioning accuracy) and 'environment manipulation' (transportation control). In addition, significant research effort has been invested in the design of swarm system for searching areas, either known or unknown, which is most relevant to our work. Specifically, in most previous works the targets, i.e., nodes with valuable information (e.g., sensor nodes) are assumed to be static. However, only a few works examine a swarm system in dynamic environments dealing with the mobility of agents [6, 7, 8] and with information validity constraints. One of the first studies in the application of PSO to dynamic environments came from [21]. The work in [9] considers dynamic targets but does not deal with certain validity issues as required in the CDP. A significant SI adaptive mechanism to detect and respond to dynamic systems is reported in [23]. The involved agents in such mechanism cannot be fully applied to mobile nodes as long as the inherent communication load and efficiency are not taken into consideration, especially when dealing with real context-aware applications. Therefore, we adopt same ideas from [23] regarding the response strategies to various changes. To the best of our knowledge, there is no prior work based on SI in order to deal with the CDP. This motivated us to define, model and propose a solution (algorithm and strategies) for the CDP.

The structure of the paper is as follows: Section 2 presents the basic idea of SI, while in Section 3 we introduce certain issues for the CDP. In Section 4 we propose an algorithm for the CDP adopting concepts from SI. We assess our algorithm in Section 5 and Section 6 concludes the article.

2 Swarm Intelligence

The *Particle Swarm Optimization* (PSO) incorporates swarm behaviors observed in flocks of birds, swarms of bees, or human social behavior, from which the idea is taken [10]. The main strength of PSO is its fast convergence, which compares favorably with many global optimization algorithms (e.g., Genetic Algorithms and Simulated Annealing). The PSO model consists of a swarm of N particles, which are initialized with a population of random candidate solutions (particles). They move iteratively through a d-dimension problem space $\mathfrak{R}^d$ to search new optima. $f: \mathfrak{R}^d \to \mathfrak{R}$ is a *fitness* function that takes a particle's solution in $\mathfrak{R}^d$ and maps it to a single decision metric; the CDP deals with the geometrical space of two dimensions, i.e., $d = 2$, as will be discussed bellow. Each particle indexed by i has a position represented by a vector $xi \in \mathfrak{R}^d$ and a velocity represented by a vector $vi \in \mathfrak{R}^d$, $i=1, ..., N$. Each particle "remembers" its own *best* position so far in a vector $xi\# = [xij\#]$. The best position vector among the swarm so far is then stored in a vector $x^* = [xj^*]$. During the iteration (time) t, the velocity update is performed as in Eq(1). The new position is then determined by the sum of the previous position and the new velocity in Eq(2).

$$u_{ij}(t + 1) = wu_{ij}(t) + c_1r_1(x_{ij}^{\#}(t) - x_{ij}(t)) + c_2r_2(x_j^{*}(t) - x_{ij}(t)) \tag{1}$$

$$x_{ij}(t + 1) = x_{ij}(t) + u_{ij}(t + 1) \tag{2}$$

w is an *inertia* factor. The r_1, r_2 random numbers are used to maintain the diversity of the population and are uniformly distributed in the interval [0, 1] for the jth dimension of the ith particle. c_1 and c_2 are positive constants called *self-recognition* and *social* component, respectively. They interpret how much the particle is directed towards good positions. That is, c_1 and c_2 indicate how much the particle's private knowledge and swarm's knowledge on the best solution is affected, respectively. The time interval between velocity updates is often taken to be unit, thus, omitted (the Equation (2) is dimensionality inconsistent). From Equation (1), a particle decides where to move at the next time considering its own experience, which is the memory of its best past position and the experience of the most successful particle in the swarm (or in a neighboring part of swarm). The inertia w regulates the trade-off between the *global* (wide-ranging) and *local* (nearby) exploration abilities of the swarm. A large inertia weight facilitates global exploration, i.e., searching new areas, while a small value facilitates local exploration, i.e., fine-tuning the current search area –exploitation [18].

The PSO algorithm is presented in Algorithm 1. The *end criterion* (line 2) may be the maximum number of iterations, the number of iterations without improvement, or the minimum objective function error between the obtained objective function and the best fitness value w.r.t. a pre-fixed anticipated threshold. Particles are started at random positions with zero initial velocities and search in parallel. What is needed is some *attraction*, if not to the absolutely best position known, at least towards a position close to the particle where the fitness is better than the fitness a particle has currently determined. All particles exploit at least one *good* position already found by some particle(s) in the swarm (line 7). Hence, particles adjust their own position and velocity based on this good position (line 9). Often, the position that is exploited is the *best* position yet found by any particle (line 5). In this case, all particles know the currently best position found and are attracted to this position. This, obviously, requires communication between particles and some sort of collective memory to the current *global best* (*gbest*). The $\mathbf{x}^*$ vector in Equation (1) represents the *gbest* position of the swarm (line 5).

Algorithm 1. Particle Swarm Optimization Algorithm

1	**Initialize** randomly the positions and zero velocities.
2	**While** (the *end criterion* is not met) **Do**
3	$t \leftarrow t + 1$;
4	**Calculate** the fitness value f of each particle;
5	$\mathbf{x}^* = \arg min^N_{i=1}\{f(\mathbf{x}^*(t-1)), f(\mathbf{x}_1(t)), ..., f(\mathbf{x}_i(t)), ..., f(\mathbf{x}_N(t))\}$;
6	**For** $i = 1: N$
7	$\mathbf{x}_i^\#() = \arg min^N_{i=1}\{f(\mathbf{x}_i^\#(t-1)), f(\mathbf{x}_i(t))\}$;
8	**For** $j = 1:d$
9	**Update** the jth dimension of $\mathbf{v}_i$ and $\mathbf{x}_i$ w.r.t. (1), (2);
	Next j
10	**Next** i
11	**End While**

Alternatively, a particle i can experience an attraction back to the best place yet found by it. The *personal best* (*pbest*) position for particle i results in its independent exploration without any input of the other particles. The *pbest* position for the ith particle is $\mathbf{x}_i^\#$ (line 7).

An idea for triggering a particle to direct to an attracted area is to balance the movement between the *gbest* and *pbest* positions by defining a *local* neighborhood around it. All N_i particles within an actual physical distance form the neighborhood of the ith particle. Each particle in N_i shares its fitness value with all other particles in that neighborhood. Hence, neighboring particles experience an attraction to the *local best* (*lbest*). The problem with *lbest* (not so critical as in *gbest*) is that, neighborhoods need to be calculated frequently and, thus, the computational cost for this operation has to be considered. The particles adjust their current velocity based on current *pbest* and prior knowledge derived from *gbest* and *lbest*. Based on *gbest*, particles have to communicate with the whole swarm for locating and maintaining information on the global best solution. In this case the best particle acts as an attractor pulling all the particles towards it. Eventually, all particles will converge to this position. Based on *lbest*, particles are required to check for any better solution appeared in adjacent particles.

In order to avoid the inherent communication cost in CDP due to the information exchange among particles for estimating *gbest* and the premature convergence obtained from *gbest*, we relate the social component c_2 in (1) to the *lbest* approach, i.e., the $\mathbf{x}^*$ vector in (1) represents the *lbest* position of a given particle. c_2 indicates the willingness of a particle to be attracted by any probable neighbor. We also adopt random relative weights for combining *lbest* and *pbest*. The continuous movement toward a position of better fitness (w.r.t. *pbest*) biases the selection of particles with even better fitness than the existing one. The discovery process, which is based on *pbest*, dramatically improves the average fitness of the positions explored. Evidently, this may result in exploration stopping at a local optimum. But, with a number of different local neighborhoods in use, there is a very good probability that the whole swarm will not get so trapped, and that any trapped particle will escape, especially if the *lbest* approach is also simultaneously in use. We adopt both approaches together with r_1, r_2 factors to set the relative influences of each.

3 The Context Discovery Problem

We firstly define the notions of *context* and *quality of context* and then map the parameters of the CDP into PSO.

3.1 Context Representation and Quality of Context

Context refers to the current values of specific parameters that represent the activity / situation of an entity and environmental state [11]. Let $\mathbf{Y} = [Y_1, \ldots, Y_m]$ be a m-dimensional vector of parameters, which assumes values y_l in the domain $\mathrm{Dom}(Y_l)$, $l = 1, \ldots, m$. A parameter Y_l is considered instantiated if at time t some y_l value is assigned to Y_l. *Context* $\mathbf{y}$ is the instantiated $\mathbf{Y}$, i.e., $\mathbf{y} = [y_1, \ldots, y_m]$. For each instantiated Y_l, a function $v: Y_l \times T \to [0, a)$, $a > 0$, is defined denoting whether the value y_l is valid at time t after the Y_l instantiation; T is the time index and a is a real positive number. The value y_l is valid at time t for a context-aware application that is executed on node i if $v(y_l, t) < \theta_{il}$ for a given threshold $\theta_{il} \in (0, a)$, which is application specific. A value of θ_{il} close to a means that y_l is not valid for the ith node. v can be any increasing function F with time t, i.e., $v = F(t)$. For simplicity reasons we can assume that F is the identity function (i.e., $v = t$). The value a is set w.r.t. application specification. For instance, in our case a is the maximum time from the sensing time of Y_l in which its value is not deprecated. A value of θ_{il} close to 0 means that y_l is of high importance. The indicator v increases over time from the sensing time of y_l. Hence, a value of v denotes the freshness of y_l, i.e., y_l refers to either an up-to-date (fresh) or obsolete measure. It should be noted that, v refers only to the temporal validity of a value. Evidently, other validity functions can be defined referring to quality indicators like spatial scope (value is usable within certain geographical boundaries), the source credibility, the reliability of the measurement, and other objective or subjective indicators [12].

We introduce the *quality of context* indicator $g: \mathbf{Y} \times T \to [0, a)$ for context $\mathbf{y}$ at time t denoting whether the values of the parameters of $\mathbf{y}$ are valid or not with respect to a certain threshold. The value of g is the minimum indicator of the values, that is $g(\mathbf{y}, t) = \min_{l=1}^{m}\{v(y_l, t)\}$ with threshold $\theta_{iy} = \min_{l=1}^{m}\{\theta_{il}\}$. A value of θ_{iy} close to a denotes invalid context, i.e., obsolete context, while a value of θ_{iy} close to 0 denotes fresh context. Context $\mathbf{y}$ turns obsolete once some parameter turns also obsolete.

Each node i attempts to maximize the time period Δt in which $g(\mathbf{y}, t + \Delta t) < \theta_{iy}$ for some t. That is, each node attempts to maintain fresh context as much time as possible. It is worth noting that a node i evaluates the quality of $\mathbf{y}$ differently from a node j, i.e., $g_i(\mathbf{y}, t) \neq g_j(\mathbf{y}, t)$. This means that, $\mathbf{y}$ may be of value for node i but not for node j at the same time. Without loss of generality we assume that all nodes evaluate the quality of context with the same θ_{iy}. That is, all nodes assess context with the same criteria / quality indicators. This does not imply that all nodes obtain context of the same quality. Instead, all nodes are interested in the same quality of context. This does not undermine the generality of the problem. In fact, if there are groups of nodes that assess context differently then groups of nodes will be formed and, consequently, each group will assess context with the same θ_{iy}.

3.2 Mapping Swarm Intelligence to Context Discovery

Let us assume discrete time and consider a square terrain of dimension L. Consider a group of N mobile nodes that maps to a swarm of particles and a set of M mobile sources (i.e., sensors that sense context) that correspond to the possible solutions in PSO. Each source regularly generates fresh context meaning that each source measures context with a given frequency-sensing rate q. Each sensed value is time stamped at the source. Every node needs to move to an area with at least a source that carries fresh context. Alternatively, a node attempts to locate areas where other nodes carry fresh context or context of better quality than the context currently available in them. In addition, a node does not know the existence of a source in a certain area and the swarm does not know the number of sources. This evidently denotes that the nodes continue searching until all sources are located or all nodes carries fresh context. However, the nodes have to adopt a mechanism in order to maintain context as fresh as possible as long as the validity fades over time.

The considered CDP is a 2-dimensional problem space in PSO ($d = 2$). It refers to the 2D location information (longitude and latitude) of the sources / nodes that carry fresh context. The exact 2D location information of a node is not known. Hence, we assume that all nodes are capable of detecting any neighboring node in a region with given transmission range equal to R. The physical presence of a node in a neighborhood can be detected thus such node is assumed to be located in the corresponding neighborhood. Moreover, a node i moves towards to a neighboring node j, which carries fresher context than node i.

The value of $g_i(\mathbf{y}, t)$ denotes the willingness of node i to seek for fresh, or at least of better quality (more up-to-date) context than the existing context. The quality of context indicator gi(y, t) resembles the fitness function f in PSO. A node i attempts to:

➢ minimize the value of $g_i(\mathbf{y}, t)$ at time t, and,
➢ maximize the duration in which it maintains fresh context,
    i.e., $g_i(\mathbf{y}, t) < \theta_{\mathbf{y}}$.

It should be noted that $g_i(\mathbf{y}, t)$ depends on time once the indicators for each parameter increase over time (w.r.t. sensing time). This means that a node i has to regularly update its fitness by dynamically adjusting its decision regarding the next movement w.r.t. *pbest* and *lbest*. Let us calculate the *pbest* and *lbest* so that node i decides in which direction to move. Let N_i be the indices of the neighboring nodes of node i at time t. The

$\mathbf{x}_i^\#$ vector at time t is the position $\mathbf{x}_j$ of the neighbor j, which carries fresher context $\mathbf{y}$ than that of node i and the freshest context among all neighbors of node i, i.e.,

$$\mathbf{x}_i^\# = \mathbf{x}_j : j = \arg min_{l \in \{Ni\}}\{g_l(\mathbf{y}, t) \wedge (g_i(\mathbf{y}, t) > g_l(\mathbf{y}, t))\}.$$

$\mathbf{x}_i^\#$ is currently the best position found at time t to which the node i adjust its next movement at time $t + 1$ assuming the *pbest* fitness value $g_i^\#(\mathbf{y}) = g_j(\mathbf{y}, t)$. The vector $(\mathbf{x}_i^\# - \mathbf{x}_i)$ refers to the self-recognition vector for node i that is attracted by the node j.

Furthermore, the node i can exploit its past knowledge. Based on *pbest* the node i locates the current best node j and moves towards it with a factor $r_1 \cdot c_1$. In addition, node i exploits the average fitness of all neighbors at time t that is

$$g_{N_i}(\mathbf{y},t) = \frac{1}{|N_i|}\sum_{i=1..k}g_i(\mathbf{y},t), k \in N_i - \{i\}$$

The proposed $g_{Ni}(\mathbf{y}, t)$ value refers to a *local fitness* of the neighborhood of node i. Node i can obtain a clear view of its neighborhood meaning that: if $g_{Ni}(\mathbf{y}, t) < \theta_{iy}$ (i.e., fresh context) then the node i might not decide to move far away from this neighborhood hoping that it will probably be within an area where nodes carry fresh context. Similarly to $g_i^\#(\mathbf{y})$, we define the *lbest* $g_i^*(\mathbf{y})$ indicator that is an estimate for the freshness of $\mathbf{y}$ at time t obtained by the neighborhood of node i. If it holds true that $g_{Ni}(\mathbf{y}, t) < g_i^*(\mathbf{y})$ then the *lbest* $\mathbf{x}_i^*$ is the current $\mathbf{x}_i$ of node i at time t and the *lbest* fitness value $g_i^*(\mathbf{y})$ equals to $g_{Ni}(\mathbf{y}, t)$. However, it may hold true that $g_{Ni}(\mathbf{y}, t) > g_i(\mathbf{y}, t)$ but this does not imply that there might not be a neighboring node j that carries more fresh context than node i. In this case, the *lbest* position is not updated contrary to the *pbest* position. Instead, the node i adjusts its next movement by combining a movement towards the current *pbest* $\mathbf{x}_i^\#$ and previous *lbest* $\mathbf{x}_i^*$. Based on the $g_i^*(\mathbf{y})$ and $g_i^\#(\mathbf{y})$ indicators, the node i self-controls its decision on the next movement at time $t + 1$. The vector $(\mathbf{x}_i^* - \mathbf{x}_i)$ refers to the social vector component for node i denoting the attraction of node i to its neighborhood. Moreover, $g_i^*(\mathbf{y})$ increases over time thus node i has to regularly update and check *lbest*. That is because as long as a previous neighborhood has maintained fresh context as a whole, at the next time the *lbest* position may not refer to the same neighborhood even with the same value of $g_{Ni}(\mathbf{y}, t)$. The node i sim ply

Table 1. Mapping Between CDP & PSO

PSO concepts	Time variant		CDP concepts
swarm of N particles	no	no	group of N mobile nodes
Particle i	-	-	node i
problem space	no	yes	context $\mathbf{y}$
global optimum solution $\mathbf{x}_i$	no	yes	source positioned at $\mathbf{x}_i$
local optimum solution $\mathbf{x}_i$	no	yes	node with fresh context positioned at $\mathbf{x}_i$
number of optima	no	no	number of sources M
Fitness f	no	yes	quality of context $g_i(\mathbf{y}, t)$
pbest $\mathbf{x}_i^\#$	no	yes	position of neighboring node j that maximizes $g_j(\mathbf{y}, t)$
lbest $\mathbf{x}_i^*$	no	yes	position of node i whose neighborhood maximizes $g_{Ni}(\mathbf{y}, t)$

stores the previously visited *lbest* position assigned to $g_i^*(\mathbf{y})$. Hence, the node i has the option to move towards to a previous visited position as a last resort.

Table 1 depicts the mapping between CDP and PSO. It should be noted that the fitness function f in PSO depends only on the solution vector $\mathbf{x}_i$ and is not time dependent. The same holds true for the *pbest* and *lbest* positions in PSO. In CDP the corresponding fitness $g_i(\mathbf{y}, t)$ depends on time t as long as the invalidity of $\mathbf{y}$ increases over time. Furthermore, the $g_i^{\#}(\mathbf{y})$ and $g_i^*(\mathbf{y})$ indicators increase over time as well.

4 The Proposed Algorithm

We propose an algorithm in which nodes search for areas where better quality of context is obtained. In other words nodes attempt to find, locate and / or follow neighboring nodes (targets) that carry context of high value. The dynamic behavior of a mobile system means that the system changes state in a repeated manner. In our case the changes occur frequently, that is, both the location of a leader and the value of the optimum (context validity) vary[1] [22]. We propose several strategies for the CDP in order to (i) experiment with the required time for finding and maintaining high quality context, (ii) reduce the inherent network load that is used to automatically detecting and tracking various changes of the context validity and (iii) effectively respond to a wide variety of changes in context validity. The network load derives from the intercommunication among nodes. In addition, several constraints that refer to the temporal validity of context are taken into account. Therefore, the best solution of CDP is time dependent (context turns obsolete over time).

The proposed behaviors indicate the intention of a node in discovering and maintaining fresh context based on its mobility and other characteristics explained below. Specifically, a node transits between three states in order to discover context. In each state, the node decides on certain actions. A state k_i of node i can be *Obsolete* (O), *Partially satisfied* (P), or *Satisfied* (S) as depicted in Figure 1. In state O, a node either carries obsolete context (or is in need of) i.e., $g_i(\mathbf{y}, t) > \theta_y$. In the S state, a node carries fresh context i.e., $g_i(\mathbf{y}, t) < \theta_y$. If context $\mathbf{y}$ turns obsolete then node i transits into O. In the P state, a node chooses to carry less obsolete context than the existing context as long as this is the current best solution it achieves (local optimum). This means that the node i has found a neighbor j with fresher context i.e., $g_i(\mathbf{y}, t) > g_j(\mathbf{y}, t) > \theta_y$. The node i escapes from the P state once another node k, which carries more fresh context, is located i.e., $g_i(\mathbf{y}, t) > \theta_y > g_k(\mathbf{y}, t)$. We assume that all nodes adopt the same threshold for assessing the quality of context ($\theta_{iy} = \theta_y = \theta$, $i = 1, \ldots, N$).

4.1 Foraging for Context

A node i in state O initiates a foraging process for context acting as follows: The node i moves randomly ($\mathbf{v}_i \sim U(\mathbf{v}_{min}, \mathbf{v}_{max})$) in the swarm and intercommunicates with neighbors till to be attracted by a neighbor j. The node j is then called *leader*. The leader j either carries objectively fresh context i.e., $g_j(\mathbf{y}, t) < \theta_y$ or carries context that is more fresh than the context carried by node i i.e., $g_i(\mathbf{y}, t) > g_j(\mathbf{y}, t) > \theta_y$ (see obsolete state in Figure 1). In the former case, the node i transits directly to state S. In the latter

---

[1] This dynamic environment refers to Type III environment ([18]).

case, the node j does not carry context of the exact quality that node i expects but such context is preferable than that of node i. Hence, node i can either follow node j hoping that it approaches areas (neighborhoods) with more fresh context -thus transiting to P state- or, alternatively, ignores such opportunity and continues moving at random - thus remaining at state O. In state P, node i settles with lower quality of context. This does not imply that node i stops communicating with other neighbors while moving. Instead, it continues exchanging information related to context quality with the purpose of locating another leader with more fresh context. The P state is an intermediary state between the O and S states (see partially satisfied state in Figure 1). The node is moving among neighborhoods carrying context of better quality and continues exploring areas. This policy reflects the idea of exploring the solution space even if a solution has already been reached (possibly a local optimum).

Node i attempts to retain fresh context for as long as possible. However, the $v(y_l, t)$ indicator for a sensed parameter Y_l increases over time t until that value turns obsolete after some Δt, i.e., $v(y_l, t + \Delta t) > \theta_l$. Hence, Y_l has to be regularly determined / sensed, with frequency at least $1/\Delta t$. In order for the node i to obtain up-to-date context $\mathbf{y}$, it follows leaders or sources that regularly generate objectively fresh context.

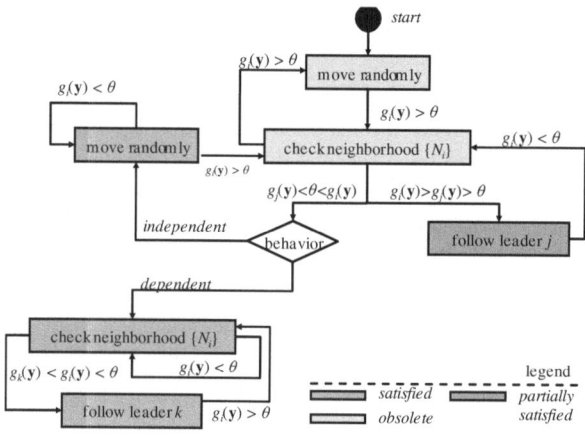

Fig. 1. A state transition in CDP

It should be noted that a localization system is needed in order to determine the solutions $\mathbf{x}_i^{\#}$ and $\mathbf{x}_i^*$, and the way node i is directed to its leader. Specifically, a node i carried by an agent (possbily a human) is directed to its leader once a GIS application displays directional information of the leader obtained, for instance, by a compass-based mechanism [16] (or other techniques, e.g., the time-of-flight technique that adopts radio frequency and ultrasound signal to determine the distance between nodes [15]). However, a non-human node i (e.g., a robot), without localization mechanisms, can "blindly" follow its leader by adjusting its direction / velocity through small improvement steps w.r.t. the signal quality [14]. Imagine for example a WLAN user trying to determine the best signal quality in a certain room by stepping around without knowing the exact location of the access point. This local-searching blind technique is not as efficient as the previously discussed method [17].

4.2 Maintaining Fresh Context

A node i, in S state, acts as follows (see the satisfied state in Figure 1): it either continues communicating with leaders (*dependent* behavior) or re-starts moving at random (*independent* behavior) with $\mathbf{v}_i \sim U(\mathbf{v}_{min}, \mathbf{v}_{max})$. In the former behavior, it is likely that node i constantly follows leaders (a.k.a. tracking optima [19]). The advantage of such behavior is that: in case node i's context turns obsolete, node i will easier find some leader provided that the latter might be yet reachable (or not far away). By adopting the independent behavior node i has no information in which direction to move towards once context turns obsolete.

Once a neighbor node k, of a node i, in S state obtains better context $\mathbf{y}$ (i.e., $\theta_y > g_j(\mathbf{y}, t) > g_k(\mathbf{y}, t)$) then node i may choose to abandon the existing leader and follow the new leader node k. Specifically, by adopting the dependent behavior, the node i communicates with neighbors with the intention of finding a node k that carries more fresh context than the objectively fresh obtained context. Hence, the node i switches constantly between among leaders. In addition, the node i never transits to state O since its leader is a source (global optimum). However, the main objective in CDP is to enable nodes to minimize the communication load and discover as many sources and leaders as possible escaping from local optima. If all nodes adopted the dependent behavior then they would attach to sources resulting in large communication effort for the sources (sources would have to communicate with a large number of nodes) but carrying objectively fresh context. It is of high importance to take into account the inherent efficiency for both behaviors.

4.3 The CDP Algorithm

A node i in state O either transits only to state S once a leader with objectively fresh context is found or transits to the immediate state P once a leader with better context is found. As long as a leader carries objectively fresh context then node i transits from state P to S. In state S node i adopts either the independent or the dependent behavior. In this paper, we present the CDP algorithm in Algorithm 2, in which node i, in state O, transits to states P and/or S, and, in state S, it adopts the dependent behavior. Initially, all N nodes in the swarm are in state O and are randomly distributed in a given terrain with random velocities in $[\mathbf{v}_{min}, \mathbf{v}_{max}]$. The inertia w is used to controlling the exploration and exploitation abilities of the swarm and eliminating the need for velocity clamping (i.e., if $|\mathbf{v}_i| > |\mathbf{v}_{max}|$ then $|\mathbf{v}_i| = |\mathbf{v}_{max}|$). The inertia is very important to ensure convergent behavior; large values for w facilitate exploration with increased diversity while small values promote local exploitation. We adopt a dynamically changing inertia values, i.e., an initially value decreases nonlinearly to a small value allowing for a shorter exploration time (due to context validity rate) with more time spent on refining optima [20]. That is,

$$w(t+1) = \frac{(w(t) - 0.4)(t_0 - t)}{t_0 + 0.4}$$

$w(0) = 0.9$, t_0 is the maximum number of iterations. In case a node transits to state O then it re-sets w to its initial value. The randomly moving M sources generate context with sensing rate q (in samples/second, Hz) and the thresholds $\theta_l = a$ for the properties Y_l are set. The c_1 and c_2 constants denote how much the *lbest* and *pbest* solutions influence the movement of the node; usually $c_1 = c_2 = 2$ ([18]). The r_1, r_2 are two random vectors with each component be a uniform random number in $[0, 1]$. In each iteration, a node i in O, or P, adjusts its movement (lines 18, 19) w.r.t. *pbest* and *lbest* (lines 20-28) once interactive communication takes place. If the node i in S adopts:

➢ dependent behavior then it adjusts its movement w.r.t lines 21-28,
➢ independent behavior then it randomly moves with v_i in $[v_{min}, v_{max}]$ (omit lines 17-28).

The end-criterion of the algorithm can be the number of iterations, the time needed to find fresh context a given portion of nodes, or energy consumption constraints. In our case the end-criterion is time dependent since the validity of context depends on the sensing rate q. Nodes adopting the independent behavior stop searching as long as they obtain fresh context and re-start foraging once context turns obsolete. The end-criterion for the dependent behavior is the minimum mean value $g_+(t)$ for the fitness function g. We require that $g_+(t)$ be as low as possible w.r.t. the a threshold that is, maximize $d(t) = (g_+(t) - a)^2$. The $d(t)$ value denotes how much fresh is context. In other words, it reflects the portion of time needed for context to turn obsolete as long as $g_+(t)$ is greater than a. For instance, let two nodes, i and j, carry context $\mathbf{y}$ with $g_i(\mathbf{y}) = a/2$ and $g_j(\mathbf{y}) = a/4$. Objectively, both nodes carry fresh context w.r.t. a. Therefore, node j carries fresher context than node i since node j will carry fresh context for longer time than node i. The convergence $g_+(t_o)$ value denotes a state in which some nodes obtain fresh context for $t \geq t_o$ and depends highly on a: a high value of a denotes a little time for context to turn obsolete. In that case, the nodes may stay for a long in S state. On the contrary, a low value of a (i.e., nodes are interested only for up-to-date context) results in values of $g_+(t)$ close to a; context turns obsolete with a high rate. It is of high interest to examine the efficiency of each behavior.

Algorithm 2. The Context Discovery Problem Algorithm

1. **Set** c_1, c_2, N, M, q	16. **For** $i = 1 : N$
2. **Set** random $\mathbf{x}_i(t)$, threshold $\theta_i \mathbf{y} = \theta \mathbf{y}$, $t \leftarrow 0$	17. **Set** random unary vectors r_1, r_2
3. **For** $i = 1 : N$	18. $\mathbf{x}_i(t) \leftarrow \mathbf{x}_i(t-1) + \mathbf{v}_i(t)$
4. $\mathbf{v}_i(t) \sim \mathrm{U}(\mathbf{v}_{min}, \mathbf{v}_{max})$, $k_i \leftarrow$ O	19. $\mathbf{v}_i(t) \leftarrow \mathbf{v}_i(t-1) + c_1 r_1 (\mathbf{x}_i^* - \mathbf{x}_i(t-1)) + c_2 r_2 (\mathbf{x}_i^\# - \mathbf{x}_i(t-1))$
5. $g_i^*(\mathbf{y}) \leftarrow (g_1(\mathbf{y}, t) + \cdots + g_{|Ni|}(\mathbf{y}, t)) / |N_i|$	20. $g_{Ni}(\mathbf{y}, t) \leftarrow (g_1(\mathbf{y}, t) + \cdots + g_{|Ni|}(\mathbf{y}, t)) / |N_i|$
6. $\mathbf{x}_i^* \leftarrow \mathbf{x}_i(t)$	21. **If** $g_i^*(\mathbf{y}) < g_{Ni}(\mathbf{y}, t)$ **Then**
7. $g_i^\#(\mathbf{y}) \leftarrow max_l\{g_l(\mathbf{y}, t)\}$, $l \in \{N_i\} \cup \{i\}$	22. $\mathbf{x}_i^* \leftarrow \mathbf{x}_i(t)$
8. $\mathbf{x}_i^\# \leftarrow \mathbf{x}_e$: $e = argmax_l\{g_l(\mathbf{y}, t)\}$,	23. $g_i^*(\mathbf{y}) \leftarrow g_{Ni}(\mathbf{y}, t)$
$l \in \{N_i\} \cup \{i\}$, $leader_i \leftarrow e$	24. **End**
9. **Next** i	25. **If** $g_i^\#(\mathbf{y}) < max_l\{g_l(\mathbf{y}, t)\}$, $l \in \{N_i\}$ **Then**
10. **While** (the *end criterion* is not met) **Do**	26. $\mathbf{x}_i^\# \leftarrow \mathbf{x}_e$: $e = argmax_l\{g_l(\mathbf{y}, t)\}$, $l \in \{N_i\}$, $leader_i \leftarrow e$
11. $t \leftarrow t + 1$;	27. $g_i^\#(\mathbf{y}) \leftarrow max_l\{g_l(\mathbf{y}, t)\}$, $l \in \{N_i\}$
12. **For** $i = 1 : N$	28. **End**
13. **Calculate** $g_i(\mathbf{y}, t)$	29. **If** $g_i^\#(\mathbf{y}) < g_i(\mathbf{y}) < \theta \mathbf{y}$ **Then** $k_i \leftarrow$ O
14. **Validate** $g_i^*(\mathbf{y})$, $g_i^\#(\mathbf{y})$ //increase validity	30. **If** $g_i(\mathbf{y}) < g_i^\#(\mathbf{y}) < \theta \mathbf{y}$ **Then** $k_i \leftarrow$ P
indicators	31. **If** $\theta \mathbf{y} < g_i^\#(\mathbf{y})$ **Then** $k_i \leftarrow$ S
15. **Next** i	32. **Next** i
	33. **End While**

5 Performance Evaluation

In this section we assess the proposed behavior for the CDP. Our objective is to enable nodes to discover and maintain fresh context. However, the fact of locating sources and leaders in an attempt to carry fresh context is at the expense of the inherent network load due to communication of nodes. We define as efficiency $e(t)$ of a certain behavior the portion of nodes $n(t)$ being in state S out of the communication load $l(t)$ among neighboring nodes exchanging information about context quality, i.e., $e(t) = n(t) / l(t)$. We require that $e(t)$ assumes high values minimizing the load $l(t)$ and maximizing $n(t)$ w.r.t. the adopted behavior.

The parameters of our simulations are: a swarm of $N = 100$ nodes, $M = 2$ sources, $a = 100$ time units, a terrain of $L = 100$ spatial units, transmission range $R = 0.01L$, the *random waypoint* model for mobility behavior in $[v_{min}, v_{max}] = [0.1, 2]$ ([13]), and 1000 runs of the algorithm. Context turns obsolete every a time units and is sensed by the sources with q ranging from $(2/a)$ Hz to 1Hz. We require that $g_+(t)$ be lower than a as time passes or, at least, lower than a between consequent intervals of a time units.

Figure 2 depicts the $g_+(t)$ value (in time units –t.u.) when all nodes in the swarm adopt the dependent behavior for different values of q. It is observed that all nodes rapidly locate leaders and then carry fresh context denoting CDP algorithm convergence. The $g_+(t)$ value converges to $g_+(t_o)$ ranging from 14.633t.u. to 40.882t.u. for q ranging from 1Hz to 0.02Hz, respectively. It is worth noting that, for $q = 1$Hz, the $g_+(t_o)$ is 14.633t.u. i.e., 14.633% of the validity threshold a indicating that most nodes can process context for 85.367% of the sensing time before it turns obsolete. Moreover, as q assumes low values (e.g., 0.02Hz), which means that the sources sense context every 50t.u., the value of $g_+(t)$ swings around the 40.882t.u. This indicates that, nodes locate sources whose context turns obsolete after 50t.u. For that reason, the $g_+(t)$ value for such nodes exhibits that behavior. On the other hand, once q assumes high values (e.g., 1Hz), the sources constantly carry up-to-date context. Consequently, nodes that locate sources carry fresh context ($g_+(t)$ converges). The achieved maximum value for $d(t_o)$ is $0.6952.10^4$ for $q = 1$Hz, as depicted in Figure 3, compared to $0.3854.10^4$ w.r.t. independent behavior, as discussed later. Evidently, by adopting the dependent behavior, a large portion of the swarm follows leaders and/or sources carrying objectively fresh context. However, such behavior requires that nodes communicate

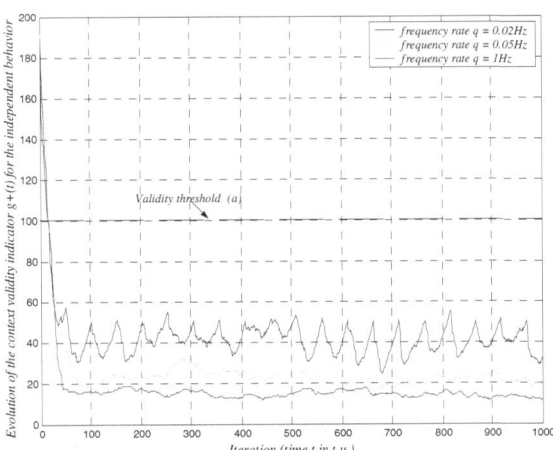

Fig. 2. The $g_+(t)$ value of the dependent behavior for sensing rate $q = 0.02$Hz, $q = 0.05$Hz and $q = 1$Hz

continuously in order to locate sources and leaders with more fresh context even if nodes are in state S for maximizing $d(t)$. That leads to additional communication load thus keeping the efficiency to 50% as depicted in Figure 4. Specifically, Figure 4 depicts the value of $e_d(t)$ for the dependent behavior for $q = 1$Hz. Obviously, the inherent communication load of such behavior is high since a large portion of nodes attempts to carry fresh context.

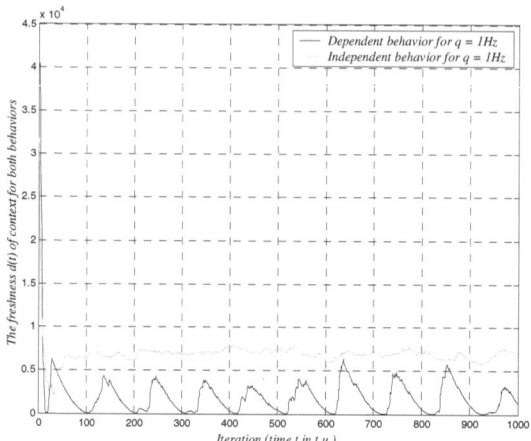

Fig. 3. The $d(t)$ value of the dependent and independent behavior with sensing rate $q = 1$Hz

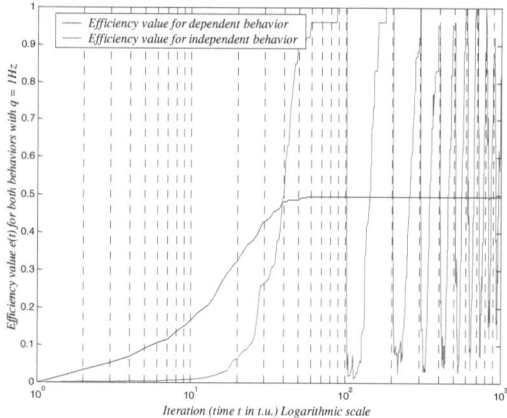

Fig. 4. The values of $e_d(t)$ and $e_i(t)$ efficiency in logarithmic scale for $q = 1$Hz

Figure 5 depicts the $g_+(t)$ value of nodes adopting the independent behavior. We illustrate $g_+(t)$ for sensing rates $q = 1$Hz, $q = 0.05$Hz and $q = 0.02$Hz. Evidently, nodes seek for fresh context only when the existing context turns obsolete. This is indicated by the sharp bend of $g_+(t)$ between intervals of a time units for $q = 1$Hz. The periodic behavior of $g_+(t)$ reflects the idea of the independent behavior denoting that a node is about to seek for context only when needed. Hence, between intervals of a time units nodes that are in S state save energy as long as they do not exchange information with others. When context turns obsolete, nodes re-start foraging but having the *pbest* solution as a candidate starting point. This means that, each time context turns obsolete nodes adjust its movement based on the last known *pbest* solution. Hence, they start moving "blindly" as long as their first direction might be the *pbest* position indicating "prolific" neighborhood. For that reason, the maximum value of $g_+(t)$ is close to a in each "period" as depicted in Figure 5. Moreover, the $g_+(t)$ value ranges from 40 t.u. to 100 t.u. compared to the convergence value of $g_+(t_0) = 14.633$t.u. in case of the dependent behavior for $q = 1$Hz. It is worth noting that the value of $g_i^*(\mathbf{y})$ for *pbest* must denote valid context, otherwise node i has to move entirely at random. Moreover, consider the $g_+(t)$ value having $q = 2/a = 0.02$Hz. Specifically, $g_+(t)$ assumes the minimum value every $(a/2) = 50$ t.u., which is greater than the minimum value of $g_+(t)$ achieved for $q = 1$Hz every at.u. In the former case nodes re-start foraging sooner than in the latter case (practically two times more), thus, the adoption of the *pbest* solution seems more prolific. For that reason, $g_+(t)$ assumes higher minimum values in the former case even though the sensing rate is lower. In such cases, the adoption of the *pbest* solution is of high importance.

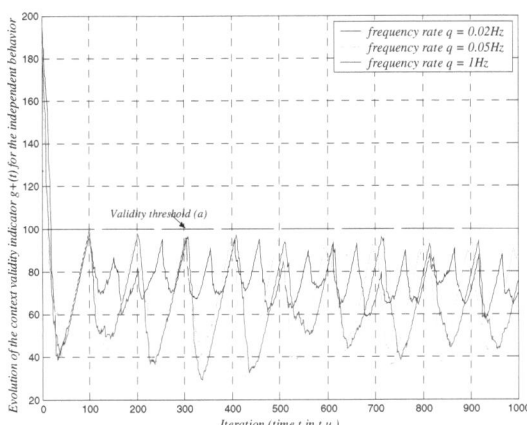

Fig. 5. The g+(t) value of the independent behavior for sensing rates q = 1Hz, q = 0.05Hz and q = 0.02Hz

Figure 3 depicts also the $d(t)$ value for the independent behavior. The $d(t)$ assumes the maximum value (therefore lower than in the case of the dependent behavior) only when a large portion of nodes carry fresh context. In addition, $d(t)$ assumes zero value regularly every a time units denoting the time that all nodes carry obsolete context. The mean value of $d(t)$ is $0.3854.10^4$, that is 44.56% lower than the convergence value of $d(t_0)$ in the case of the dependent behavior (for the same sensing rate $q = 1$Hz). Hence, the adoption of the independent behavior for CDP results in 44.56% lower quality of context than that achieved by dependent behavior.

By adopting the independent behavior we can achieve high values of efficiency $e_i(t)$ during intervals in which nodes carry fresh context. This is due to the fact that in such intervals nodes stop communicating with each other thus reducing the load $l(t)$.

However, when context turns obsolete then $e_i(t)$ assumes a very low value (mean value lower than 0.1) as long as a large portion of nodes do not carry fresh context thus reducing $n(t)$. In Figure 4 the behavior of $e_i(t)$ for sensing rate $q = 1Hz$ is also illustrated. We can observe that $e_i(t)$ ranges from 0.069 (mean value) to 0.93 (mean value) compared to the convergence value of $e_d(t) = 0.5$.

Each behavior can be applied on a context-aware application considering the specific requirements of the application. Once the application needs critically up-to-date context then the adoption of the dependent behavior is preferable. On the other hand, once we are interested in saving energy then nodes can adopt the independent behavior. However, a hybrid scheme combining both behaviors can be adopted. For instance, a portion of nodes can adopt the independent behavior for reducing energy consumption and the rest nodes adopt the dependent behavior maintaining up-to-date information. Another combination refers to the adoption of the dependent behavior for rapidly locating sources and leaders followed by the adoption of the independent behavior till context turns obsolete.

6 Conclusions

PSO is a simple algorithm with a wide application range on different optimization problems. We deal with the CDP by adopting the decentralized control and self-organization of SI. We provide the mapping between PSO and CDP, and study how SI-inspired computing can facilitate context discovery. We introduce the time-variant context quality indicator g that refers to the fitness function f in PSO. Hence, each particle-node attempts to carry and maintain fresh context w.r.t. the g indicator. We propose the independent and dependent foraging behaviors (strategies) for mobile nodes The use of such behaviors in conjunction to the local fitness of the neighborhood enables node to discover sources and/or leaders that provide up-to-date contextual information. The proposed algorithm for the CDP supports such behaviors provided that context turns obsolete over time in a dynamic environment. We evaluated the efficiency and the effectiveness of each behavior. The adoption of each behavior relies on the context-aware application itself: for critically up-to-date context constrained applications the dependent behavior is preferable while, once energy savings are of high importance then the independent behavior exhibits satisfactory results. Our simulation results indicate the applicability of SI in context discovery and the proposed foraging behaviors provide useful tools in mobile computing. In addition, the adoption of SI for transmission range / power adjustment, so that context-aware nodes control their energy consumption, is a future work that can be considered in the CDP.

References

1. Deneubourg, J., Goss, S., Sandini, G., Ferrari, F., Dario, P.: Self-Organizing Collection and Transport of Objects in Unpredictable Environments. In: Symposium on Flexible Automation, pp. 1093–1098 (1990)
2. Drogoul, A., Ferber, J.: From Tom Thumb to the Dockers: Some Experiments With Foraging Robots. In: 2nd Int. Conference on Simulation of Adaptive Behavior, pp. 451–459 (1992)

3. Parker, L.E.: ALLIANCE: An Architecture for Fault-Tolerant Multi-Robot Cooperation. IEEE Transactions on Robotics and Automation 14(2), 220–240 (1998)

4. LaVille, S.M., Lin, D., Guibas, L.J., Latombe, J.C., Motwani, R.: Finding an Unpredictable Target in a Workspace with Objects. In: IEEE Int. Conf. on Robotics and Automation, pp. 737–742 (1997)

5. Dorigo, M., Gambardella, L.M.: Ant Colony Systems: A Cooperative Learning Approach to the Traveling Salesman Problem. IEEE Trans. on Evolutionary Computation 1(1), 53–66 (1997)

6. Pasino, K., Polycarpou, M., Jacques, D., Pachter, M., Liu, Y., Yang, Y., Flint, M., Baum, M.: Cooperative Control for Autonomous Air Vehicles. In: Murphy, R., Pardalos, P. (eds.) Cooperative Control and Optimization. Kluwer Academics Publishers, Boston (2002)

7. Polycarpou, M., Yang, Y., Pasino, K.: A Cooperative Search Framework for Distributed Agents. In: IEEE Int. Symposium on Intelligent Control, pp. 1–6 (2001)

8. Stone, L.D.: Theory of Optimal Search. Academic Press, New York (1975)

9. Vincent, P., Rubin, I.: A Framework and Analysis for Cooperative Search Using UAV Swarms. In: ACM Symposium on Applied Computing (2004)

10. Kennedy, J., Eberhart, R.: Swarm Intelligence. Morgan Kaufmann Publishers, Inc., San Francisco (2001)

11. Anagnostopoulos, C., Tsounis, A., Hadjiefthymiades, S.: Context Awareness in Mobile Computing Environments, Special Issue on Advances in Wireless Communications Enabling Technologies for 4G. Wireless Personal Communication Journal 2(3), 454–464 (2007)

12. Anagnostopoulos, C., Sekkas, O., Hadjiefthymiades, S.: Context Fusion: Dealing with Sensor Reliability. In: IEEE Int. Workshop on Information Fusion and Dissemination in Wireless Sensor Networks - IEEE Int. Conference on Mobile Ad-hoc and Sensor Systems, pp. 1–7 (2007)

13. Bettstetter, C., Hartenstein, H., Pérez-Costa, X.: Stochastic properties of the random waypoint mobility model. ACM/Kluwer Wireless Networks: Special Issue on Modeling and Analysis of Mobile Networks 10(5) (September 2004)

14. Matthias, O., Hanspeter, A.: Biomimetic robot navigation. Robotics and Autonomous Systems 30(1-2), 133–153 (2000)

15. Savvides, A., Han, C., Srivastava, M.: Dynamic Fine-grained Localization in Ad-hoc Networks of Sensors. In: ACM Mobicom, July 2001, pp. 166–179 (2001)

16. Anjum, F., Mouchtaris, P.: Security for Wireless Ad Hoc Networks. Wiley-Interscience, Hoboken (2007)

17. Zhao, F., Guibas, L.: Wireless Sensor Networks - An Information Processing Approach. Elsevier Science, Amsterdam (2004)

18. Engelbrecht, A.P.: Computational Intelligence: An Introduction, 2nd edn. Wiley, Chichester (2007)

19. Eberhart, R.C., Shi, Y.: Tracking and Optimizing Dynamic Systems with Particle Swarms. In: IEEE Cong. on Evolutionary Computation, vol. 1, pp. 94–100 (2001)

20. Peram, T., Veeramachaneni, K., Mohan, C.K.: Fitness-Distance-Ration based Particle Swarm Optimization. In: IEEE Symp. Swarm Intelligence, pp. 174–181 (2003)

21. Carlishe, A., Dozier, G.: Adapting Particle Swarm Optimization to Dynamic Environments. In: Intl. Conf. Artificial Intelligence, pp. 429–434 (2000)

22. Hu, X., Eberhart, R.C.: Tracking Dynamic Systems with PSO: Where's the Cheese? In: Proc. Workshop on Particle Swarm Optimization, pp. 80–83 (2001)

23. Hu, X., Eberhart, R.: Adaptive Particle Swarm Optimization: Detection and Response to Dynamic Systems. In: IEEE Congress on Evolutionary Computation, USA, pp. 1666–1670 (2002)

Consequences of Social and Institutional Setups for Occurrence Reporting in Air Traffic Organizations

Alexei Sharpanskykh

Vrije Universiteit Amsterdam, Department of Artificial Intelligence
De Boelelaan 1081a, 1081 HV Amsterdam, The Netherlands
sharp@cs.vu.nl

Abstract. Deficient safety occurrence reporting by air traffic controllers is an important issue in many air traffic organizations. To understand the reasons for not reporting, practitioners formulated a number of hypotheses, which are difficult to verify manually. To perform automated, formally-based verification of the hypotheses an agent-based modeling and simulation approach is proposed in this paper. This approach allows modeling both institutional (prescriptive) aspects of the formal organization and social behavior of organizational actors. To our knowledge, agent-based organization modeling has not been attempted in air traffic previously. Using such an approach four hypotheses related to consequences of controller team composition in particular organizational contexts were examined.

Keywords: Agent-based simulation, organization modeling, formal analysis, air traffic.

1 Introduction

One of the safety problems, which air navigation service providers (ANSP) face, is that many safety occurrences happened during air and ground operations are not reported by air traffic controllers. An example of a ground occurrence is 'taxiing aircraft initiates to cross due to misunderstanding in communication'. Knowledge about occurrences is particularly useful for timely identification of safety problems.

To understand the reasons for such a behavior of controllers a number of hypotheses have been formulated by professionals in air traffic control that concern particular controller types. In [6] the following types of controllers that prevail in controller teams are distinguished: (1) *rule-dependent*: controllers who show strict adherence to formal regulations; (2) *peer-dependent*: controllers whose behaviour depends strongly on the behaviour and opinions of their peers. Following the discussions from [6,1] and based on the interviews with safety professionals from an existing ANSP, the following four hypotheses related to occurrence reporting and to the considered types of controllers have been identified:

A.V. Vasilakos et al. (Eds.): AUTONOMICS 2009, LNICST 23, pp. 176–191, 2010.

Hypothesis 1: Reprimands provided to controllers for safety occurrences, in which they were involved, serve the purpose of improvement of the reporting quality.

Hypothesis 2: The rule-dependent controllers demonstrate more uniform reporting behavior over time than the peer-dependent controllers of the same team.

Hypothesis 3: Teams with majority of peer-dependent members report poorly in the absence of reprimands in ANSPs with low actual commitment to safety.

Hypothesis 4: To neutralize negative effects of peer influence on reporting, a mixed composition of teams with comparable numbers of controllers of both types is useful.

Hypotheses over safety occurrence reporting were attempted to be verified using conventional analysis techniques in air traffic control, which are based predominantly on fault/event trees used for sequential cause-effect reasoning for accident causation [2]. However, such trees do not capture complex, non-linear dependencies and dynamics inherent in ANSPs. Agent-based modeling has been proposed as a means to assess safety risks of and identify safety issues in air traffic operations in a complex ANSP [8,12,13]. However, existing agent-based approaches known to us model air traffic systems without considering the organizational layer, often with a simplified representation of agents (i.e., without or with a very simple internal (or cognitive) structure), cf. [12,13]. Disregarding significant knowledge about formal and informal organization structures of an ANSP may lead to mediocre analysis results, when actual causes of issues remain unidentified. Furthermore, a large number of existing agent-based approaches aim at efficient air traffic management (planning, scheduling), which is not the type of research questions pursued in this research.

To incorporate organizational aspects in agent-based safety analysis of an ANSP, an approach is proposed in this paper that allows modeling both institutional (prescriptive) aspects of an ANSP and proactive social behavior of organizational agents. To define the prescriptive aspects the general organization modeling framework from [10] was used, which has formal foundations precisely defined based on the order-sorted predicate logic. In this framework formal organizations are considered from three interrelated perspectives: the performance-oriented, the process-oriented, and the organization-oriented. The behavior of organizational agents was modeled from external and internal perspectives. From the external perspective interaction of an agent with other agents and with the environment by observation, communication and performing actions was modeled. From the internal perspective the behavior of an agent was modeled by direct causal relations between internal (or cognitive) agents states, based on which an externally observable behavioral pattern is generated. In particular, the internal dynamics of a decision making process of a controller agent whether to report an observed occurrence is considered in the paper.

The developed model of the formal organization extended with a specification of the agents was used to perform simulation of safety occurrence reporting in an ANSP. The four hypotheses formulated above were tested on the obtained simulation results. Previously an approach for validation of models using the framework from [10] has been developed [11]. Using this validation approach, models developed for verifying hypotheses can be validated.

The paper is organized as follows. In Section 2 the developed model for formal reporting is given. A specification of the organizational agents is described in Section 3. The simulation setup and the hypotheses verification results are described in Section 4. One of the steps of statistical validation of the model sensitivity analysis -is considered in Section 5. Finally, Section 6 concludes the paper.

2   Modeling Formal Reporting in an ANSP

For modeling the formal reporting in an ANSP the modeling framework and methodology from [10] was used, which comprises a sequence of organization design steps. To design the model, data obtained from a real ANSP were used. For a more detailed description with formal details see [15].

Step 1. The identification of the organizational roles. A role is a (sub-)set of functionalities of an organization, which are abstracted from specific agents who fulfill them. Each role can be composed by several other roles, until the necessary detailed level of aggregation is achieved. The environment is modeled as a special role. In this study roles are identified at three aggregation levels, among them (see Fig. 1 and 2): ANSP (level 1), Tower Control Unit (level 2), Controller (level 3), Controller Supervisor (level 3), Safety Investigation Unit (level 2), Safety Investigator (level 3). Furthermore, role instances may be specified, which besides the inherited characteristics and behavior of the role may possess additional characteristics. For example, two instances of Controller role were defined for each sector of the airport with the characteristics and behavior of Controller role.

Step 2. The specification of the interactions between the roles. Relations between roles are represented by interaction and interlevel links. An interaction link is an information channel between two roles at the same aggregation level. An interlevel link connects a composite role with one of its subroles to enable information transfer between aggregation levels. For the considered example some of the identified interaction relations are given in Fig. 1 and 2. To formalize interactions, for each role an interaction ontology is introduced. An ontology is a signature or a vocabulary that comprises sets of sorts (or types), sorted constants, functions and predicates. In particular, to specify communications, interaction ontologies of roles include the predicate:

$$communicated\_from\_to : ROLE \times ROLE \times MSG\_TYPE \times CONTENT$$

Here the first argument denotes the role-source of information, the second the role-recipient of information, the third argument denoted the types of the communication (which may be one of the following *observe, inform, request, decision, readback*) and the fourth the content of the communication. The sort *ROLE*

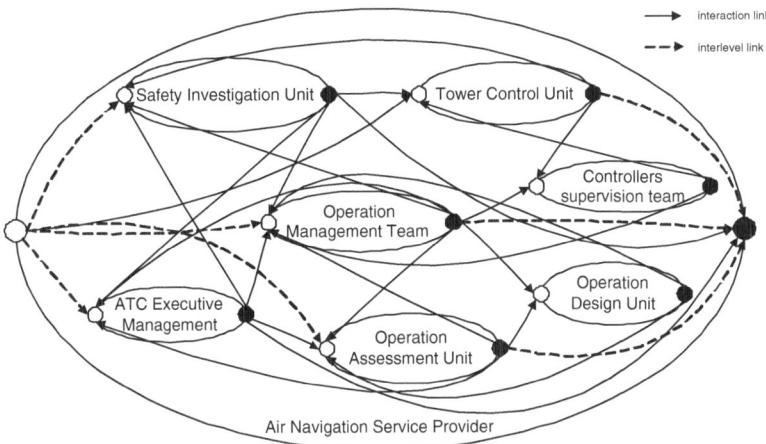

Fig. 1. Interaction relations in ANSP role considered at the aggregation level 2

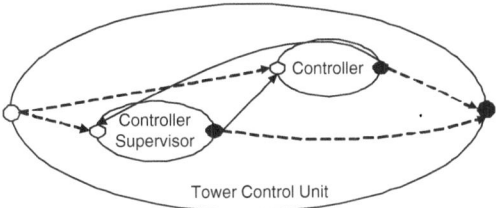

Fig. 2. Interaction relations in Tower Control Unit role considered at the aggregation level 3

is a composite sort that comprises all subsorts of the roles of particular types (e.g., $CONTROLLER$). The sort $CONTENT$ is also the composite sort that comprises all names of terms that are used as the communication content. Such terms are constructed from sorted constants, variables and functions in the standard predicate logic way. For example, communication by role $controller1$ to role $controller\_supervisor$ about $occurrence1$ is formalized as $communicated\_from\_to(controller1, controller\_supervisor, inform, occurrence1)$.

Note that an agent who eventually will be allocated to a role will take over all interaction relations defined for this role. Moreover, an agent may be involved in other (informal) interaction relations with other agents defined in a specification of the agents behaviour (considered in Section 3).

Step 3. The identification of the requirements for the roles. The requirements on knowledge, skills and personal traits of the agent implementing a role at the lowest aggregation level are identified.

Step 4. The identification of the organizational performance indicators and goals. A performance indicator (PI) is a quantitative or qualitative indicator that reflects the state/progress of the company or individual. Goals are objectives that describe a desired state or development and are defined as expressions over PIs. PI evaluated in this paper is the reporting quality (ratio reported/observed occurrences) and the corresponding goal is G1 'It is required to maintain reporting quality > 0.75'. A goal can be refined into subgoals forming a hierarchy. Goals are related to roles: e.g., G1 is attributed to ANSPs Tower Control Unit role.

Step 5. The specification of the resources. In this step organisational resource types and resources are identified, and characteristics for them are provided, such as: name, category: discrete or continuous, measurement unit, expiration duration: the time interval during which a resource type can be used; location; sharing: some processes may share resources. Examples of resource types are: airport's diagram, aircraft, incident classification database, clearance to cross a runway, an incident investigation report.

Step 6. The identification of the tasks and workflows. A *task* represents a function performed in an organization and is characterized by name, maximal and minimal duration. Each task should contribute to the satisfaction of one or more organizational goals. For example, task 'Create a notification report' contributes to goal G1 defined at step 4. Tasks use, produce and consume resources: e.g., task 'Investigation of an occurrence' uses a notification report and produces a final occurrence assessment report. *Workflows* describe temporal ordering of tasks in particular scenarios. Fig.3 describes formal occurrence reporting initiated by a controller. For each task from the workflow responsibility relations on roles were defined. In the following the workflow is considered briefly.

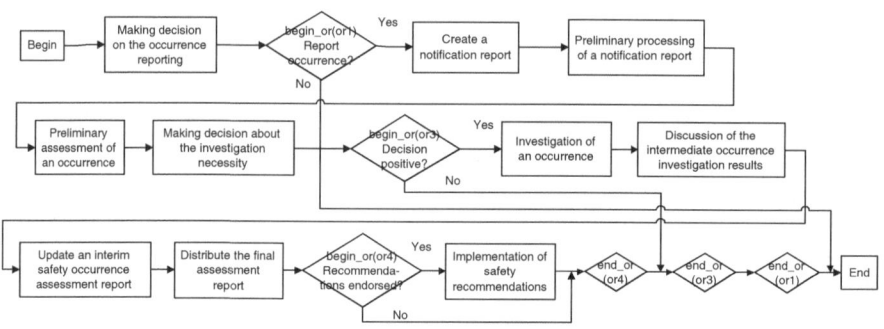

Fig. 3. The workflow for the formal occurrence reporting

After a controller decides to report an observed occurrence, s/he creates a notification report, which is provided to the Safety Investigation Unit (SIU). Different aspects of responsibility relations are distinguished: e.g., Controller role is responsible for execution of and decision making with respect to task Create a notification report, Controller Supervisor is responsible for monitoring and consulting for this task. Depending on the occurrence severity and the collected information about similar occurrences, SIU makes the decision whether

Table 1. Organizational reprimand policies used in simulation

Low severity	$repr(1, A) = 1$
Average severity	$repr(1, A) = 1; repr(1, B) = 0.5$
High severity	$repr(1, A) = 1; repr(1, B) = 0.5;$
	$repr(2, C) = 0.2; repr(4, other) = 0.1$

to initiate a detailed investigation. During the investigation accumulated organizational knowledge about safety related issues is used. As the investigation
result, a final occurrence assessment report is produced, which is provided to the
controller-reporter as a feedback. Furthermore, often final reports contain recommendations for safety improvement, which are required to be implemented
by ANSP (e.g., provision of training, improvement of procedures).

Step 7. The identification of domain-specific constraints. Constraints restrain the
allocation and behavior of agents. In particular, a prerequisite for the allocation
of an agent to a role is the existence of a mapping between the capabilities and
traits of the agent and the role requirements. Furthermore, the ANSPs reprimand
policies related to reporting were formalized as constraints using function repr
that maps the number of occurrences of some type to a reprimand value $[0, 1]$.
Table 1 lists three reprimand policies with the increasing severity of personal
consequences used in simulation.

3   Modeling of Agents

First general agent modeling aspects are presented in Section 3.1, then a decision
making model of an agent is considered in Section 3.2.

3.1   Modeling Internal States and Interaction

Agent models are formally grounded in order-sorted predicate logic with finite
sorts. More specifically, the static properties of a model are expressed using
the traditional sorted first-order predicate logic, whereas dynamic aspects are
specified using the Temporal Trace Language (TTL) [10], a variant of the order-
sorted predicate logic. In TTL, the dynamics of a system are represented by a
temporally ordered sequence of states. Each state is characterized by a unique
time point and a set of state properties that hold, specified using the predicate
$at : STATE\_PROPERTY \times TIME$. Dynamic properties are defined in TTL
as transition relations between state properties. For example, the property that
for all time points if an agent ag believes that action a is rewarded with r, then
ag will eventually perform a, is formalized in TTL as:

$$\forall t : TIME\ [\ at(internal(ag, belief(reward\_for\_action(r, a))), t)$$
$$\rightarrow \exists t1\ \&\ t1 > t\ \&\ at(output(ag, performed\_action(a)), t1)\ ]$$

The behavior of an agent can be considered from external and internal perspectives. From the external perspective the behavior can be specified by temporal

correlations between agents input and output states, corresponding to interaction with other agents and with the environment. An agent perceives information by observation and generates output in the form of communication or actions.

From the internal perspective the behavior is characterized by a specification of direct causal relations between internal states of the agent, based on which an externally observable behavioral pattern is generated. Such types of specification are called causal networks. In the following different types of internal states of agents are considered that form such causal networks, used further in decision making.

It is assumed that agents create time-labeled internal representations (beliefs) about their input and output states, which may persist over time:

$$\forall ag : AGENT \; \forall p : STATE\_PROPERTY \; \forall t : TIME \; at(input(ag, p), t)$$

$$\rightarrow at(internal(ag, belief(p, t), t + 1))$$

Information about observed safety occurrences is stored by agents as beliefs: e.g., $belief(observed\_occurrence\_with(ot \; : \; OCCURRENCE\_TYPE, ag \; : \; AGENT)), t : TIME)$. Besides beliefs about single states, an agent forms beliefs about dependencies between its own states, observed states of the environment, and observed states of other agents (such as expectancies and instrumentalities from the following section):

$belief(occurs\_after(p1 : STATE\_PROPERTY, p2 : STATE\_PROPERTY, t1 : TIME, t2 : TIME), t : TIME)$, which expresses that state property $p2$ holds t' $(t1 < t' < t2)$ time points after $p1$ holds.

In social science behavior of individuals is considered as goal-driven. It is also recognized that individual goals are based on needs. Different types of needs are distinguished: (1) *extrinsic needs* (n1) associated with biological comfort and material rewards; (2) *social interaction needs* that refer to the desire for social approval and affiliation; in particular own group approval (n2) and management approval (n3); (3) *intrinsic needs* that concern the desires for self-development and self-actualization; in particular contribution to organizational safety-related goals (n4) and self-esteem, self-confidence and self-actualization needs (n5). Different needs have different priorities and minimal acceptable satisfaction levels for individuals in different cultures. To distinguish different types of controllers investigated in this paper, the cultural classification framework by Hofstede [4] was used. The following indexes from the framework were considered: *individualism*(IDV) is the degree to which individuals are integrated into groups; *power distance index* (PDI) is the extent to which the less powerful members of an organization accept and expect that power is distributed unequally; and *uncertainty avoidance index* (UAI) deals with individuals tolerance for uncertainty and ambiguity. The indexes for individuals from the Western European culture adapted from [4] were changed to reflect the features of peer-dependent (low IDV) and rule-dependent (high UAI) agents (see Table 2).

The knowledge of an agent w.r.t. the ATC task is dependent on the adequacy of the mental models for this task, which depends on the sufficiency and timeliness

Table 2. The ranges for the uniformly distributed individual cultural characteristics and minimal acceptable satisfaction values of needs used in simulation

Agent type	IDV	PDI	UAI	$min(n1)$	$min(n2)$	$min(n3)$	$min(n4)$	$min(n5)$
peer-dependent	[0.3, 0.5]	[0.3, 0.5]	[0.4, 0.6]	1	0.8	0.5	0.7	0.9
rule-dependent	[0.7, 0.9]	[0.3, 0.5]	[0.7, 0.9]	1	0.5	0.7	0.7	0.9

of training provided to the controller and the adequacy of knowledge about safety-related issues. Such knowledge is contained in reports resulted from safety-related activities: final occurrence assessment reports resulted from occurrence investigations and monthly safety overview reports. Many factors influence the quality of such reports, for specific details we refer to [15]. Thus, the maturity level of a controller agent ($e5$) is calculated as:

$$e5 = w22 \cdot e19 + w23 \cdot e20 + w24 \cdot e21 + w25 \cdot e10 + w26 \cdot e42 + w27 \cdot e43,$$

here $e19 \in [0, 1]$ is the agent's self-confidence w.r.t. the ATC task (depends on the number of occurrences with the controller); $e20 \in [0, 1]$ is the agent's commitment to perform the ATC task; $e21 \in [0, 1]$ is the agents development level of skills for the ATC task; $e10 \in [0, 1]$ is the indicator for sufficiency and timeliness of training for changes; $e42 \in [0, 1]$ is the average quality of the final occurrence assessment reports received by the agent; $e43 \in [0, 1]$ is the average quality of the received monthly safety overview reports, $w22 - w27$ are the weights (sum up to 1).

The agent's commitment to safety is also influenced by the perceived commitment to safety of other team members and by how much the priority of safety in enforced and supported by management. An agent evaluates the managements commitment to safety by considering factors that reflect the managements effort in contribution to safety (investment in personnel and technical systems, training, safety arrangements).

In such a way, the commitment value is calculated based on a feedback loop: the agent's commitment influences the team commitment, but also the commitment of the team members and of the management influence the agents commitment:

$$e6 = w1 \cdot e1 + w2 \cdot e2 + w3 \cdot e3 + w4 \cdot e5,$$

here $e1 \in [0, 1]$ is the priority of safety-related goals in the role description, $e2 \in [0, 1]$ is the perception of the commitment to safety of management, $e3 \in [0, 1]$ is the perception of the average commitment to safety of the team, $e5 \in [0, 1]$ is the controller's maturity level w.r.t. the task; $w1 - w4$ are the weights (1 in total). For rule-dependent agents $w1 > w3$ and $w2 > w3$ and for peer-dependent agents $w3 > w2$ and $w3 > w1$.

3.2 Modeling Decision Making of a Controller Agent

Reporting quality analyzed in this paper is determined based on the decisions of controllers agents whether to report observed occurrences. To model decision making of agents a refined version of the expectancy theory by Vroom [7] has been used. Some advantages of the expectancy theory are: (a) it can be formalized; (b) it allows incorporating the organizational context; (c) it has received good empirical support. According to this theory, when a human evaluates alternative possibilities to act, s/he explicitly or implicitly makes estimations for the following factors: *valence, expectancy* and *instrumentality*. In Fig. 4 and 5 the decision making models for reporting and not reporting an occurrence are shown.

Expectancy refers to the individual's belief about the likelihood that a particular act will be followed by a particular outcome (called a first-level outcome). For example, E12 refers to the agent's belief of how likely that reporting of an occurrence will be followed by an administrative reprimand. *Instrumentality* is a belief concerning the likelihood of a first level outcome resulting into a particular second level outcome; its value varies between -1 and +1. Instrumentality takes negative values when a second-level outcome does not follow a first-level outcome. A second level outcome represents a desired (or avoided) by an agent state of affairs that is reflected in the agent's needs. For example, I32 refers to the belief about the likelihood that own group appreciation of the action results in own group approval. In the proposed approach the original expectancy model is refined by considering specific types of individual needs, described in section 3.1 *Valence* refers to the strength of the individual's desire for an outcome or state of affairs. Values of expectancies, instrumentalities and valences change over time, in particular due to individual and organizational learning.

In the Vrooms model the force on an individual to perform an act is defined as:

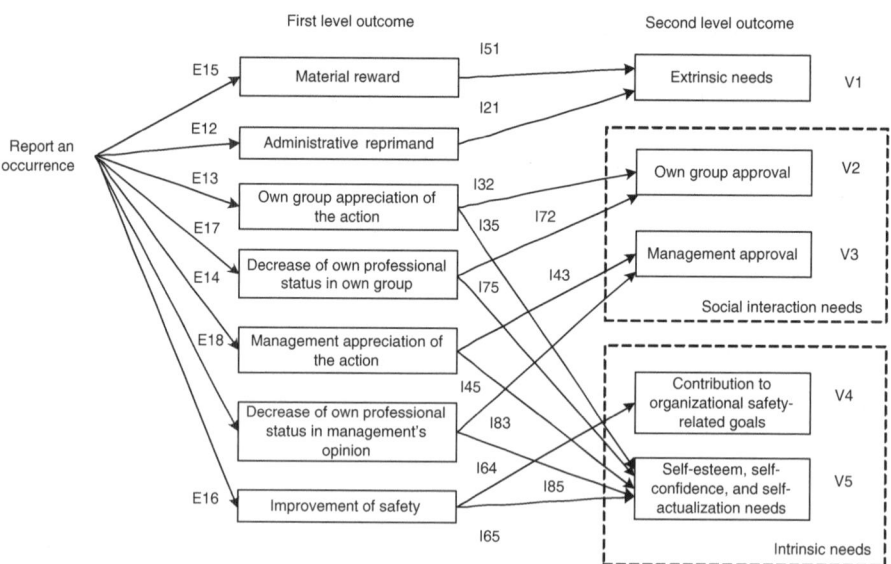

Fig. 4. Decision making model for reporting an occurrence. Here E's are expectancies, I's are instrumentalities and V's are valences.

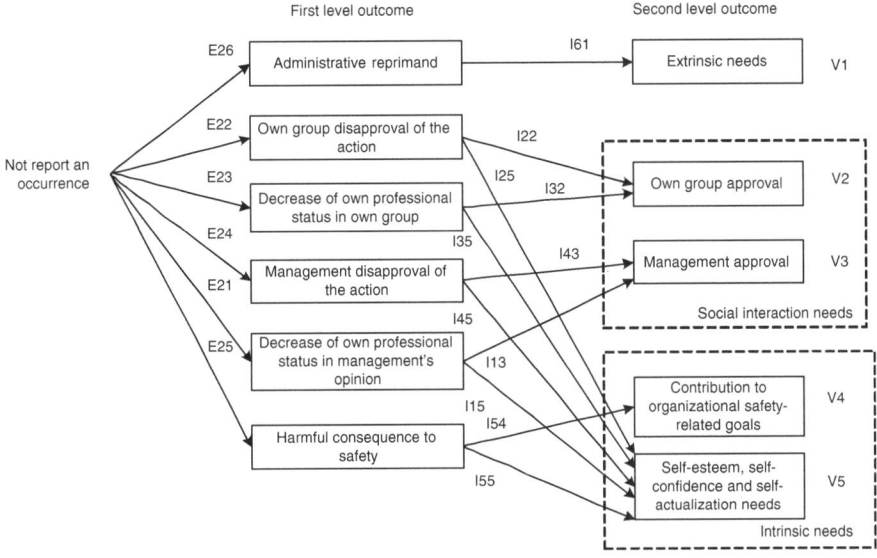

Fig. 5. Decision making model for not reporting an occurrence. Here E's are expectancies, I's are instrumentalities and V's are valences.

$$F_i = \sum_{j=1}^{n} E_{ij} \cdot \sum_{k=1}^{m} V_{ik} \cdot I_{jk}$$

Here E_{ij} is the strength of the expectancy that act i will be followed by outcome j; V_{ik} is the valence of the second level outcome k; I_{jk} is perceived instrumentality of outcome j for the attainment of outcome k.

The agent's decision making consists in the evaluation of the forces for two alternatives: to report and to not report. The agent chooses to perform the alternative with a greater force. In the following the basis for calculation of the variables of the decision making model for reporting is discussed. The precise, elaborated details of the mathematical model can be found in [15].

The factors $E15$, $E12$, $I51$ and $I21$ are defined based on the ANSP's formal reprimand/reward policies (see Table 1). In particular, $E12 = 1$ for an observed occurrence, which completes a set of occurrences, for which a reprimand is defined; $E12 = 0$ for all other observed occurrences. The values of $E13$ and $I32$ depend largely on the average commitment of the team of controllers to safety, and $E18$ and $I43$ depend on the management commitment to safety (considered in section 3.1).

With each set of occurrences, in which a controller agent was involved during an evaluation period (e.g., a month), the measure of severity is associated, calculated as the sum of the severities of the occurrences from the set. The factors $E17$, $E18$, $I72$, $I43$ depend mostly on the severity of the set of occurrences of the controller known to his/her team and known to the management. $E16$ is based on the agent's beliefs about the dependencies between previous reporting of similar occurrences and improvement of safety that followed.

$I35$ and $I75$ are based on the agent's IDV index, which indicates the degree of importance of team's opinions for the agent (e.g., high for peer-dependent agents, low for rule-dependent agents). $I45$ and $I85$ are based on the agent's PDI index. Furthermore, also the values of the basis valences (the degrees of importance of particular needs taken alone, see Fig.2) of a controller agent depend on its indexes:

$$v1_b = 1 \quad v2_b = 1 - IDV \quad v3_b = 0.7 \cdot PDI + 0.3 \cdot UAI \quad v4_b = 0.3 + 0.7 \cdot UAI$$

The values of valences change over time depending on the degree of satisfaction of the agent's needs: the more a need is satisfied, the less its valence:

$$v(need) = \begin{cases} v_b \cdot \frac{min\_accept(need)}{sat(need)}, & sat(need) \geq min\_accept(need) \\ v_b + v_b \cdot \frac{min\_accept(need) - sat(need)}{min_a ccept(need)} & sat(need) < min\_accept(need) \end{cases}$$

here $sat(need)$ is the current satisfaction value of a need.

4　Simulation Results

To test the hypotheses formulated previously 6 types of ANSPs have been considered (see Table 3). The informal descriptions of the ANSPs were formalized using the modeling framework from Section 2. The simulated organizations were populated with 48 controller agents distributed over 6 airport sectors, working in 4 shifts, 12 hours per day (12 controllers per shift; 2 per sector). The simulation has been done in the Matlab environment.

Many evidences exist (cf [1]) that due to a strict selection procedure and similarity of training, controllers have highly developed ATC skills which was also specified in the simulation model. Three types of controller teams were considered for each ANSP type: (a) with majority of peer-dependent members (75%); (b) with equal numbers of peer- and rule-dependent members; (c) with majority

Table 3. ANSP types used in simulation

Organizational aspect	Settings 1/2	Settings 3/4	Settings 5/6
Formal commitment to safety	high	high	low
Investment in personnel average	high	low	
Quality of technical systems	average	high	low
Formal support for confidentiality of reporting	average	high	low
Quality of management of safety activities	low	high	low
Personal consequences of occurrences	high/low	high/low	high/low
Influence of a controller on organizational safety arrangements	low	high	low
Quality of identification of occurrences	high/average	high/average	high/average

Table 4. The average reporting quality obtained from the simulations for each ANSP setting

Setting #	1	2	3	4	5	6
more rule-dependent	0.78	0.7	0.78	0.86	0.43	0.35
more peer-dependent	0.74	0.48	0.77	0.87	0.34	0.22
equal number	0.78	0.64	0.78	0.88	0.4	0.27

Table 5. The variances of reporting quality obtained from the simulations for each ANSP setting

Setting #	1	2	3	4	5	6
more rule-dependent	5e-3	2e-3	5e-3	5e-3	3e-3	7e-3
more peer-dependent	4e-3	3e-3	7e-3	3e-3	3e-3	2e-3
equal number	6e-3	3e-3	6e-3	8e-3	3e-3	4e-3

of rule-dependent members (75%). Different types of occurrences happened randomly in the environment with the frequencies provided by a real ANSP. 1000 simulations of each type have been performed. The obtained average reporting quality is given in Tables 4 and 5. As follows from the obtained results, the hypothesis 1 which states that reprimands serve the purpose of improvement of reporting quality was confirmed for settings 1 (in comparison with 2) and 5 (in comparison with 6). However, in setting 3 quite an opposite effect was observed: reprimands and close control in the ANSPs committed to safety cause a notable decrease in the reporting quality (in comparison with 4).

To verify the hypothesis 2 that rule-dependent controllers demonstrate more uniform reporting behavior over time than peer-dependent controllers, the mean and standard deviation values of the reporting force for the teams of types (a) and (c) were calculated. The obtained results show that the difference between the standard deviation values of the forces for the teams of types (a) and (c) for all settings was 7% (of the team's (a) value) at most. This finding may be explained by a high coherence of the teams of type (a), in which the attitude towards reporting (i.e., reporting force) stabilizes quickly due to intensive observation/interaction of/between the team members. In the teams of type (c) the homogeneous reporting behavior is achieved by rule adherence of most of the team members. Thus, although the standard deviation was less for the team of type (c) in all settings, the hypothesis 2 is supported weakly.

The hypothesis 3, which states that in ANSPs with low actual commitment to safety in the absence of reprimands, teams of type (a) may not report often, has been confirmed strongly by the simulation results. As can be seen from Table 4 the reporting quality dropped from 0.74 in the setting 1 to 0.48 in the setting 2 and from 0.34 in the setting 5 to 0.22 in the setting 6.

The hypothesis 4 that to neutralize negative effects of peer influence, mixing composition of teams may be proposed is also supported by the simulation results. From Table 4 it can be seen that the reporting quality of the teams of type

(b) is never worse and for some settings is much better than of the teams of type (a). Furthermore, as can be seen from Table 4 such an increase in reporting depends non-linearly on the number of rule-dependent agents; this is a joint effect of the organizational context and the non-linear behavior of the agents situated in this context.

5 Sensitivity Analysis

The validity of the results of automated checking of hypotheses depends on the validity of the model used. One of the tools used commonly for statistical validation of simulation models is sensitivity analysis [5,9,14]. By sensitivity analysis one can identify the most important factors of a model that influence particular outputs of the model. Then, the validity of the significance of the identified factors for the models outputs may be checked by performing face validation with domain experts and/or based on available domain knowledge.

The simulation model considered in this paper has one measured output the average occurrence reporting quality in the ANSP. Using sensitivity analysis the degree of influence of the input factors of the model given in Table 6 on the average occurrence reporting quality in the ANSP was investigated. To this end two sensitivity analysis techniques were used: Monte-Carlo filtering [14] and factor fixing [9].

Table 6. The input factors of the ANSP model

Factor	Description
e1	Priority of safety-related goals in the role description
e4	Influence of a controller on safety activities
e7	Sufficiency of the amount of safety investigators
e8	Sufficiency of the amount of controllers
e9	Availability of up-to-date technical systems for controllers
e10	Sufficiency and timeliness of training for changes
e11	Regularity of safety meetings
e12	Developed and implemented SMS
e14	Level of development of managerial skills of the controller supervisor
e19	Initial value of the self-confidence of a controller
e20	Commitment to perform ATC task
e21	Development level of skills for ATC task
e25	Sufficiency of the number of maintenance personal
e26	Quality of formal procedures for system checks and repairs
e35	Intensity of informal interactions in the team of controllers
e36	Quality of the formal safety occurrence assessment procedure
e40	Quality of the communication channel between controllers and safety investigators
e44	Average commitment of the agents involved in the safety analysis
e71	Formal support for confidentiality of reporting

Monte-Carlo filtering is often applied if a definition for 'good' or 'acceptable' model outcome can be given, e.g., through a set of constraints. In the considered model, the acceptable reporting quality is considered to be > 0.8. The aim of the Monte Carlo filtering is to perform multiple model evaluations with the input factors randomly chosen from suitable ranges and then split the output values into two subsets: those considered as 'acceptable' and those considered as 'unacceptable', depending on whether they lead to acceptable or unacceptable outputs. All factors in Table 6 have range $(0, 1]$. The Smirnov test is applied to each input factor to test whether the distributions of the 'acceptable' and 'unacceptable' values can be regarded as significantly different [9]. The higher the Smirnov test value for an input factor, the higher its influence on the model output, and hence the higher the sensitivity of output due to changes in the input. In detail, the Monte Carlo filtering method is implemented by the following two steps.

Step 1: MC simulations: 1000 Monte Carlo simulations were performed. For each input factor x_i two sets of values were determined: $x_i|B$, containing all values of x_i from the simulations that produced the desired organizational behaviour, and $x_i|\underline{B}$, containing all x_i values that did not produce the desired behaviour.

Step 2: Smirnov test: The Smirnov two sample test was performed for each input factor independently. The test statistics are defined by

$$d(x_i) = sup_Y||F_B(x_i|B) - F_{\underline{B}}(x_i|\underline{B})||,$$

where F_B and $F_{\underline{B}}$ are marginal cumulative probability distribution functions calculated for the sets $x_i|B$ and $x_i|\underline{B}$, respectively, and where Y is the output.

A low level of $d(x_i)$ supports the null-hypothesis $H_0 : F_B(x_i|B) = F_{\underline{B}}(x_i|\underline{B})$, meaning that the input factor x_i is not important, whereas a high level of $d(x_i)$ implies the rejection of H_0 meaning that x_i is a key factor.

It is determined at what significance level α, the value of $d(x_i)$ implies the rejection of H_0, where α is the probability of rejecting H_0 when it is true. In the sensitivity analysis, we used the classification High / Medium / Low for the importance of each factor:

- If $\alpha \leq 0.01$, then the importance of the corresponding factor x_i is considered High;
- If $0.01 < \alpha \leq 0.1$, then the importance of the corresponding factor is considered Medium;
- If $\alpha > 0.1$, then the importance of the corresponding factor is considered Low.

The Monte Carlo filtering method provides a measure of the sensitivity of the model output with respect to variations in the input factors. A limitation is that it captures only first-order effects and it does not detect interactions among factors. To solve this problem, variance-based global sensitivity analysis techniques can be used. Such techniques are able to capture interaction (correlation) between input factors by decomposing the variance of the output. One of such

Table 7. Importance of input factors classified by categories High and Medium for three types of controller teams

Importance	High	Medium
more rule-dependent	$e1, e4, e7, e8, e9, e10, e12, e14, e71$	$e11, e20, e21$
more peer-dependent	$e1, e4, e7, e8, e9, e10, e12, e14, e35$	$e11, e20, e21$
equal number	$e1, e4, e7, e8, e9, e10, e12, e14$	$e11, e20, e21$

techniques - the factor fixing [9] was used in this study. By this technique one is able to identify input factors recognized as insignificant by the Monte Carlo filtering approach, but which nevertheless should be considered as significant due to their interaction with other input factors.

The results of the sensitivity analysis for the simulation model considered in this paper are given in Table 7.

The factors $e1, e7, e8, e9, e10, e12$ were identified as highly influential for the quality of occurrence reporting by domain experts. The factor $e4$ is particularly important for high-quality reporting in a Western European ANSP, as argued in the literature [1,6]. Although the factor $e14$ was recognized as relevant, the degree of its influence on occurrence reporting was difficult to judge for the experts. A high importance of $e71$ for occurrence reporting in teams with most rule-dependent members can be explained by the rule adherence of the members. The factor $e35$ gains a high importance for teams with most peer-dependent members due to high importance of informal interactions in such teams.

Thus, none of the identified factors of high importance was identified as irrelevant or incorrect by domain experts and in the literature.

6   Conclusions

Many existing ANSPs face the problem that many safety occurrences observed by controllers are not reported. Practitioners in air traffic formulated hypotheses in the attempt to understand the reasons for such behavior. However, most of these hypotheses are difficult to verify manually due to a high complexity and temporal interdependency of institutional and social factors that should be taken into account. To address this issue an approach based on formal agent-based modeling and simulation has been proposed. Four hypotheses related to consequences of team composition in particular organizational contexts were examined. Two of these hypotheses were supported strongly by the simulation results, for one hypothesis only a weak support was found, and one hypothesis was partially supported, for particular types of organizational contexts only.

The validity of the results of automated checking of hypotheses depends on the validity of the model used. In general, to prove that a developed simulation model is valid, a number of validation steps should be performed [5]. In this paper the results of an important statistical validation step - sensitivity analysis are presented. The identified important factors influencing the average quality of occurrence reporting in an ANSP were recognized as highly relevant by domain experts and the literature.

However, sensitivity analysis alone is not sufficient to ensure the validity of a model. Previously, an approach for validation of agent-based organization models in air traffic based on questionnaires was developed [11]. Such an approach can be followed for simulation models similar to the one considered in the paper, when relevant questionnaire data are available.

References

1. Ek, A., Akselsson, R., Arvidsson, M., Johansson, C.R.: Safety culture in Swedish air traffic control. Safety Science 45(7), 791–811 (2007)
2. Eurocontrol: Air navigation system safety assessment methodology. SAF.ET1.ST03.1000-MAN-01, edition 2.0 (2004)
3. Hersey, P., Blanchard, K.H., Johnson, D.E.: Management of Organizational Behavior: Leading Human Resources (2001)
4. Hofstede, G.: Cultures and Organizations. McGraw-Hill, New York (2005)
5. Kleijnen, J.P.C.: Verification and validation of simulation models. European Journal of Operational Research 82(1), 145–162 (1995)
6. Patankar, M.S., Brown, J.P., Treadwell, M.D.: Safety ethics. Ashgate (2005)
7. Pinder, C.C.: Work motivation in organizational behavior. Prentice-Hall, NJ (1998)
8. Pritchett, A.R., Lee, S., Goldsman, D.: Hybrid-System Simulation for National Airspace System Safety Analysis. AIAA Journal of Aircraft 38(5), 835–840 (2001)
9. Saltelli, A., Ratto, M., Andres, T., Campolongo, F., Cariboni, J., Gatelli, D., Saisana, M., Tarantola, S.: Global Sensitivity Analysis: The Primer. Wiley-Interscience, Hoboken (2008)
10. Sharpanskykh, A.: On Computer-Aided Methods for Modeling and Analysis of Organizations. PhD thesis, Vrije Universiteit Amsterdam (2008)
11. Sharpanskykh, A., Stroeve, S.H.: Safety modelling and analysis of organizational processes in air traffic – Deliverable D5: Validation plan. NLR, report NLR-CR-2008-653 (2008)
12. Stroeve, S.H., Blom, H.A.P., Van der Park, M.N.J.: Multi-agent situation awareness error evolution in accident risk modelling. In: 5th ATM R & D Seminar (2003)
13. Tumer, K., Agogino, A.: Distributed agent-based air traffic flow management. In: AAMAS 2007, pp. 342–349. ACM Press, New York (2007)
14. Young, P.C.: Data-based mechanistic modelling, generalised sensitivity and dominant mode analysis. Comput. Phys. Commun. 117, 113–129 (1999)
15. Appendix, http://www.few.vu.nl/~sharp/app.pdf

Can Space Applications Benefit from Intelligent Agents?

Blesson Varghese[1] and Gerard McKee[2]

[1] Active Robotics Laboratory, School of Systems Engineering, University of Reading, Whiteknights Campus, Reading, Berkshire, UK, RG6 6AY
b.varghese@student.reading.ac.uk
[2] School of Systems Engineering, University of Reading, Whiteknights Campus, Reading, Berkshire, UK, RG6 6AY
g.t.mckee@reading.ac.uk

Abstract. The work reported in this paper proposes a Swarm-Array computing approach based on 'Intelligent Agents' to apply autonomic computing concepts to parallel computing systems and build reliable systems for space applications. Swarm-array computing is a swarm robotics inspired, novel computing approach considered as a path to achieve autonomy in parallel computing systems. In the intelligent agent approach, a task to be executed on parallel computing cores is considered as a swarm of autonomous agents. A task is carried to a computing core by carrier agents and can be seamlessly transferred between cores in the event of a predicted failure, thereby achieving self-* objectives of autonomic computing. The approach is validated on a multi-agent simulator.

1   Introduction

Autonomic computing has recently emerged as a domain of interest to computing researchers worldwide. What is autonomic computing, and what are its inspiration and vision? What are its distinct perspectives? What are the autonomic approaches? What needs to be focused ahead? These are the few questions answered in this section, before commencing discussions on Intelligent Agents and their feasibility in Swarm-Array Computing, the primary focus of this paper.

What is autonomic computing, and what are its inspiration and vision? With the advancements of computing techniques, biologically inspired computing has emerged as a major domain in computing. Many computing paradigms, namely amorphous computing, evolutionary computing and organic computing have emerged as a result of being inspired from natural phenomenon. Autonomic computing is one such biologically inspired computing paradigm based on the autonomic human nervous system [1].

Autonomic computing is a visionary paradigm for developing large scale distributed systems, reducing cost of ownership and reallocating management responsibilities from administrators to the computing system itself [2] - [9]. Autonomic computing paves the necessary foundation autonomic computing principles have paved necessary foundations towards self-managing systems.

Self-managing [10] systems are characterized by four objectives and four attributes. The objectives and attributes that contribute to self-management are not independent

A.V. Vasilakos et al. (Eds.): AUTONOMICS 2009, LNICST 23, pp. 192–202, 2010.

functions. The objectives considered are [1, 11, 12]: (a) Self-configuration, (b) Self-healing, (c) Self-optimizing and (d) Self-protecting. The attributes considered are [1, 11, 12]: (a) Self-awareness, (b) Self-situated, (c) Self-monitoring and (d) Self-adjusting.

What are the perspectives of autonomic computing? There are mainly two perspectives, namely business and research oriented perspectives that provide a bird's-eye view of the paradigm. Firstly, from a business oriented perspective, autonomic computing was proposed by IBM for better management of increasingly complex computing systems and reduce the total cost of ownership of systems today, hence aiming to reallocate management responsibilities from administrators to the computing systems itself based on high-level policies [2] - [9].

Secondly, the research oriented perspective primarily focuses on the worms-eye view, laying necessary foundations for the newly emerging computing paradigm. There are two categories of ongoing research in the area of autonomic computing. Firstly, research describing approaches and technologies related to autonomic computing [10]. The aim of the approaches is to achieve autonomy without specifying the technology to be implemented [2]. Any existing technology capable of achieving autonomy (in any degree) can be used in the approaches. Secondly, research attempting to develop autonomic computing as a unified project [10]. The research lays emphasis on the means to achieve autonomy and initiatives are taken to define a set of standard practices and methods as the path towards autonomy.

What are the autonomic computing approaches? Autonomic computing researchers have adopted six different approaches, namely emergence-based, component/service-based, control theoretic based, artificial intelligence, swarm intelligence and agent based approaches to achieve self-managing systems.

The emergence based approach for distributed systems considers complex behaviours of simple entities with simple behaviours without global knowledge [13]. Intelligent behaviour is thus repercussions of interactions and coordination between entities. One major challenge in emergence based approaches is on how to achieve global coherent behaviour [14]. Autonomic computing research on emergence based approaches is reported in [13] - [16].

The component/service based approach for distributed systems employ service-oriented architectures. With advancements in software engineering practices, component/service based approaches are also implemented in many web based services. The autonomic element of the autonomic system is a component whose interfaces, behaviours and design patterns aim to achieve self-management. These approaches are being developed for large scale networked systems including grids. Autonomic computing research on component/service based approaches is reported in [17] - [20].

The control theoretic based approach aims to apply control theory for developing autonomic computing systems. The building blocks of control theory such as reference input, control input, control error, controller, disturbance input, measured output, noise input, target system and transducer are used to model computing systems and further used to study properties like stability, short settling times, and accurate regulation. Using a defined set of control theory methodologies, the objectives of a control system namely regulatory control, disturbance rejection and optimization can be achieved. These objectives are closely associated with the objectives of autonomic computing.

Research on control theoretic based approaches applied to autonomic computing is reported in [21] - [23].

The artificial intelligence based approaches aim for automated decision making and the design of rational agents. The concept of autonomy is realized by maximizing an agent's objective based on perception and action in the agent's environment with the aid of information from sensors and in-built knowledge. Work on artificial intelligence approaches for autonomic computing is reported in [24, 25].

The swarm intelligence based approaches focuses on designing algorithms and distributed problem solving devices inspired by collective behaviour of swarm units that arise from local interactions with their environment [26, 27]. The algorithms considered are population-based stochastic methods executed on distributed processors. Autonomic computing research on swarm intelligence approaches is reported in [28] - [30].

The agent based approaches for distributed systems is a generic technique adopted to implement emergence, component/service, artificial intelligence or swarm intelligence based approaches. The agents act as autonomic elements or entities that perform distributed task. The domain of software engineering considers agents to facilitate autonomy and hence have a profound impact on achieving the objectives of autonomic computing. Research work based on multi-agents supporting autonomic computing are reported in [6] [31] - [36].

What needs to be focused ahead? The focus of researchers in autonomic computing should be towards two directions. Firstly, researchers ought to aim towards applying autonomic computing concepts to parallel computing systems. This focus is essential since most distributed computing systems are closely associated with the parallel computing paradigm. The benefits of autonomy in computing systems, namely reducing cost of ownership and reallocating management responsibilities to the system itself are also relevant to parallel computing systems. It is surprising that only few researchers have applied autonomic computing concepts to parallel computing systems in the approaches above.

Secondly, autonomic computing researchers ought to focus towards implementing the approaches for building reliable systems. One potential area of application that demands reliable systems is space applications. Space crafts employ FPGAs, a special purpose parallel computing system that are subject to malfunctioning or failures of hardware due to 'Single Event Upsets' (SEUs), caused by radiation on moving out of the protection of the atmosphere [37] - [39]. One solution to overcome this problem is to employ reconfigurable FPGAs. However, there are many overheads in using such technology and hardware reconfiguration is challenging in space environments. In other words, replacement or servicing of hardware is an extremely limited option in space environments. On the other hand software changes can be accomplished. In such cases, autonomic computing approaches can come to play.

How can a bridge be built between autonomic computing approaches and parallel computing systems? How can autonomic computing approaches be extended towards building reliable systems for space applications? The work reported in this paper is motivated towards bridging this gap by proposing swarm-array computing, a novel technique to achieve autonomy for distributed parallel computing systems and experimenting the feasibility of a proposed approach on FPGAs that can be useful for space applications.

The remainder of the paper is organized as follows. Section 2 introduces swarm-array computing. Section 3 investigates the feasibility of the proposed approach by simulations. Section 4 concludes the paper and considers future work.

2   Swarm-Array Computing

Swarm-array computing is a swarm robotics inspired approach and is proposed as a path to achieve autonomy. The development of the swarm-array computing approach is from the foundations of parallel and autonomic computing paradigms. The constitution of the swarm-array computing approach can be separated into four constituents. Three approaches are proposed that bind the swarm-array computing constituents together. The four constituents and the three approaches are considered in the following sub sections.

2.1   Constituents

There are four prime constituents that make up the constitution of swarm-array computing. They are the computing system, the problem/task, the swarms and the landscape considered in this section.

Firstly, the computing systems which are available for parallel computing are multicore processors, clusters, grids, field programmable gate arrays (FPGA), general purpose graphics processing units (GPGPU), application-specific integrated circuit (ASIC) and vector processors. With the objective of exploring swarm-array computing, FPGAs are selected as an experimental platform for the proposed approaches.

FPGAs are a technology under investigation in which the cores of the computing system are not geographically distributed. The cores in close proximity can be configured to achieve a regular grid or a two dimensional lattice structure. Another reason of choice to look into FPGAs is its flexibility for implementing reconfigurable computing.

The cores of the computing system can be considered as a set of autonomous agents, interacting with each other and coordinating the execution of tasks. In this case, a processing core is similar to an organism whose function is to execute a task. The focus towards autonomy is laid on the parallel computing cores abstracted onto intelligent cores. The set of intelligent cores hence transform the parallel computing system into an intelligent swarm. The intelligent cores hence form a swarm-array. A parallel task to be executed resides within a queue and is scheduled onto different cores by the scheduler. The swarm of cores collectively executes the task.

The intelligent cores described above are an abstract view of the hardware cores. But then the question on what intelligence can be achieved on the set of cores needs to be addressed. Intelligence of the cores is achieved in two different ways. Firstly, by monitoring local neighbours. Independent of what the cores are executing, the cores can monitor each other. Each core can ask the question of 'are you alive' to its neighbours and gain information. Secondly, by adjusting to core failures. If a core fails, the process which was executed on the core needs to be shifted to another core where resources previously accessed can be utilized. Once a process has been shifted, all data dependencies need to be re-established.

To shift a process from one core to another, there is a requirement of storing data associated and state of the executing process, referred to as checkpointing. This can be achieved by a process monitoring each core or by swarm carrier agents that can store the state of an executing process. The checkpointing method suggested is decentralized and distributed across the computing system. Hence, though a core failure may occur, a process can seamlessly be transferred onto another core. In effect, awareness and optimizing features of the self-ware properties are achieved.

Secondly, the problem/task to be executed on the parallel computing cores that can be considered as a swarm of autonomous agents. To achieve this, a single task needs to be decomposed and the sub tasks need to be mapped onto swarm agents. The agent and the sub-problems are independent of each other; in other words, the swarm agents are only carriers of the sub-tasks or are a wrapper around the sub-tasks. The swarm displaces itself across the parallel computing cores or the environment. The goal would be to find an area accessible to resources required for executing the sub tasks within the environment. In this case, a swarm agent is similar to an organism whose function is to execute on a core. The focus towards autonomy is laid on the executing task abstracted onto intelligent agents. The intelligent agents hence form a swarm-array.

The intelligent agents described above are an abstract view of the sub-tasks to be executed on the hardware cores. Intelligence of the carrier agents is demonstrated in two ways. Firstly, the capabilities of the carrier swarm agents to identify and move to the right location to execute a task. In this case, the agents need to be aware of their environments and which cores can execute the task. Secondly, the prediction of some type of core failures can be inferred by consistent monitoring of power consumption and heat dissipation of the cores. If the core on which a sub-task being executed is predicted to fail, then the carrier agents shift from one core to another gracefully without causing an interruption to execution, hence making the system more fault-tolerant and reliable. An agent can shift from one core to another by being aware of which cores in the nearest vicinity of the currently executing core are available.

Thirdly, a combination of the intelligent cores and intelligent swarm agents leads to intelligent swarms. The intelligent cores and intelligent agents form a multi-dimensional swarm-array. The arena in which the swarms interact with each other is termed as landscape.

Fourthly, the landscape that is a representation of the arena of cores and agents that are interacting with each other in the parallel computing system. At any given instance, the landscape can define the current state of the computing system. Computing cores that have failed and are predicted to fail are holes in the environment and obstacles to be avoided by the swarms.

A landscape is modelled from three different perspectives which is the basis for the swarm-array computing approaches discussed in the next section. Firstly, a landscape comprising dynamic cores (are autonomous) and static agents (are not autonomous) can be considered. In this case, the landscape is affected by the intelligent cores. Secondly, a landscape comprising of static cores and dynamic agents can be considered. In this case, the landscape is affected by the mobility of the intelligent agents. Thirdly, a landscape comprising of dynamic cores and dynamic agents can be considered. In this case, the landscape is affected by the intelligent cores and mobility of the carrier agents.

2.2   Approaches

At this point it is appropriate to consider how the constitution of swarm-array computing fits together? To answer this question, three approaches that combine the constituents of swarm-array computing are proposed.

In the first approach, only the intelligent cores are considered to be autonomous swarm agents and form the landscape. A parallel task to be executed resides within a queue and is scheduled onto the cores by a scheduler. The intelligent cores interact with each other as considered in section 2.1 to transfer tasks from one core to another at the event of a hardware failure.

In the second approach, only the intelligent swarm agents are considered to be autonomous and form the landscape. A parallel task to be executed resides in a queue, which is mapped onto carrier swarm agents by the scheduler. The carrier swarm displace through the cores to find an appropriate area to cluster and execute the task. The intelligent agents interact with each other as considered in Section 2.1 to achieve mobility and successful execution of a task. Figure 1 describes the approach diagrammatically. In the third approach, both the intelligent cores and intelligent agents are considered to form the landscape. Hence, the approach is called a combinative approach. A parallel task to be executed resides in a queue, which is mapped onto swarm agents by a scheduler. The swarm agents can shift through the landscape utilizing their own intelligence, or the swarm of cores could transfer tasks from core to core in the landscape. The landscape is affected by the mobility of intelligent agents on the cores and intelligent cores collectively executing a task by accommodating the intelligent agent.

However, in this paper the major focus is the second approach and is only considered for experimental studies. The feasibility of the first method is reported in [40].

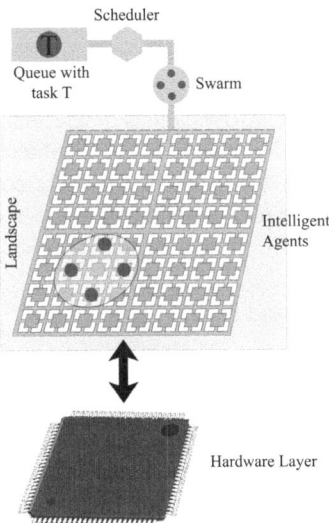

Fig. 1. Second Approach in Swarm-array computing

3 Simulation Studies

Simulation studies were pursued to validate and visualize the proposed approach in swarm-array Computing. Since FPGA cores are considered in this paper and the approach proposed in this paper considers executing cores as agents; hence a multi-agent simulator is employed. This section is organized into describing the simulation environment, experimental platform and model and simulation results.

3.1 Simulation Environment

The feasibility of the proposed swarm-array computing approach was validated on the SeSAm (Shell for Simulated Agent Systems) simulator. The SeSAm simulator environment supports the modelling of complex agent-based models and their visualization [41, 42].

The environment has provisions for modelling agents, the world and simulation runs. Agents are characterized by a reasoning engine and a set of state variables. The reasoning engine defines the behaviour of the agent, and is implemented in the form of an activity diagram, similar to a UML-based activity diagram. The state variables of the agent specify the state of an agent. Rules that define activities and conditions can be visually modelled without the knowledge of a programming language. The building block of such rules is primitives that are pre-defined. Complex constructs such as functions and data-types can be user-defined.

The world provides knowledge about the surroundings the agent is thriving. A world is also characterized by variables and behaviours. The modelling of the world defines the external influences that can affect the agent Hence, variables associated with a world class can be used as parameters that define global behaviours. This in turn leads to the control over agent generation, distribution and destruction.

Simulation runs are defined by simulation elements that contribute to the agent-based model being constructed. The simulation elements include situations, analysis lists, simulations and experiments. Situations are configurations of the world with pre-positioned agents to start a simulation run. Analysis lists define means to study agents and their behaviour with respect to time. Simulations are combinations of a situation, a set of analysis items and a simulation run; or in other words a complete definition of a single simulation run. Experiments are used when a combination of single simulation runs are required to be defined.

3.2 Experimental Platform and Model

As considered in Section 2.1 and 2.2, the swarm-array computing approach needs to consider the computing platform, the problem/task and the landscapes. The parallel computing platform considered in the studies reported in this paper is FPGAs and is modelled in SeSAm. The hardware cores are arranged in a 5 X 5 regular grid structure. The model assumes serial bus connectivity between individual cores. Hence, a task scheduled on a core can be transferred onto any other core in the regular grid.

The breakdown of any given task to subtasks is not considered within the problem domain of swarm-array computing. The simulation is initialized with sub-tasks scheduled to a few cores in the grid. Each subtask carrying agent consistently monitors the

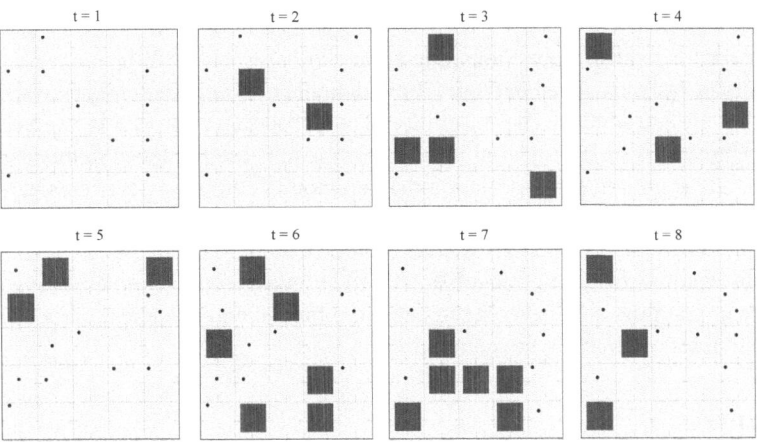

Fig. 2. Sequence of eight simulation screenshots (a) - (h) of a simulation run from initialization on the SeSAm multi-agent simulator. Figure shows how the carrier agents carrying sub-tasks are seamlessly transferred to a new core when executing cores fail.

hardware cores. This is possible by sensory information (in our model, temperature is sensed consistently) passed onto the carrier agent. In the event of a predicted failure, the carrier agent displaces itself to another core in the computing system. The behaviour of the individual cores varies randomly in the simulation. For example, the temperature of the FPGA core changes during simulation. If the temperature of a core exceeds a predefined threshold, the subtask being executed on the core is transferred by the carrier agent to another available core that is not predicted to fail. During the event of a transfer or reassignment, a record of the status of execution of the subtask maintained by the carrier agent also gets transferred to the new core. If more than one sub-task is executed on a core predicted to fail, each sub-task may be transferred to different cores.

3.3 Simulation Results

Figure 2 is a series of screenshots of a random simulation run developed on SeSAm for eight consecutive time steps from initialization. The figure shows the executing cores as rectangular blocks in pale blue colour. When a core is predicted to fail, i.e., temperature increases beyond a threshold, the core is displayed in red. The subtasks wrapped by the carrier agents are shown as blue filled circles that occupy a random position on a core. As discussed above, when a core is predicted to fail, the subtask executing on the core predicted to fail gets seamlessly transferred to a core capable of processing at that instant.

The simulation studies are in accordance with the expectation and hence are a preliminary confirmation of the feasibility of the proposed approach in swarm-array computing. Though some assumptions and minor approximations are made, the approach is an opening for applying autonomic concepts to parallel computing platforms.

4   Conclusion

In this paper, a swarm-array computing approach based on intelligent agents that act as carriers of tasks has been explored. Foundational concepts that define swarm-array computing are introduced. The feasibility of the proposed approach is validated on a multi-agent simulator. Though only preliminary results are produced in this paper, the approach gives ground for expectation that autonomic computing concepts can be applied to parallel computing systems and build reliable systems for space applications.

Future work will aim to study the third proposed approach or the combinative approach in swarm-array computing. Efforts will be made towards implementing the approaches in real time and exploring in depth the fundamental concepts associated with the constituents of swarm-array computing.

References

1. Hinchey, M.G., Sterritt, R.: 99% (Biological) Inspiration. In: Proceedings of the 4th IEEE International Workshop on Engineering of Autonomic and Autonomous Systems, pp. 187–195 (2007)
2. Lin, P., MacArthur, A., et al.: Defining Autonomic Computing: A Software Engineering Perspective. In: Proceedings of the Australian Software Engineering Conference, pp. 88–97 (2005)
3. Sterritt, R., Hinchey, M.: Autonomic Computing - Panacea or Poppycock? In: Proceedings of the 12th IEEE International Conference and Workshops on the Engineering of Computer-Based Systems, pp. 535–539 (2005)
4. Sterritt, R., Bustard, D.: Autonomic Computing - a Means of Achieving Dependability? In: Proceedings of the 10th IEEE International Conference and Workshop on the Engineering of Computer-Based Systems, pp. 247–251 (2003)
5. Nami, M.R., Sharifi, M.: Autonomic Computing a New Approach. In: Proceedings of the 1st Asia International Conference on Modelling and Simulation, pp. 352–357 (2007)
6. Jarrett, M., Seviora, R.: Constructing an Autonomic Computing Infrastructure using Cougaar. In: Proceedings of the 3rd IEEE International Workshop on Engineering of Autonomic and Autonomous Systems, pp. 119–128 (2006)
7. Lightstone, S.: Foundations of Autonomic Computing Development. In: Proceedings of the 4th IEEE Workshop on Engineering of Autonomic and Autonomous Systems (2007)
8. Gentsch, W., Iano, K., et al.: Self-Adaptable Autonomic Computing Systems: An Industry View. In: Proceedings of the 16th IEEE International Workshop on Database and Expert Systems Applications (2005)
9. Cybenko, G., Berk, V.H., et al.: Practical Autonomic Computing. In: Proceedings of the 30th IEEE Annual International Computer Software and Applications Conference (2006)
10. Nami, M.R., Bertels, K.: A Survey of Autonomic Computing Systems. In: Proceedinbgs of the 3rd International Conference on Autonomic and Autonomous Systems, pp. 26–30 (2007)
11. Marshall, T., Dai, Y.S.: Reliability Improvement and Models in Autonomic Computing. In: Proceedings of the 11th International Conference on Parallel and Distributed Systems, pp. 468–472 (2005)
12. King, T.M., Babich, D., et al.: Towards Self-Testing in Autonomic Computing Systems. In: Proceedings of the 8th International Symposium on Autonomous Decentralized Systems, pp. 51–58 (2007)
13. Anthony, R.J.: Emergence: a Paradigm for Robust and Scalable distributed applications. In: Proceedings of the International Conference on Autonomic Computing, pp. 132–139 (2004)

14. De Wolf, T., Holvoet, T.: Emergence as a general architecture for distributed autonomic computing. K. U. Leuven, Department of Computer Science, Report CW 384 (2004)
15. Saffre, F., Halloy, J., et al.: Self-Organized Service Orchestration Through Collective Differentiation. IEEE Transactions on Systems, Man and Cybernetics, Part B, 1237–1246 (2006)
16. Champrasert, P., Lee, C., et al.: SymbioticSphere: Towards an Autonomic Grid Network System. In: Proceedings of the IEEE International Conference on Cluster Computing, pp. 1–2 (2005)
17. Zeid, A., Gurguis, S.: Towards Autonomic Web Services. In: Proceedings of the 3rd ACS/IEEE International Conference on Computer Systems and Applications (2005)
18. Almeida, J., Almeida, V., et al.: Resource Management in the Service Oriented Architecture. In: Proceedings of the IEEE International Conference on Autonomic Computing, pp. 84–92 (2006)
19. White, S.R., Hanson, J.E., et al.: An Architectural Approach to Autonomic Computing. In: Proceedings of the IEEE International Conference on Autonomic Computing (2004)
20. Parashar, M., Li, Z., et al.: Enabling Autonomic Grid Applications: Requirements, Models and Infrastructure. In: Babaoğlu, Ö., Jelasity, M., Montresor, A., Fetzer, C., Leonardi, S., van Moorsel, A., van Steen, M. (eds.) SELF-STAR 2004. LNCS, vol. 3460, pp. 273–290. Springer, Heidelberg (2005)
21. Diao, Y., Hellerstein, J.L., et al.: Self-Managing Systems: A Control Theory Foundation. In: Proceedings of the 12th IEEE International Conference and Workshops on the Engineering of Computer-Based Systems, pp. 441–448 (2005)
22. Abdelwahed, S., Kandasamy, N., et al.: Online Control for Self-Management in Computing Systems. In: Proceedings of the 10th IEEE Real-Time and Embedded Technology and Applications Symposium, Toronto, Canada (2004)
23. Zhu, Q., Lin, L., et al.: Characterizing Maintainability concerns in Autonomic Element Design. In: Proceedings of the IEEE International Conference on Software Maintenance, pp. 197–206 (2008)
24. Kephart, J.O., Walsh, W.E.: An Artificial Intelligence Perspective on Autonomic Computing Policies. In: Proceedings of the 5th IEEE International Workshop on Policies for Distributed Systems and Networks, pp. 3–12 (2004)
25. Peddemors, A., Niemegeers, I., et al.: A System Perspective on Cognition for Autonomic Computing and Communication. In: Proceedings of the 16th International Workshop on Database and Expert Systems Application (2005)
26. Hinchey, M.G., Sterritt, R., et al.: Swarms and Swarm Intelligence. IEEE Computer 40(4), 111–113 (2007)
27. Kennedy, J., Eberhart, R.C., et al.: Swarm Intelligence. Morgan Kaufmann Publishers, San Francisco (2001)
28. Wang, J., d'Auriol, B.J., et al.: A Swarm Intelligence inspired Autonomic Routing Scenario in Ubiquitous Sensor Networks. In: Proceedings of the International Conference on Multimedia and Ubiquitous Engineering, pp. 745–750 (2007)
29. Hinchey, M., Dai, Y.S., et al.: Modeling for NASA Autonomous Nano-Technology Swarm Missions and Model-Driven Autonomic Computing. In: Proceedings of the 21st International Conference on Advanced Information Networking and Applications, pp. 250–257 (2007)
30. Carrasco, L.M.F., Marin, H.T., et al.: On the Path Towards Autonomic Computing: Combining Swarm Intelligence and Excitable Media Models. In: Proceedings of the 7th Mexican International Conference on Artificial Intelligence, pp. 192–198 (2008)
31. De Wolf, T., Holovet, T., et al.: Towards Autonomic Computing: Agent-Based Modelling, Dynamical Systems Analysis, and Decentralised Control. In: Proceedings of the IEEE International Conference on Industrial Informatics, pp. 470–479 (2003)
32. Bonino, D., Bosca, A., et al.: An Agent based Autonomic Semantic Platform. In: Proceedings of the International Conference on Autonomic Computing, pp. 189–196 (2004)

33. Tianfield, H.: Multi-agent Autonomic Architecture and its Application in e-Medicine. In: Proceedings of the IEEE/WIC International Conference on Intelligent Agent Technology (2003)
34. Pour, G.: Prospects for Expanding Telehealth: Multi-Agent Autonomic Architectures. In: Proceedings of the International Conference on Computational Intelligence for Modelling and Automation, and International Conference on Intelligent Agents, Web Technologies and Internet Commerce (2006)
35. Guo, H., Gao, J., et al.: A Self-Organized Model of Agent-Enabling Autonomic Computing for Grid Environment. In: Proceedings of the 6th World Congress on Intelligent Control and Automation, pp. 2623–2627 (2006)
36. Hu, J., Gao, J., et al.: Multi-Agent System based Autonomic Computing Environment. In: Proceedings of the International Conference on Machine Learning and Cybernetics, pp. 105–110 (2004)
37. O'Bryan, M.V., Poivey, C., et al.: Compendium of Single Event Effects Results for Candidate Spacecraft Electronics for NASA. In: Proceedings of the IEEE Radiation Effects Data Workshop, pp. 19–25 (2006)
38. Johnson, E., Wirthlin, M.J., et al.: Single-Event Upset Simulation on an FPGA. In: Proceedings of the International Conference on Engineering of Reconfigurable Systems and Algorithms, USA (2002)
39. Habinc, S.: Suitability of Reprogrammable FPGAs in Space Applications. Feasibility Report for the European Space Agency by Gaisler Research under ESA contract No. 15102/01/NL/FM(SC) CCN-3 (2002)
40. Varghese, B., McKee, G.T.: Towards Self-ware via Swarm-Array Computing. In: Proceedings of the International Conference on Computational Intelligence and Cognitive Informatics, Paris, France (2009)
41. Klugl, F., Herrler, R., et al.: SeSAm: Implementation of Agent-Based Simulation Using Visual Programming. In: Proceedings of the Fifth International Joint Conference on Autonomous Agents and Multi-Agent Systems, Japan, pp. 1439–1440 (2006)
42. SeSAm website, http://www.simsesam.de

A Generic Agent Organisation Framework for Autonomic Systems

Ramachandra Kota, Nicholas Gibbins, and Nicholas R. Jennings

School of Electronics and Computer Science
University of Southampton, Southampton, UK
{rck05r,nmg,nrj}@ecs.soton.ac.uk

Abstract. Autonomic computing is being advocated as a tool for managing large, complex computing systems. Specifically, self-organisation provides a suitable approach for developing such autonomic systems by incorporating self-management and adaptation properties into large-scale distributed systems. To aid in this development, this paper details a generic problem-solving agent organisation framework that can act as a modelling and simulation platform for autonomic systems. Our framework describes a set of service-providing agents accomplishing tasks through social interactions in dynamically changing organisations. We particularly focus on the organisational structure as it can be used as the basis for the design, development and evaluation of generic algorithms for self-organisation and other approaches towards autonomic systems.

Keywords: Organisation, Autonomic Systems, Organisation Model.

1 Introduction

Autonomic systems are envisaged as self-managing, distributed computing systems containing several service-providing components interacting with each other over large networks to accomplish complex tasks. The features of such systems are that they are robust, decentralised, adapting to changing environments and self-organising. Within this, a central concern that needs to be focused on is the interactions between the various computing entities involved. In particular, the interactions within the system are critical for it to achieve its system-wide goals as the tasks tend to be too complex to be accomplished by any single component or entity alone. Given this, and taking inspiration from self-organisation principles, the development of effective autonomic systems involves, to a significant extent, adapting local interactions towards achieving a better performance globally [18]. By so doing, the system can robustly reconfigure itself to the changing requirements and environmental conditions. Therefore, the self-management aspect of the system requires that the individual components of the system are allowed the freedom to adapt their local interactions with other components. In particular, adapting these interactions is necessary because, purely changing the internal characteristics of the components will not be sufficient for improving performance as most of the tasks and goals involve multiple components and interactions across them.

A.V. Vasilakos et al. (Eds.): AUTONOMICS 2009, LNICST 23, pp. 203–219, 2010.

For example, consider the interconnected network of a university as a form of an autonomic grid system. Being a university, it contains various labs with their own specialised computing systems, as part of the overlaying network of the university. That is, there might be a graphics lab containing computers with some high end graphics cards for rendering rich intensive images. Similarly, some computers in the geography lab might contain various GIS maps and related software. Also, there will be complex computing tasks that need several computers (possibly situated in different labs) providing specialised services for their accomplishment. A task might need statistical analysers from the mathematics department for analysing data available from the sociology department in order to predict natural resource, like water and wood, usage as needed by the institute on environmental conservation. Thus, the computers on the university network, need to interact with one another to perform these complex tasks. Moreover, as these individual computers are controlled by different people in different labs, the respective loads on them, at any time, cannot be known or predicted. Also, some might go offline, some might be upgraded and so on. Hence, the computers need to continuously adapt their interactions with others in the network to keep up with the changes and, also optimise the overall performance.

Now, these social interactions of the components can be quite reactive and not guided by definite regulations or they can be structured using an explicitly depicted network or organisation. That is, the individual components of the system will be modelled as autonomous agents participating in an organisation and the interactions between the components are governed by the structure of this organisation. In such a context, regulating the interactions in the system through the organisation structure will aid in the design of adaptation techniques by suitably representing the recurring interactions between the components. For example, consider the autonomic system being used to maintain the computing systems in a university, as discussed earlier. Now, given the large number of computers or components in the system, one computer can hope to maintain links with just a limited number of those in the network. Now, say a computer in the geography lab regularly needs computers with good graphics capabilities for rendering its maps. It has to choose between maintaining links with just one computer or with many in the graphics lab. The former case will lead to less processing at the geography computer during allocation but might lead to delays in task completion when that particular graphics computer is busy. In contrast, the latter case will require more processing at the geography computer every time it has to allocate a task,but might help in getting quicker outputs once the task is allocated. Now, if provided with the structure, the geography computer can smartly choose how many graphics computers to maintain links with by evaluating the possible delays that might occur when accessing most of the graphics computers indirectly and compare that with the resources saved at itself in terms of processing cycles per each allocation task. Once the social interactions are explicitly depicted by the organisation structure, designers seeking to embed adaptation into the system can then use and focus on this organisation as a whole rather than working on each of the individual components separately. Thus, the organisation model will provide a better overview of the global performance of the system without compromising on the individuality of the constituent entities.

In summary, we argue that a formally modelled organisational representation of the components will help in managing their social interactions [22], and at the same time, provide insights into possible avenues for self-organisation and adaptation towards improving global behaviour. More specifically, we contend that depicting the distributed computing systems, including the service providers, their social interactions and the task environment, using an abstract organisation framework will provide a suitable platform to develop and test techniques attempting to bring about autonomicity into the system.

Against this background, we seek to develop a problem-solving agent organisation model that serves as a fitting abstract representation of such distributed computing systems. Our model will provide an appropriate simulation framework for distributed systems by modelling the task environment, the computational entities and their interactions along with their performance evaluation measures. Such a platform can then be used for developing and evaluating generic approaches designed for autonomic systems. In this context, by problem-solving agent organisations we refer to those containing some service-providing agents that receive inputs, perform tasks and return results. We chose to use problem-solving agent organisations because they can be decentralised with autonomous and independent agents which accomplish tasks by providing services and collaborating with each other through social interactions governed by the organisation's structure. Thus, it models the salient features of distributed computing systems and, at the same time, contains the flexibility required to make them autonomic. Following the reasons detailed above, the focus of the model is on the inter-agent interactions; that is, the organisation structure and its effect on the system. Moreover, we also present an evaluation mechanism for the performance of the organisation based on the tasks or goals achieved by it. This method is developed such that the critical role played by the organisation structure on the performance is made explicit and clear. Therefore, any designer of self-organisation techniques, especially those focusing on the structure or network, will be able to see their method in action and evaluate its performance before being transported and put in the actual domain specific autonomic systems.

In more detail, our organisation models a set of service-providing, resource constrained agents facing a dynamic stream of tasks requiring combinations of services with some ordering. The agents only posses local views and need to interact with each other to accomplish the tasks. These interactions are governed by the relations existing between the agents (organisation structure) and affect task allocation and organisation's performance. Finally, keeping in mind the development of adaptation techniques based on the structure, we also provide a method of representing the costs and resources involved in reorganisation.

In the next section (Sec. 2), we discuss the current literature in our context. Then in Sec. 3, we present our organisation framework together with the task environment, agents, organisational characteristics and performance measures. We illustrate our model with an example in Sec. 4 and conclude in Sec. 5.

2   Related Work

As we are seeking to develop an organisation framework that suitably represents distributed computing systems, it should provide an abstract representation of the components of the system, their social interactions and the tasks that they perform along with the environment that they are based in. Correspondingly, organisation modelling involves modelling the task environment, the organisational characteristics (structure and norms), the agents and the performance measures. In the following, we study the current literature in each of these aspects.

2.1   Modelling Tasks

The tasks faced by the organisation can be atomic or made up of two or more tasks (or subtasks) which, in turn, may be composed of other tasks. The tasks may have dependencies among them, resulting in a temporal ordering of the tasks in the organisation. In this context, [19] identifies three kinds of such dependencies— pooled, sequential and reciprocal. Two or more tasks whose results are jointly required to execute another task are said to be in a pooled dependency relation with each other. A sequential dependency exists between tasks if they have to be performed in a particular sequence. Finally, a reciprocal dependency exists if the tasks are mutually dependent on each other and have to be executed at the same time. However, the tasks dependencies as suggested by Thompson have subsequently been interpreted in different ways in different models. In particular, [11] model the task dependencies in their 'Virtual Design Team (VDT)' closely following Thompson's model. In fact, they even extend the sequential and reciprocal dependencies by classifying each of them into different types. In contrast, in the PCANS model, [13] demonstrate that both pooled and reciprocal dependencies, as described by Thompson, can be derived from sequential dependencies. Thus, their representation enables the designer to model just a single dependency type. For our present requirements, we just require a simple task model containing dependencies, and hence we will use the PCANS model.

2.2   Modelling Organisational Characteristics

Approaches towards organisational design in multi-agent systems can be considered to be either agent-centric or organisation-centric [14]. The former focus on the social characteristics of agents like joint intentions, social commitment, collective goals and so on. Therefore, the organisation is a result of the social behaviour of the agents and is not created explicitly by the designer. On the other hand, in organisation-centric approaches, the focus of design is the organisation which has some rules or norms which the agents must follow. Thus, the organisational characteristics are *imposed* on the agents. As we are primarily interested in problem-solving agent organisations, we only study organisations in multi-agent systems whose design is modelled explicitly.

In this context, the OperA and OMII frameworks [2,21] formally specify agent societies on the basis of social, interaction, normative and communication structures. However, in both of these frameworks, the agents are not permitted to modify the pre-designed organisational characteristics. Hence, they do not provide a suitable platform for self-organisation. Islander [16] also uses a similar approach by expecting the designer to pre-design the roles and the possible interactions between them thus delivering fixed patterns of behaviour for the agents to follow. Thus, this too is not flexible enough to incorporate reorganisation.

A more useful and simpler model developed by [3] provides a meta-model to describe organisations based on agents, groups and roles (AGR). While their model mainly pertains to groups of agents and intra-group dynamics (which does not apply to our requirements), they model organisation structure as defining the possible interactions between the groups. This interpretation of the structure matches our purpose and lies behind our model as well. A somewhat similar approach is followed by Moise [8], which considers an organisation structure as a graph defined by a set of roles, a set of different types of links and a set of groups. An agent playing a role must obey some permitted behaviours comprising the role. Organisation links are arcs between roles and represent the interactions between them. These links can be of three types— communication, authority and acquaintance. However, the links have a context associated with them and are valid within that context only. We seek an organisation structure that is not so specific or bounded. Nevertheless, some of the ideas used in this model, especially those relating to the organisation structure will be used while developing our model. A slightly different approach is followed by the Virtual Design Team (VDT) framework [11]. Its purpose is to develop a computational model of real-life project organisations. It does not use the agent-role paradigm. Instead, the agents are fixed to their duties and are called actors. The organisation structure is composed of a control structure and a communication structure. Evidently, VDT attempts to model a problem-solving organisation, and therefore, very relevant for our requirements. However, it lacks flexibility in the organisation structure, as it only permits purely hierarchical organisations. Therefore, we do not directly use the whole VDT model but only some parts of it.

In contrast to the above models, mathematical approaches have been developed for creating organisations [10,17]. However, they produce an instantiated organisation according to complex and elaborate specification of organisational requirements but not the generic model we need. In a more relevant work, [15] aim at an organisation framework that is flexible enough for self-organisation. However, they take a strictly emergent view of self-organisation and focus mainly on the social delegation aspects in agent organisations. Furthermore, their method specifies a set of organisation models, and the participating agents choose, whether or not, to join such organisations. Therefore, it does not inherently aid the development of problem-solving agent organisations. Another work [20] follows a norm based approach for modelling hierarchical organisations in which every role has a *position profile* associated with it. This profile is specified by positional norms and an agent can take up a role by changing its own set of norms to conform to these positional norms. However, the model requires that all positions and norms are specified at the outset itself thereby not allowing for the flexibility in the interactions as sought by us.

2.3 Modelling Agents

An overview of modelling agents in the context of organisations is presented by [1]. From this, it is apparent that the modelling of agents varies across different organisation models. In particular, agents may be homogeneous or belong to different classes, be cooperative or competitive. Their abilities may be represented as a simple vector or as a complex combination of skills, strategies, preferences and so on. Against this background, while all the organisation design approaches described above, with the exception of VDT, leave the agent development to the designer, VDT models the members of the organisation called actors in great detail. The main characteristics of the actors are attention allocation (determines the decision making behaviour of how the actor chooses among several task alternatives) and information processing (determines the skills, capacity and other processing characteristics). This design of agents will be partly used in our organisation model as it meets our requirements for modelling agents in the context of problem-solving organisations. Another concept that we will use is obtained from [6] where the agents are required to perform task assignment but can only address one request per time-step. Thus, we will also make use of this concept of agents possessing limited computational capacities so that the efficiency of the agents plays a prominent role in the performance of the organisation, thereby, reflecting the real-life scenarios where the components of the autonomous systems often possess small and limited computational power.

2.4 Evaluating an Organisation's Effectiveness

Organisation characteristics play a major role in the performance of the organisation [5]. Therefore, there are a number of existing methods for evaluating an organisation's characteristics based on parameters like robustness of the structure, connectivity and degree of decentralisation [9,7]. However, these measures are independent of the tasks being handled by the system and thus, fail to capture the suitability of the organisation according to the environment it is situated in. A contrasting criterion is to measure the performance of the organisation on the basis of how well it performs its tasks [4]. We believe this provides a good indication of the organisation's efficiency during run-time. In this context, in VDT, the measure of the performance of the organisation is on the basis of the load on the organisation. The load on the organisation is represented in units of *work volume*, thereby providing a common calibration for different tasks. The total work volume of a task depends partly on the task specification and partly on the organisational characteristics. Therefore, the resultant load on the organisation is a function of the tasks and the organisational characteristics and acts as a performance indicator. Therefore, the approach chosen by VDT is more suitable for our requirements and will be taken into account.

3 The Agent Organisation Framework

We describe our organisation framework by first detailing the task environment. Then we describe the agents and the organisation structure, before discussing the performance evaluation mechanism.

3.1 Task Representation

The task environment contains a continuous dynamic stream of tasks that are to be executed by the organisation. A task can be presented to the organisation at any point of time and the processing of the task must start immediately from that time-step. Thus, the organisation of agents is presented with a dynamic incoming stream of tasks that they should accomplish. In detail, the organisation of agents provides a set of services which is denoted by S. Every task requires a subset of this set of services. Services are the skills or specialised actions that the agents are capable of. We model the tasks as work flows composed of a set of several service instances (SIs) in a precedence order, thereby representing a complex job as expected in autonomic systems. We define a service instance si_i to be a 2-tuple: $\langle s_i, p_i \rangle$ where $s_i \in S$ (i.e. s_i is a member of the services set S), $p_i \in \mathbb{N}$ denotes the amount of computation required.

Following the PCANS model of task representation (see Sec. 2.1), we only consider sequential dependencies between the service instances. Thus, the SIs of a task need to be executed following a precedence order or dependency structure. This dependency structure is modelled as a tree in which the task execution begins at the root node and flows to the subsequent nodes. The task is deemed complete when all its SIs have been executed in the order, terminating at the leaf nodes. The complete set of tasks is denoted by W and contains individual tasks w_i which are presented to the organisation over time.

3.2 Organisation Representation

Since, we aim to model the agent organisation to represent a distributed computing system, our organisation framework consists of a set of computational agents representing the individual components. An agent is an independent computational entity that can provide one or more services. We model our agents by simplifying the agent model used by VDT (see Sec. 2.3) and consider only the information processing characteristics of the agents by overlooking the attention allocation characteristic. The attention allocation characteristic enables an agent to schedule its allocated tasks. The task scheduling algorithms at an agent will depend on the system that is being represented. However, this aspect is internal to an agent and independent of the organisational dynamics which is our primary focus. Therefore, we do not need to model this aspect.

In more detail, the agents are associated with particular sets of services (like say, in the example home-management system, a controller manages the heating system and can also access the internet for communication, thus containing two services in its service set). These sets can be overlapping, that is two or more agents may provide the same service. Also, building on the agent model used by Gershenson (see Sec. 2.3), every agent also has a computational capacity associated with it. The computational load on an agent (explained later), in a time-step, cannot exceed this capacity. This modelling of resource constrained agents is necessary because, generally the components of an autonomic system are small embedded devices with low computational power. Formally, let A be the set of agents in the organisation. Every element in this set is a tuple of the form:- $a_x = \langle S_x, L_x \rangle$ where the first field, $S_x \in S$ denotes a set of services that belong to the complete service set S and $L_x \in \mathbb{N}$ denotes the capacity. The agents, their service sets and their capacities may change during the lifetime of the organisation.

The other features of an agent organisation, in general, are its structure and norms. The structure of an organisation represents the relationships between the agents in the organisation, while the norms govern the kind of interactions and messages possible between the agents. However, since we are developing a problem-solving organisation, the agents are all internal to the organisation and share the same goals. Moreover, all the agents will be designed in the same way, and therefore, their interaction protocol will be similar and can be internally specified. Therefore, an explicit definition of norms is not required to regulate their interactions. Thus, in our model, the relationships between the agents (denoted by the structure) also determine the interactions between the agents. Formally, an organisation is defined as consisting of a set of agents and a set of organisational links. It can be represented by a 2-tuple of the form:- $ORG = \langle A, G \rangle$ where A, as stated above, is the set of agents, G is the set of directed links between the agents (will be described later in this section).

As mentioned previously, the organisation is presented with a continuous stream of tasks which are completed by the agents through their services. Tasks come in at random time-steps and the processing of a task starts as soon as it enters the system. Task processing begins with the assignment of the first SI (root node). The agent that executes a particular SI is, then, also responsible for the allocation of the subsequent dependent SIs (as specified by the task structure) to agents capable of those services. Thus, the agents have to perform two kinds of actions: (i) execution and (ii) allocation. Moreover, every action has a load associated with it. The load incurred for the execution of a SI is equal to the computational amount specified in its description, while the load due to allocation (called management load) depends on the relations of that agent (will be explained later). As every agent has a limited computational capacity, an agent will perform the required actions in a single time-step, as long as the cumulative load (for the time-step) on the agent is less than its capacity. If the load reaches the capacity and there are actions still to be performed, these remaining actions will be deferred to the next time-step and so on. We allow the agents to perform more than one action in a time-step to de-couple the time-step of the simulation with the real-time aspect of the actual computing systems. Thus, the time-step of the model places no restrictions whatsoever and can represent one or several processor cycles in the actual system.

As stated earlier, agents need to interact with one another for the allocation of SIs. The interactions between the agents are regulated by the structure of the organisation. Inspired from the Moise approach (see Sec. 2.2), we adopt the organisational links paradigm to represent the structure. However, unlike in Moise, the links in our case are not task-specific because we do not assume that the agents will be aware of all the tasks at the outset itself. Moreover, instead of using several graphs to represent particular aspects, we use an *organisation graph* (G) to represent the structure. The nodes in the graph represent the agents of the organisation while the links represent the relations existing between them. Thus, the structure of the organisation is based on the relations between the agents that influence their interactions.

In more detail, we classify the relationships that can exist between agents into four types — (i) `stranger` (not knowing the presence), (ii) `acquaintance` (knowing the presence, but no interaction), (iii) `peer` (low frequency of interaction) and (iv) `superior-subordinate` (high frequency of interaction). The superior-subordinate relation can also be called the authority relation and depict the authority held by the *superior* agent over the *subordinate* agent in the context of the organisation. The peer relation will be present between agents who are considered equal in authority with respect to each other and is useful to cut across the hierarchy. Also, the relations are mutual between the agents, that is for any relation existing between two agents, both the concerned agents respect it. The type of relation present between two agents determines the information that they hold about each other and the interactions possible between them. The information that an agent holds about its relations is:

1. The set of services provided by each of its peers (S_y of each peer a_y)
2. The accumulated set of services provided by each of its subordinates. The *accumulated service set* of an agent is the union of its own service set and the accumulated service sets of its subordinates, recursively. Thus, the agent is aware of the services that can be obtained from the sub-graph of agents rooted at its subordinates though it might not know exactly which agent is capable of the service. AS_x denotes the accumulated service set of agent a_x.

Whenever an agent finishes the execution of a particular SI, it has to allocate each of the subsequent dependent SIs to other agents (this may include itself). The mechanism for allocating SIs to other agents is also mainly influenced by the agents' relations. The decision mechanism of an agent is as follows:

- When an agent needs to allocate a SI, it will execute the SI if it is capable of the service and has no waiting tasks (capacity is not completely used up)
- Otherwise, it will try to assign it to one of its subordinates containing the service in its accumulated service set. This involves the agent traversing through the accumulated service sets (AS_x) of all its subordinates and then choosing one subordinate from among the suitable ones. If the agent finds no suitable subordinate (no subordinate or their subordinates are capable of the service) and it is capable of the service itself (but did not initially assign to self because its capacity is filled), then it will add the SI to its waiting queue for execution.

- If neither itself nor its subordinates are capable of the service, then the agent tries to assign it one of its peers by traversing through their service sets and choosing from among the suitable ones (those capable of the service).
- If none of the peers are capable of the service either, then the agent will pass it back to one of its superiors (who will then have to find some other subordinates or peers to execute the service).

Therefore, an agent mostly delegates SIs to its subordinates and seldom to its peers. Thus, the structure of the organisation influences the allocation of the SIs among the agents. Moreover, the number of relations of an agent contributes to the management load that it incurs for its allocation actions, since an agent will have to sift through its relations while allocating a SI. One unit of management load is added to the load on the agent every time it considers an agent for an allocation (mathematically modelled in Sec. 3.3). Therefore, an agent with many relations will incur more management load per allocation action than an agent with fewer relations. Also, a subordinate will tend to cause management load more often than a peer because an agent will search its peers only after searching through its subordinates and not finding a capable agent. Generally, it is expected that an agent will interact more frequently with its subordinates and superiors than its peers. This process of assigning a SI to an agent requires sending and receiving messages to/from that agent. Thus, task allocation also requires inter-agent communication which adds to the cost of the organisation.

In summary, the authority relations impose the hierarchical structure in the organisation, while the peer relations enable the otherwise disconnected agents to interact directly. It is important to note that while we present only these kinds of relations, the model allows the flexibility to depict more relation types in a similar fashion. Thus, the set of the relation types presented here can be expanded or contracted depending on the domain that is to be represented by the organisation model. Using this model, we abstract away the complex interaction problems relating to issues like service negotiation, trust and coordination. We do so, so that the model keeps the focus on the essence of self-organisation and autonomicity and isolates its impact on system performance.

Formally denoting the structure, every link g_i belonging to G is of form:-
$g_i = \langle a_x, a_y, type_i \rangle$ where a_x and a_y are agents that the link originates and terminates at respectively and $type_i$ denotes the type of link and can take any of the values in the set $\{Acqt, Supr, Peer\}$ to denote the type of relation existing between the two agents. The absence of a link between two agents means that they are strangers.

3.3   Organisation Performance Evaluation

The performance of a computing system denotes how well it performs its tasks. In terms of an agent organisation, the performance measure can be abstracted to the profit obtained by it. In our model, the profit is simply the sum of the rewards gained from the completion of tasks when the incurred costs have been subtracted. In more detail, the cost of the organisation is based on the amount of resources consumed by the agents. In our case, this translates to the cost of sending messages (communication) and the cost of any reorganisation taking place within the organisation. Thus, the cost of the organisation is:

$$cost_{ORG} = C. \sum_{a_x \in A} c_x + D.d \qquad (1)$$

where C is the communication cost coefficient representing the cost of sending one message between two agents and D is the reorganisation cost coefficient representing the cost of changing a relation. c_x is the number of messages sent by agent a_x and d is the number of relations changed in the organisation.

As stated earlier, agents have limited capacities and their computational load cannot increase beyond this capacity. Since, an agent might perform three kinds of actions in a time-step (task execution, task allocation, adaptation), the load on an agent is the summation of the computational resources used by the three actions and can be represented by three terms. Thus, the load l_x on agent a_x in a given time-step is:

$$l_x = \sum_{si_i \in W_{x_E}} p_i + M \sum_{si_j \in W_{x_F}} m_{j,x} + R.r_x \qquad (2)$$

- p_i is the amount of computation expended by a_x for executing SI si_i.
- $m_{j,x}$ is the number of relations considered by a_x while allocating SI si_j.
- W_{x_E} is the set of SIs (possibly belonging to many tasks) executed by a_x.
- W_{x_F} is the set of SIs being allocated by a_x.
- M is the 'management load' coefficient denoting the computation expended by an agent to consider one of its relations while allocating a single SI.
- R is the 'reorganisation load' coefficient, denoting the amount of computational units consumed by an agent while reasoning about adapting a relation.
- r_x is the number of agents considered by a_x for adaptation, in that time-step.

In this way, M and $m_{j,x}$, together represent the computational load for task allocation that is affected by the relations possessed by the agent, thereby providing a simple and explicit method of denoting the effect of the organisation structure on the individual agents. Similarly, R and r_x are used to represent the load caused by reasoning about adaptation (if any). Thus, the coefficient R denotes the amount of resources at the agent that are diverted for adaptation rather than performing tasks and help in deciding about when to reason about adaptation (meta-reasoning).

Since, the load l_x of a_x cannot exceed its capacity L_x, any excess SIs will be waiting for their turn, thus delaying the completion time of the tasks. The rewards obtained by the organisation depend on the speed of completion of tasks. In particular, a task w completed on time accrues the maximum reward b_w which is the sum of the computation amounts of all its SIs:

$$b_w = \sum_{i=0}^{|si_w|} p_i \qquad (3)$$

where si_w is its set of SIs. For delayed tasks, this reward degrades linearly with the extra time taken until it reaches 0:

$$reward_w = b_w - (t_w^{taken} - t_w^{reqd}) \tag{4}$$

where t_w^{taken} represents the actual time taken for completing the task, while t_w^{reqd} is the minimum time needed. Thus, the total reward obtained by the organisation is the sum of the rewards of the individual tasks completed by the organisation:

$$reward_{ORG} = \sum_{w \in W} reward_w \tag{5}$$

where W is the set of all tasks. The organisation's performance is measured as:

$$profit_{ORG} = reward_{ORG} - cost_{ORG} \tag{6}$$

Thus, for higher profits, the reward should be maximised, while the cost needs to be minimised. Both of these are affected by the allocation of tasks between the agents which, in turn, is influenced by the organisation structure. It is important to note that the agents only have a local view of the organisation. They are not aware of all the tasks coming in to the organisation (only those SIs allocated to them and the immediately dependent SIs of those allocated SIs) and neither are they aware of the load on the other agents. In spite of this incomplete information, they need to cooperate with each other to maximise the organisation profit through faster allocation and execution of tasks. Therefore, by modelling both the decentralisation and individual agent load along with inter-agent dependence and global profit, this evaluation mechanism suitably models the requirements faced by a designer while developing autonomic systems. In the same vein, reasoning and adapting the organisation also take up resources (as denoted by R and D) in our model, thus reflecting real-life scenarios.

4   Applying the Agent Organisation Framework

We illustrate our framework by using it to depict an autonomic system in charge of the university grid network system as outlined in Sec. 1. First, to illustrate our task model, consider a sample task possibly faced by this system. Assume that a project involves producing a predictive model of a given city. Such a task will involve analysing the GIS data of the required city, obtaining the population density of the city over the past years and then using some kind of statistical analysers on this data to estimate the population distribution in the future. It will also involve predicting the changes to the city transport system using the GIS information on this estimated population, and alongside render the map of the city graphically. In more detail, let us assume that the first part of this task will be to obtain the geographical data of the city and analyse it. In terms of our model, this can be designated as SI geo_map needing service gis-analyser with computation 20 (very intensive job). After this, let us say that the subsequent sub-tasks are obtaining the historical population data of the city and rendering the city-map graphically. These will form SIs get_census and draw_city requiring services census-data (getting and cleaning the census information from the archives) and graphics (graphically modelling to result in an image). Finally, execution of get_census might reveal that further statistical analysis is required to properly predict the population growth in the future and also that the growth caused by immigration depends on the transport incoming to the city. These can be designated as analyse_census and transport_flow requiring services stat-analyser and gis-analyser (as the transport network of the city can be obtained by analysing the GIS data) respectively. Also, note that the computation required for geo_map is much higher than that required for transport_flow even though both SIs need the same service. The task structure for this scenario, including the SIs and the dependencies is shown in Fig. 1(a). Representing this task formally:

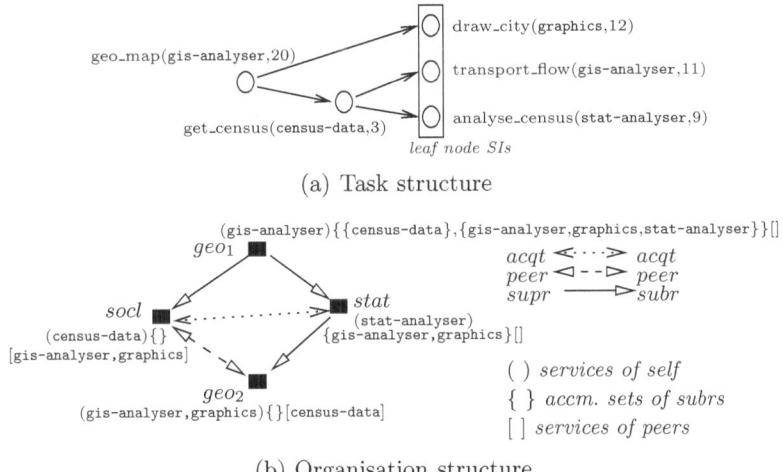

(a) Task structure

(b) Organisation structure

Fig. 1. Representation of an example task and organisation

In the same vein, consider a sample agent organisation to represent the autonomic system. Taking a limited view, let us focus on only four agents in this organisation— geo_1 and geo_2(two computers in the geography department), *socl* (a computer in the sociology department) and *stat* (an analyser in the statistics department). The services provided by the agents are basically their capabilities in terms of hardware, software and data accessible to them. Therefore, let us assume that geo_1 provides service gis-analyser. Similarly *socl* provides census-data, which is the population data of various places in all the past years, and *stat* provides stat-analyser service. However, geo_2 is capable of providing both gis-analyser (just like geo_1) and graphics (because it also contains high end graphics cards for rendering maps). Given this, let us look at the possible structure of the organisation. Let *socl* and geo_2 have a peer relationship. Also, assume geo_1 has two subordinates — *socl* and *stat* (because, say, often GIS based jobs are followed by either census information or statistical analysis).

stat, in turn, has *geo₂* as a subordinate. Moreover, while *socl* and *stat* are acquaintances of each other, *geo₂* and *geo₁* are not aware of each other. The G for this organisation contains 5 organisational links:

$$G = \{\langle geo_1, socl, Supr \rangle, \langle geo_1, stat, Supr \rangle, \langle stat, geo_2, Supr \rangle,$$
$$\langle socl, geo_2, Peer \rangle, \langle socl, stat, Acqt \rangle\}$$

For this organisation, the organisation graph is shown in Fig. 1(b). The absence of an arrow between two agents means that they are strangers. In addition, the information possessed by the agents about the services provided by their relations is also shown. For example, the accumulated service set (*AccmSet*) of agent *geo₁*, in turn, contains three sets representing its own service (`gis-analyser`), *AccmSet* of its subordinate *socl* (`census-data`) and of its other subordinate *stat* (`gis-analyser`,`graphics`,`stat-analyser`).

As an illustration of the allocation process, consider the sample organisation in Figure 1(b) executing the task shown in Figure 1(a). The allocation of SIs across the agents occurs as shown in Figure 2(a). In detail, we assume that the task arrives at agent *geo₁*. Hence, *geo₁* checks that it is capable of `geo_map` (as it is capable of service `gis-analyser` and has available capacity) and therefore, allocates `geo_map` to itself. After execution, *geo₁* needs to allocate the two dependencies of `geo_map` which are `get_census` and `draw_city` to capable agents. For allocating `get_census`, it checks the accumulated service sets of its two subordinates (*socl* and *stat*) and allocates to *socl* (because it is the only one capable of service `census-data`). Similarly, it allocates `draw_city` to *stat* because this subordinate contains service `graphics` in its accumulated service set. However, *stat* has to reallocate `draw_city` to its subordinate *geo₂* which is actually capable of that service. Similarly, after *socl* executes `get_census`, it needs to allocate the two dependencies (`transport_flow` and `analyse_census`) to appropriate agents. So, *socl* allocates `transport_flow` to its peer *geo₂* as it has no subordinates. It also

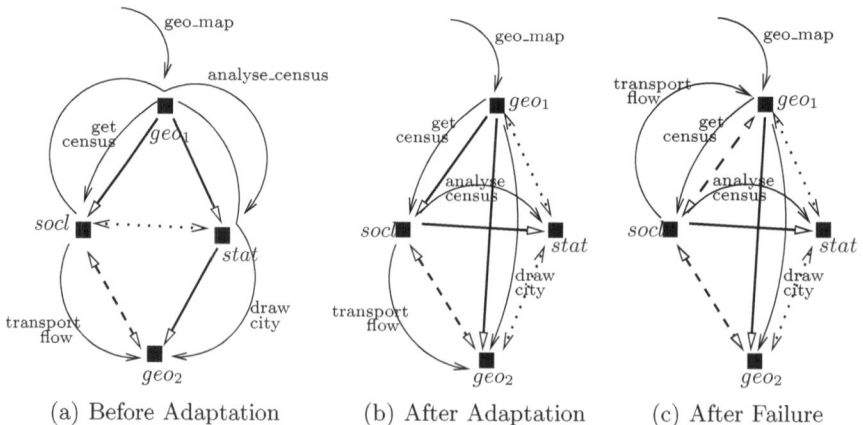

(a) Before Adaptation   (b) After Adaptation   (c) After Failure

Fig. 2. Allocation of the task in Fig. 1(a) in the organisation

hands back analyse_census to its superior geo_1 as it has found no subordinates or peers with that service (stat-analyser). geo_1 then assigns analyse_census to its subordinate *stat* (capable of stat-analyser) which then proceeds to execute it.

Thus, the structure of the organisation influences the allocation of service instances among the agents. Therefore, an efficient structure can lead to better and faster allocation of tasks. We see that in Figure 2(a), the allocation of draw_city and analyse_census was indirect and needed intermediary agents (*stat* and geo_1 respectively). Now, suppose on the basis of some adaptation method (such as that detailed in [12]), the agents modify their relations to form the structure as shown in Figure 2(b). That is, geo_1 and geo_2 decide to form a superior-subordinate relation and so do *socl* and *stat*. Meanwhile *stat* ends up becoming only an acquaintance of geo_1 and geo_2 as they decide to change the previously existing authority relations. With this new structure, the allocation of the SIs turns out to be much more efficient as all the allocations end up being direct one-step process. Therefore, they take shorter time because intermediary agents are not involved. Moreover, this allocation process requires less computation and communication because, for any SI, only a single agent performs the allocation and sends only one message. Compared to previous structure, this decreases the load on geo_1 and *stat* without putting extra load on other agents.

Now, let us suppose that after some time has passed, geo_2 is reconfigured (perhaps, the OS is reinstalled) such that it is no longer able to to provide gis-analyser. In such a scenario, *socl* will no longer be able to delegate transport_flow to geo_2 and will be handing back the SI to its superior geo_1. *socl* does so only after unsuccessfully considering its own subordinates and peers for allocation, thus causing more load onto itself and also taking more time. Under these changed circumstances, *socl* and geo_1 should realise that it is better to change their current relation into a peer relation so that *socl* can delegate to geo_1 quicker. Reversing the existing superior-subordinate relation will not be as useful because geo_1 also needs to continue delegating SIs like get_census to *socl*. Hence, these two agents change their relation as shown in Fig. 2(c).

In this way, the performance of the organisation can be improved by modifying the organisation structure through changes to the agent relationships. This will involve changes to the organisation graph G.

With this example and the sample adaptation scenarios, we see that adaptation of the structure plays an important role in the performance of the organisation. Furthermore, we illustrated that our framework not only provides a well-suited platform to represent autonomic systems, but also gives insights into possible avenues for self-organisation and permits the agents to perform the required adaptation. Here, while we showed how a more efficient structure can lead to the better allocation of a task, we should note that the organisation is performing several tasks at any given time and that the structure is common to all these possibly dissimilar tasks. Given this, the adaptation method should be such that the agents are able to identify which set of relations will be most suitable for their current context on the basis of the kind of tasks facing them in addition to their own service sets and allocation patterns.

5   Conclusions

In this paper, we introduced an abstract organisation framework for depicting distributed computing systems to aid in the development of autonomic systems. We presented our model by detailing our representation of the task environment and the organisation along with a performance evaluation system. The tasks are made up of service instances, each of which specifies the particular service and the computation required. The organisation consists of agents providing services and having computational capacities. The structure of the organisation manifests the relationships between the agents and regulates their interactions. Any two agents in the organisation could be strangers, acquaintances, peers or superior-subordinate. The relations of the agents determine what service information is held by the agents about the other agents and how to allocate service instances to them. We also presented the coefficients that affect the environment (communication cost, management load, reorganisation load) and the functions for calculating the organisation's cost and reward, thus enabling us to evaluate the profit obtained by it when placed in a dynamic task environment.

Our organisation framework provides a simulation platform that can be used by designers to implement and test their adaptation techniques before porting them to real and domain-specific systems. In particular, we designed our model such that the agents, though generic, realistically represent the components that would compose autonomic systems. The organisation is decentralised and agents possess local views and limited capacities like any large distributed computing system. Nevertheless, the agents interact with each other based on the organisation structure, which also influences the task allocations and thereby the organisational performance. This presents a suitable environment for self-organisation, which we have illustrated by using it to represent an intelligent and adapting, autonomous home-management system. In this context, our framework provides sufficient flexibility for the agents to modify their characteristics and social interactions, that is, manage themselves, just as expected in autonomic systems. Furthermore, we also provided the reorganisation cost (D) and load coefficients (R) to represent the price of adaptation. Thus, we have presented a suitable organisation framework that can be used as a platform for developing adaptation techniques, especially focusing on the agents' social interactions.

References

1. Carley, K.M., Gasser, L.: Computational organization theory. In: Multiagent Systems: A Modern Approach to Distributed Artificial Intelligence, pp. 299–330. MIT Press, Cambridge (1999)
2. Dignum, V.: A model for organizational interaction: based on agents, founded in logic. Ph.D. thesis, Proefschrift Universiteit Utrecht (2003)
3. Ferber, J., Gutknecht, O., Michel, F.: From agents to organizations: An organizational view of multiagent systems. In: Proc. of 4th Intl. Workshop on Agent Oriented Software Engineering, Melbourne, Australia, pp. 214–230 (2003)
4. Fox, M.S.: An organizational view of distributed systems, pp. 140–150. Morgan Kaufmann Publishers Inc., San Francisco (1988)

5. Galbraith, J.R.: Organization Design. Addison-Wesley, Reading (1977)
6. Gershenson, C.: Design and control of self-organizing systems. Ph.D. thesis, Vrije Universiteit Brussel (2007)
7. Grossi, D., Dignum, F., Dignum, V., Dastani, M., Royakkers, L.: Structural evaluation of agent organizations. In: Proc. of the 5th AAMAS (2006)
8. Hannoun, M., Boissier, O., Sichman, J.S., Sayettat, C.: Moise: An organizational model for multi-agent systems. In: Proc. of the 7th Ibero-American Conf. on AI (IBERAMIA-SBIA 2000), pp. 156–165 (2000)
9. Horling, B., Lesser, V.: A survey of multi-agent organizational paradigms. The Knowledge Engineering Review 19(4), 281–316 (2005)
10. Horling, B., Lesser, V.: Using quantitative models to search for appropriate organizational designs. In: Autonomous Agents and Multi-Agent Systems (2008)
11. Jin, Y., Levitt, R.E.: The virtual design team: A computational model of project organizations. Computational & Mathematical Organization Theory (1996)
12. Kota, R., Gibbins, N., Jennings, N.R.: Self-organising agent organisations. In: The Eighth International Conference on Autonomous Agents and Multiagent Systems (AAMAS 2009), pp. 797–804 (2009)
13. Krackhardt, D., Carley, K.M.: A pcans model of structure in organizations. In: Proc. of Intl. Symp. on Command and Control Research and Technology (1998)
14. Lematre, C., Excelente, C.B.: Multi-agent organization approach. In: Proc. of 2nd Ibero-American Workshop on DAI and MAS, Toledo, Spain (1998)
15. Schillo, M., Bettina Fley, M.F., Hillebrandt, F., Hinck, D.: Self-organization in multiagent systems: from agent interaction to agent organization. In: Proc. of Intl. Workshop on Modeling Artificial Societies and Hybrid Organizations (2002)
16. Sierra, C., Rodriguez-Aguilar, J.A., Noriega, P., Esteva, M., Arcos, J.L.: Engineering multi-agent systems as electronic institutions. UPGRADE The European Journal for the Informatics Professional V(4), 33–39 (2004)
17. Sims, M., Corkill, D., Lesser, V.: Automated organization design for multi-agent systems. Autonomous Agents and Multi-Agent Systems 16(2), 151–185 (2008)
18. Tesauro, G., Chess, D.M., Walsh, W.E., Das, R., Segal, A., Whalley, I., Kephart, J.O., White, S.R.: A multi-agent systems approach to autonomic computing. In: Proc. of 3rd AAMAS (2004)
19. Thompson, J.D.: Organizations in Action: Social Science Bases in Administrative Theory. McGraw-Hill, New York (1967)
20. Montealegre Vázquez, L.E., López y López, F.: An agent-based model for hierarchical organizations. In: Noriega, P., Vázquez-Salceda, J., Boella, G., Boissier, O., Dignum, V., Fornara, N., Matson, E. (eds.) COIN 2006. LNCS (LNAI), vol. 4386, pp. 194–211. Springer, Heidelberg (2007)
21. Vazquez-Salceda, J., Dignum, V., Dignum, F.: Organizing multiagent systems. Autonomous Agents and Multi-Agent Systems 11(3), 307–360 (2005)
22. Zambonelli, F., Jennings, N.R., Wooldridge, M.: Developing multiagent systems: The gaia methodology. ACM Trans. Softw. Eng. Methdol. 12(3) (2003)

Metareasoning and Social Evaluations in Cognitive Agents

Isaac Pinyol and Jordi Sabater-Mir

IIIA - Artificial Intelligence Research Institute
CSIC - Spanish National Research Council
Bellaterra, Barcelona, Spain
{ipinyol,jsabater}@iiia.csic.es

Abstract. Reputation mechanisms have been recognized one of the key technologies when designing multi-agent systems. They are specially relevant in complex open environments, becoming a non-centralized mechanism to control interactions among agents. Cognitive agents tackling such complex societies must use reputation information not only for selecting partners to interact with, but also in metareasoning processes to change reasoning rules. This is the focus of this paper. We argue about the necessity to allow, as a cognitive systems designers, certain degree of freedom in the reasoning rules of the agents. We also describes cognitive approaches of agency that support this idea. Furthermore, taking as a base the computational reputation model Repage, and its integration in a BDI architecture, we use the previous ideas to specify metarules and processes to modify at run-time the reasoning paths of the agent. In concrete we propose a metarule to update the link between Repage and the belief base, and a metarule and a process to update an axiom incorporated in the belief logic of the agent. Regarding this last issue we also provide empirical results that show the evolution of agents that use it.

Keywords: Reputation, Trust, Cognitive Agents, Metareasoning, BDI agents.

1 Introduction

Reputation mechanisms have been recognized one of the key technologies when designing multi-agent systems (MAS) [1]. In this relatively new paradigm, reputation models have been adapted to confront the increasing complexity that open multi-agent environments bring. Thus, the figure of agents endowed with their own private reputation model takes special relevance as a non-centralized mechanism to control interactions among agents. Following this line, cognitive agents using cognitive reputation models arise as one of the most complete and generic approaches when facing very complex societies. Usually, cognitive agent's architectures, like BDI (*Belief, Desire, Intention*), follow logic-based reasoning mechanisms, providing then a high flexibility and theoretically well-founded reasoning.

Repage [2] is a reputation system based on a cognitive theory of reputation that has been used in logical BDI reasoning processes [3] (BDI+Repage), offering then an integrated reasoning framework. Even when this work faces the field of computational

A.V. Vasilakos et al. (Eds.): AUTONOMICS 2009, LNICST 23, pp. 220–235, 2010.

reputation models, the focus is not on the model itself, but on the integration of the information that it provides with the other elements of the agent. Following this line, a very important aspect of cognitive agents is the capacity to reason about their own reasoning processes. These metareasoning processes act at all levels in the agents' mind. However, we are interested in the aspects related to reputation information.

In this concrete work we justify the use of metarules and metaprocesses in cognitive agents and provide mechanisms to specify them. We apply these ideas to some rules and axioms of the BDI+Repage model. In concrete, we propose a specification that allows the modification at run-time of the rule that relates reputation information with logic belief formulas, and also of an axiom rule integrated in the logical belief based of the agent. Regarding this last point we provide both a metareasoning process to update such axiom, and empirical results of an implementation we develop of the BDI+Repage model placed in a replication of a simple market, populated by buyers, sellers and informant agents. To detail our work, in section 2 we introduce a cognitive theory of agency to justify the use of metarules and metareasoning when modeling cognitive agents. In section 3 we explain the agent model, and in section 4 we provide the tools to specify metareasoning rules and processes. In the same section we apply them to some reasoning rules of the BDI+Repage model. In section 5 we present empirical results to show how the dynamic modification of axioms can produce good results in the agents' level of satisfaction. Finally in section 6 we conclude our analysis and present the future work.

2 Reasoning and Metareasoning: A Cognitive Approach

In this section we briefly get in touch with the cognitive theory that supports the work done in this paper. The theory developed by Castelfranchi and Paglieri [4] is quite generic and focuses on the dynamics of goals processing and its relation with the beliefs of the agent. Although the specific topic of the paper relies on describing which *typology* of beliefs participates in each stage of goal dynamics, we are very interested in the concepts of belief-supporting goals and belief-supporting beliefs that are pointed out by the theory. The authors argue that goals and beliefs have a supported structure of beliefs, i.e., beliefs from which a given goal or belief is activated. Moreover, such structure can be explicitly represented also as beliefs, achieving then a metabelief level.

In our work we are specially interested in the idea of belief-supporting beliefs which has been deeply studied in [5]. When such belief structures are explicitly represented as beliefs, the authors named them *reasons*. Thus, a given belief has a set of *reasons* that continuously supports it. Because of this explicit representation, agents can also reason about them, achieving then a *metareasoning* process that also relies on beliefs.

Reasons are important because allow the agent to *justify* herself *why* the set of beliefs and goals are activated, but also to justify her beliefs to other agents. This last issue has been extensively studied in the field of argumentation [6]. Reasons are also important because they allow the agents to review their own reasoning process, by starting metareasoning processes that *change* the reasoning paths.

A logical perspective is a good starting point for the explanation of more technical details. Lets consider a propositional logic in which proposition p holds. If formula $p \rightarrow q$ also holds, then for sure q will hold. From the logical view this reasoning has been produced by *modus ponents* by two explicit formulas (p and $p \rightarrow q$). If we are talking about the beliefs of an agent, the belief on q (Bq) is justified by the beliefs Bp and $B(p \rightarrow q)$. Notice that if the agent realizes that p was not the case, then q must be withdrawn. This is the typical belief revision process. However, the agent may also realize that it is not the case that $p \rightarrow q$. Thus, q should be also withdrawn. However, from a cognitive point of view this situation is quite different than the first one, since the formula $p \rightarrow q$ is a *reason*, an explicit belief saying that from p can be deduced q. Therefore, a reasoning concerning the truth of $p \rightarrow q$ could be seen as a metareasoning. It could be argued that the truth of p also affects the reasoning process, but the object of the reasoning is not a representation of an explicit reasoning step, as it is in the other case. See figure Figure 4 for a graphical representation of the structure.

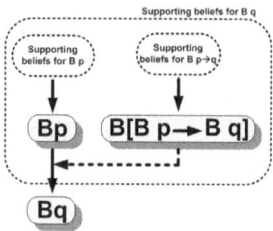

Fig. 1. The generic belief-supporing beliefs structure with explicit representation of a deduction step

The point of the discussion is that logics offers a nice way to construct trees of supporting formulas, through logical reasoning, but in the general case, the links can be modified by the same agent due to other beliefs. As a designers of cognitive agents architectures, we must deal with these concepts to consider real autonomous agents. In this paper we extend a BDI agent architecture that incorporates reputation information [3] by allowing a partial update of the rules that govern the reasoning process, focusing on the belief-supporting beliefs, letting for future work the relationship between desires and intentions. Next section provides the description of the BDI+Repage model.

3   A Multicontext BDI Agent with Repage System

The model of agent we present in this section is a BDI model in which Repage reputation system is also incorporated. To explain it is necessary first to get in touch with the Repage reputation model, and the cognitive theory of reputation that supports it.

3.1   Preliminaries: Social Evaluations, Image and Reputation

Repage [2] is a computational system designed to be part of the agents architecture and based on a cognitive theory of reputation [7]. It provides social evaluations as image and reputation. A social evaluation is a generic term used to encapsulate the information resulting from the evaluation that an agent (evaluator) might make about another agent's (target's) performance regarding some skill, standard or norm (object of the evaluation). The object of the evaluation relies on which property of the target agents is evaluated. The value of the evaluation indicates how *good* or *bad* the performance resulted to be.

A social evaluation in Repage has three elements: a target agent, a role and a probability distribution over a set of labels. The target agent is the agent being evaluated. The role is the object of the evaluation and the probability distribution the value of the evaluation. The evaluator is the agent making the social evaluation.

The role uniquely identifies a kind of transaction and the classification of the possible outcomes. The current implementation of Repage considers five abstract linguistic labels for this classification: *Very bad, Bad, Neutral, Good, Very good*(*VB, B, N, G, VG* from now on), and assigns a probabilistic value to each label, however, we generalize it considering a finite number of labels $w_1, w_2 \ldots$. The *meaning* of each label must be contextualized depending on the role. For instance, we can represent a Repage image predicate as $img_i(j, seller, [0.4, 0.2, 0.2, 0.1, 0.1])$. This indicates that agent i holds an image predicate about agent j in the role of $seller$, and the value of the evaluation is $[0.4, 0.2, 0.2, 0.1, 0.1]$. This value reflects a probability distribution over the labels *VB, B, N, G, VG*. Then, it means that agent i believes that in the transaction of buying, when agent j acts as a seller, there is a probability of 0.4 to achieve a VB result (in the context of this transaction, this may mean a very low quality of the product), with a probability of 0.2 a B result, etc. For reputation predicates, it is the same as image, but instead, the agent believes that the evaluation is said by all or most of the agents in the group. We refer to [2] for details on the calculus and the internal architecture.

In the next subsection we detail the BDI model, starting from the basic framework of multicontext systems.

3.2 The Multi-Context BDI Model

Multi-context systems (MCS) provide a framework that allows several distinct theoretical components to be specified together with the mechanisms that link them together [8]. These systems are composed of a set of contexts (or units), and a set of bridge rules. Each context can be seen as a logic and a set of formulas written in that logic. Bridge rules are the mechanisms with which to infer information from one context to another. Each bridge rule has a set of antecedents (preconditions) and a consequent. The consequent is a formula that becomes true in the specific context when each antecedent holds in its respective context.

The specification of our BDI agent as a multi-context system is inspired by the models presented in [9,10]. It is formalized with the tuple $Ag = \langle \{BC, DC, IC, PC, CC, RC\}, \triangle_{br} \rangle$. These correspond to Belief, Desire, Intention, Planner, Communication and Repage contexts respectively. The set of bridge rules $\triangle_{br}$ incorporates the rules $1, 2, 3, 4, P, Q$ and B, the bridge rules A_I and A_R shown in Figure 3, and rule B. Figure 2 shows a graphical representation of this multi-context specification. In the next subsections we briefly explain each context and bridge rule.

3.3 Belief Context (BC)

This context contains the beliefs of the agent. For this we use BC-logic [11], a probabilistic dynamic belief logic with a set of special modal operators. We are specially

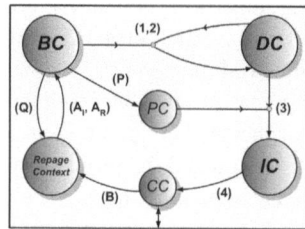

Fig. 2. The Repage context embedded in a multi-context BDI agent. Circles represent context and arrows represent bridge rules.

interested in the operators B_i and S, the first expressing what is believed by agent i, and the latter, what has been said by all the agents in the group respectively. The dynamic aspect of this logic is introduced by defining a set Π of actions. Then, for $\alpha \in \Pi$ and $\varphi \in BC$, formulas like $[\alpha]\varphi$ indicate that after the execution of α, the formula φ holds.

This logic incorporates specific axioms to reason about the probabilities of formulas by means of the operator Pr and constants $\bar{p}$ such that $p \in [0,1] \bigcap \mathbb{Q}$. It follows that for formulas $\varphi \in BC$, the expression $\bar{p} \leq Pr\varphi$ indicates that the probability of holding φ is higher or equal to p. This logic is based on the Logic of Knowledge and Probability introduced by Fagin and Halpern in [12].

BC-logic allows expressions like $B_i(\bar{p} \leq Pr([\alpha]\varphi))$. This indicates that agent i believes that the probability of holding φ after the execution of action α is at least p. Thereby, the formula $S(\bar{p} \leq Pr([\alpha]\varphi))$ expresses the same but in terms of what all agents have said. To simplify the notation, we will write expressions like $B_i(\bar{p} \leq Pr\varphi)$ as $(B_i\varphi, p)$, and $S(\bar{p} \leq Pr\varphi)$ as $(S\varphi, p)$.

This logic allows us to express image information in terms of beliefs $B_i\varphi$, and reputation information in terms of beliefs about what is said, $B_iS\varphi$ (see section 3.8). By grounding image and reputation into simple elements, we endow the agent with a powerful tool to reason over these concepts.

The complete syntax, semantics and axiomatization of BC-logic can be found at [11]. The belief operator follows the standard K, D, 4 and 5 axioms of modal logic, while operator S has its owns. The most interesting axioms are those that describe the interaction between S and B_i. These are closely related to the concept of *trust* that Demolombe in [13] defined regarding agents as information sources. The relationship of the two operators implies a relation between image and reputation at the belief level [11]. For instance, if for every φ the formula $B_i((S\varphi \rightarrow \varphi), p)$ holds (trust axiom), then agent i believes that what all agents say is really true with a probability p. The trust axiom has big implications in the relation between image and reputation information at the belief level [11].

3.4   Desire Context (DC)

This context deals with the desires of the agent. Like the BDI model described by Rao and Georgeff in [14], they are attitudes that are explicitly represented and that reflect the general objectives of the agent. We consider that desires are graded, and for that, we use the multi-valued logic (DC-logic) based on the Lukasiewicz logic described in [10].

DC-logic includes two fuzzy modal operators[1]: D_i^+ and D_i^-. The intended meaning of $D_i^+\varphi$ is that the formula φ is desired by agent i, and its truth degree, from 0 (minimum) to 1 (maximum), represents the level of satisfaction if φ holds. The intended meaning of $D_i^-\varphi$ is that φ is negatively desired, and the truth degree represents de level of disgust if φ holds. Also, DC-logic includes truth constants $\bar{r}$ where $r \in [0,1] \cap \mathbb{Q}$, and the connectives $\&$ and $\Rightarrow$ corresponding to the Lukasiewicz conjunction and implication respectively.

3.5 Intention Context (IC)

This context describes the intentions of the agent. Like in the Rao and Georgeff's BDI model [14], intentions are explicitly represented, but in our case generated from beliefs and desires. Also, we consider that intentions are graded, and for this we use the IC-logic defined in [10].

Similar to DC-logic, IC-logic defines the fuzzy modal operator $I_i\varphi$, indicating that agent i has the intention to achieve φ, and its truth degree (from 0 to 1) represents a measure of the trade-off between the benefit and counter-effects of achieving φ. Furthermore, IC-logic is defined in terms of a Lukasiewicz logic in the same way as DC-logic. Also, formulas like $\bar{r} \Rightarrow I_i\varphi$ will be written as $(I_i\varphi, r)$.

3.6 Planner Context (PC) and Communication Context (CC)

The logic in the Planner context is a first-order logic restricted to Horn clauses. In this first approach, this context only holds the special predicate $action$, which defines a primitive action together with its precondition. We look forward to introducing plans as a set of actions in the future. Communication context is a functional context as well, and its logic is also a first-order logic restricted to Horn clauses with the special predicates $does$ (to perform actions), and rec_{ij} (to notify that agent i has received a communication from agent j).

3.7 Repage Context (RC)

The Repage context contains the Repage model. We can assume that Repage predicates are specified in first-order logic restricted to Horn clauses, where the special predicates Img and Rep are defined. We write them as img_i $(j, r, [V_{w_1}, V_{w_2}, \ldots])$ and rep_i $(j, r, [V_{w_1}, V_{w_2}, \ldots])$, corresponding to the Image and Reputation of agent j playing the role r, from the point of view of i.

When in Repage the role and its labeled weights are defined, the role uniquely identifies which kind of transaction is part of, and each w_k identifies a predicate. To simplify, we can assume that the transaction identified by a role is summarized in a single action. To state this, we presuppose the definition of a mapping $\mathcal{R}_r$ between each role r and its

---

[1] The original logic in [10] does not contain the reference to the agent. We include it to remark the desires of agent i.

action. In a similar way, we assume a mapping $\mathcal{T}_{r,w_k}$ between each role r and label w_k to a predicate.

We illustrate this with an example: In a typical market, the transaction of buying certain product involves two agents, one playing the role of buyer and the other playing the role of seller. From the point of view of the buyer, if she wants to evaluate other agents that play the role of seller, she knows that the associated action is buy. So, $\mathcal{R}_{seller}$ maps to buy. In the same way, the agent must know the meaning of each label w_k of Repage. Then, we can define that $\mathcal{T}_{seller,w_1}$ is $veryBadProduct$, $\mathcal{T}_{seller,w_2}$ is $okProduct$, etc.

In this mapping, the Repage predicate $img_i(j, seller, [0.2, 0.3, \dots])$ indicates that agent i believes that there is a probability of 0.2 that after executing the action $\mathcal{R}_{seller}$ (buy) with agent j as a seller, she will obtain a $\mathcal{T}_{seller,w_1}$ ($veryBadProduct$); with 0.3 that she will obtain $\mathcal{T}_{seller,w_2}$ ($OKproduct$), etc. With reputation predicates it is similar, but the concept is quite different. In this case it indicates that agent i believes that the corresponding evaluation is said by the agents in the group.

3.8 Bridge Rules

Bridge rules A_I and A_R (see Figure 3) are in charge of generating the corresponding beliefs from images and reputations respectively. Notice that given a Repage social evaluation, these bridge rules generate one belief for each weight w_k. Both bridge rules use the belief operator (B_i) over certain formula, but meanwhile rule A_I states a knowledge that agent i believes as true, A_R states a knowledge that agent i believes to be said. They follow the definition of image and reputation we have given in the Repage context in section 3.7.

The detail of the following rules can be found at [3]. Rules 1,2,3,4 perform the actual BDI reasoning. Bridge rules 1 and 2 transform generic desires to more concrete and realistic desires. To do this, these bridge rules merge generic desires from DC (with absolute values of satisfaction or disgust) with the information contained in BC, which includes the probability to achieve the desire by executing certain action. The result is a desire whose gradation has changed, becoming more realistic. This is calculated by the function g. If we define it as the product of both values, we obtain an expected level of satisfaction/disgust. Notice that we require that the belief information implies the achievement of the desired predicate.

$$A_I: \quad \frac{RC : img_i(j, r, [V_{w_1}, V_{w_2}, \dots])}{BC : (B_i([\mathcal{R}_r(j)]\mathcal{T}_{r,w_1}, V_{w_1}))}$$
$$BC : (B_i([\mathcal{R}_r(j)]\mathcal{T}_{r,w_2}, V_{w_2}))$$
$$\dots$$

$$A_R: \quad \frac{RC : rep_i(j, r, [V_{w_1}, V_{w_2}, \dots])}{BC : (B_i(S([\mathcal{R}_r(j)]\mathcal{T}_{r,w_1}, V_{w_1})))}$$
$$BC : (B_i(S([\mathcal{R}_r(j)]\mathcal{T}_{r,w_2}, V_{w_2})))$$
$$\dots$$

Fig. 3. The bridge rules A_I and A_R (see Figure 2). They translate Image and Reputation predicates respectively into beliefs expressions in BC.

Bridge rule 3 generates intentions. It takes into account both the expected level of satisfaction and the cost of the action. At the same time, executing an action to achieve certain formula can generate undesirable counter-effects. Thus, bridge rule 3 also takes into account the possible negative desires that can be reached by executing this action. In this bridge rule, for each positive realistic desire (D^+), we must include all negative desires (D^-) that can result from the same action. In this way we have the value of the positive desire (δ^+) and the sum of all negative desires (δ^-) that can be achieved by executing the same action. The strength of the intention that is created is defined by a function f. Different f functions would model different behaviors. In our examples we use the following definition: $f(\delta^+, \delta^-) = max(0, \delta^+ - \delta^-)$.

Finally, bridge rule 4 instantiates a unique intention (the one with maximum degree) and generates the corresponding action in the communication context.

4   The Metalevel Specification

In this section we specify a possible metalevel reasoning regarding the trust axiom of the BC-logic and the bridge rules A_I and A_R. For this task we take ideas from the specification of dynamic protocols [15] in the frame of open multiagent organizations. Here, the specification of interaction protocols is described as a set of rules specified at the design time. However, when facing open systems, often environmental or social conditions for instance, may carry the necessity to modify such protocols at run-time. These modifications must be product of a dialog, as a metaprotocol, among the participants. In [15], the author presents an infrastructure to allow agents the modification of a subset of rules. It considers a k-level infrastructure, where at level 0, the main rules of the protocol are specified with certain *degrees of freedom* (Dof). At level 1, a metaprotocol can be specified to allow the discussion about how to change the protocol of level 0. More levels can be specified following the idea that at level i the protocol allows the discussion of the degrees of freedom of level $i - 1$.

4.1   DoF for Reasoning Rules

We apply the same DoF principle to some axioms and bridge rules of our BDI+Repage architecture. Instead of using belief revisions techniques, we encourage the use of DoF to update parts of rules that govern a reasoning process, not only for preserving consistencies, but also for adaptation. Belief revision processes rely on crisp logic and look for the smallest subset of formulas to keep a logical theory consistent when a formula is added in the theory. For our needs this vision is limited because only faces logical theories and because is used to avoid inconsistencies.

By using degrees of freedom, we bound the space of states by constraining what can be modified and what not. Thus, a main reasoning structure remains constant, but not static. In the case of logic-based BDI agents, this is very clear. For instance, the original model that Rao and Georgeff presented [14] states some basic and untouchable axioms to ensure several properties of the logical reasoning, but also considers other set of axioms that when included in the logic, model totally different behaviors. A clear example is the relation between the three main attitudes: beliefs, desires and intentions.

They define a typology of agents, the main ones being realist, strong realist and weak realist agents.

This is one of the advantages of using BDI models. The flexibility they achieve. By simply adding or erasing some axioms we can model an infinity of agents. However, when facing autonomous agents that must deal with open environments, we need some more flexibility. In Rao and Georgeff's BDI model, could an agent move from a strong realism to weak realism at run-time? From a technical point of view it is just a matter of changing two axioms. From a logical point of view, this process is outside the logic, and must be done at a meareasoning level. The possibility to update or modify some axioms is supported by the cognitive theory presented in the introduction of this paper, in which real autonomous agents should be aware of the way they reason. Due to that, agents can *think* about how they *think* and act in consequence (see section 2 for more details).

Notice that in the model of Rao and Georgeff the switch between strong realism and weak realism implies the totally substitution of a set of axioms for another set. This is the most extreme scenario in which the DoF involves the whole rule, because of the nature of the logic, which is crisp. More complex and expressive logics, like the BC-logic presented in the previous section, can deal with probabilities, which can be incorporated in the axioms to somehow tune their strength in the reasoning process, for instance. In a similar way, brige rules, which in fact are outside the logic, can be also tune by similar elements.

In the following sections we show how a similar formalism used for the DoF of dynamic protocols can be used to specify a metareasoning model for our BDI+Repage model.

4.2   A Metalevel Specification for the Rules A_I and A_R

In this subsection we focus on the relationship between the Repage model and the Belief context. This relation is statically specified by the rules A_I and A_R. As we mentioned, these rules are responsible for translating Image and Reputation predicates into atomic beliefs. Following the specifications of the Repage reputation model and the underlying theory, these rules are a very accurate formalism to generate the belief that in a more atomic way represent the information provided by the reputation model. However such transformation may carry out the logical inconsistency on the belief theory. This is the case when the Trust axiom is present and we have very contradictory information between an image and a reputation predicate of a given agent in a given role.

These inconsistencies always refer to probabilistic issues. To illustrate this, we can assume that the trust predicate is present in the belief context as $B_i(S\varphi \rightarrow \varphi)$. Then, it may occur that rules A_I and A_R have generated the following beliefs:

$$B_i([buy(alice)]VeryBad, 0.9) \tag{1}$$

$$B_i([buy(alice)]VeryGood, 0.1) \tag{2}$$

$$B_iS([buy(alice)]VeryBad, 0.5) \tag{3}$$

$$B_iS([buy(alice)]VeryGood, 0.5) \tag{4}$$

Due to the trust axiom, the formulas 3 and 4 imply the following formulas:

$$B_i([buy(alice)]VeryBad, 0.5) \tag{5}$$

$$B_i([buy(alice)]VeryGood, 0.5) \tag{6}$$

Notice that formula 1 implies formula 5, and formula 6 implies formula 2. The inconsistency relies on that propositions $veryBad$ and $veryGood$ should be mutually disjoint. Then, how is it possible to belief that with a probability higher that 0.9 after execution of the action $buy(alice)$ we will obtain a very bad product, and with a probability higher that 0.5 we will obtain a very good product?

To solve this kind of situations we provide the agent with the capability to modify its bridge rules. To do so, we define one degree of freedom at each one of the rules A_I and A_R. By doing this we are specifying metarules $(M(A_I), M(A_R))$, which map to a family of different A_I and A_R rules:

$$\mathbf{M(A_I):} \quad \frac{RC : img_i(j, r, [V_{w_1}, V_{w_2}, \dots])}{\begin{array}{l} BC : (B_i(pr([\mathcal{R}_r(j)]\mathcal{T}_{r,w_1}, V_{w_1})) = \mathcal{X}_{j,r,1}) \\ BC : (B_i(pr([\mathcal{R}_r(j)]\mathcal{T}_{r,w_2}, V_{w_2})) = \mathcal{X}_{j,r,2}) \end{array}}$$

$$\dots$$

$$\mathbf{M(A_R):} \quad \frac{RC : rep_i(j, r, [V_{w_1}, V_{w_2}, \dots])}{\begin{array}{l} BC : (B_i(pr(S([\mathcal{R}_r(j)]\mathcal{T}_{r,w_1}, V_{w_1})) = \mathcal{Y}_{j,r,1})) \\ BC : (B_i(pr(S([\mathcal{R}_r(j)]\mathcal{T}_{r,w_2}, V_{w_2})) = \mathcal{Y}_{j,r,2})) \end{array}}$$

$$\dots$$

Notice that if we set the default value of $\mathcal{X}_{j,r,1}$ and $\mathcal{Y}_{j,r,1}$ to 1, we have exactly the original rules A_I and A_R, since in BC-logic, $\varphi \leftrightarrow pr(\varphi) = 1$. The process by which the agent decides which value to take is a metareasoning process. Different heuristics can be used to perform such task. What it is clear is that such heuristics is a process that depends on a set of beliefs (Trust axiom and the beliefs that refer to the same agent and role must be part of the inputs of such process).

4.3   A Metalevel Specification for the Trust Axiom

In a similar way, the generic trust axiom that relates what is said with what is believed can be modified, and in fact, it is crucial for the adaptation of the agent. On the one side, if no trust axiom is present in the theory, formulas like $B_i S\varphi$ will never become $B_i\varphi$. On the other side, if the trust axiom is present, $B_i S\varphi$ would imply $B_i\varphi$. However, we also talk about graded trust, $B_i(pr(S\varphi \rightarrow \varphi) = g)$, and the effects on the formulas like $B_i S\varphi$ and $B_i(S(pr(\varphi) \geq p))^2$. Different values of g model different behaviors of the agent. Thus, we can consider this g as a DoF of the axiom. We can write the meta axiom $M(Trust)$ as

$$B_i(pr(S\varphi \rightarrow \varphi) = \mathcal{Z})$$

---

[2] It can be proved that if the graded trust axiom is present, and the formula $B_i(S(pr(\varphi) \geq p))$ holds, then it can be deduced $B_i(pr(\varphi) \geq p \cdot g)$ [11].

As well, different heuristics as metaprocesses can be considered for the update of the DoF. In section 5 we consider an heuristics for this axiom and show how agents behavior changes during time.

Notice that the inclusion of such axiom with a $\mathcal{Z}$ higher that 0 may cause inconsistencies in the theory, as we explained in the previous subsection. Because of that, this process can start another process to update the values of $\mathcal{X}_{j,r,p}$ and $\mathcal{Y}_{j,r,p}$ agent j, role r and weight p.

4.4 Processes Description

Figure 4 shows a graphical representation of the metarules dependences. Circles $M1$ and $M2$ represent processes:

- **M1:** This process is in charge for deciding the DoF $\mathcal{Z}$, which belongs to $M(Trust)$. In the experimental section we state how this concrete process could be performed. In any case, the graphical representation shows that this process is fed by the actual instantiation of the trust axiom (which is only characterized by the DoF variable $\mathcal{Z}$), and information from the Repage context. The output of the process is a new value for $\mathcal{Z}$ (in the graphic is shown as $\mathcal{Z}'$).
- **M2:** This process only receives the current instantiation of the Trust predicate, characterized by $\mathcal{Z}$. As we argued, this value is the only that can produced inconsistencies in the theory of BC-Context. As output, it provides new values from the DoF $\mathcal{X}_{j,r,p}$ and $\mathcal{Y}_{j,r,p}$, for each different agent j, role r and weight p. Thus, the number of different instantiated rules is computed by $|Ag| \cdot |R| \cdot |W|$. This means that potentially, each agent in each role can have a different inference rule from Repage to BC-Context.

We leave the exploration of the metaprocess $M2$ for future work. Instead, we focus on $M1$, providing a possible mechanism to compute $\mathcal{Z}$.

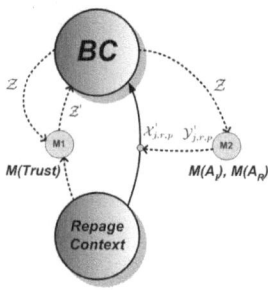

Fig. 4. The metalevel specification. Dot lines specify the metalevel reasoning. Notice that all them come from the belief context.

5   Experimentation

In this section we propose a concrete solution for the $M1$ process, to update the trust axiom represented by the metarule

$$B_i(pr(S\varphi \rightarrow \varphi) = \mathcal{Z})$$

As previously shown, this axiom plays a crucial role in the relationship between image and reputation predicates. Different values of the degree of freedom produce a typology of agents. When $\mathcal{Z}$ is 0, the agent only takes into account image information. When $\mathcal{Z}$ is 1, reputation information is as valuable as image information in terms of the impact that the information has in the mind of the agent. Previous work on cognitive theories and simulation of image and reputation dynamics [16,7] reveals that the amount of reputation information that circulates in a society is a lot higher than image-based information, due to the implicit commitment that sending image information carries out.

However, even when reputation information is mostly inaccurate, open societies perform *better* when reputation information is allowed in the system, and also are more robust with respect to certain level of cheating information[3]. This indicates that agents face mostly inaccurate information but that they need to use it to face real uncertain and unpredictable scenarios.

These studies are very helpful when defining a process to decide $\mathcal{Z}$. Our trust axiom is in fact a predicate that indicates how much information that circulates in the society can be considered true. In the way we have defined rules A_I and A_R, settings of $\mathcal{Z}$ tending to 0 could be useful when the number of cheaters is considerably big, meanwhile settings of $\mathcal{Z}$ close to 1 would be helpful in the opposite way.

5.1   Scenario and Simulation Settings

We replicate a simple market where in the society we have a set of buyer, seller and informant agents. In this scenario, all sellers offer the same product, which has a certain quality going from 0 (minimum) to 100 (maximum). Also, a delivery time expressed in weeks is associated with the seller. These agents are completely reactive and sell the products on demand. Buyers are BDI agents following the model described in this paper. Therefore, the main goals of the agents are described in terms of graded desires.

The set of informant agents send out reputation information about the sellers. We control the experiment by setting a percentage of informants that spread *bad* reputation, the number of sellers and the distribution of qualities and delivery times.

The performance of the buyers is evaluated by the level of satisfaction obtained after their decision. As we mentioned, buyer agents state their preferences by a set of graded desires. These desires can be positive or negative, and each one of them has a grade. After an action is performed, the agent receives the fulfillment of the interaction, obtaining the real quality of the product and the delivery time. This information is compared with the objectives of the agent. The *level of satisfaction* of the agent is calculated by summing the grades of positive achieved desires and subtracting the grades of the achieved negative desires.

At each turn buyers need to perform an action. In this case, they need to buy a product to some of the available sellers. To simulate the fact that reputation information is more present than image information, at each turn all informants send reputation information

---

[3] More that 50% of cheaters in a society still produces a benefit in the overall performance when reputation is allowed.

to the buyers. In this experiments we do not consider image communications. Therefore, image information is only calculated through direct experience. In this sense, at each turn one direct experience is contrasted with N reputation communications from the informants (where $N > 1$).

In the specification, we are considering the evolution of a single buyer with 10 sellers and 5 informants. We execute 10 times each experiment and consider the average level of satisfaction for each turn. We state a distribution of qualities and delivery times in such a way that the best qualities and best delivery times are very scarce. If these properties are the norm, the society does not need the exchange of information, since a random choice from the buyer would get already a very good seller[4]

5.2 Static Experiments

It is easy to show the effects of a fix trust axiom in different situations. Figure 5 shows the accumulated average level of satisfaction obtained by a buyer at each turn in an environment where all informants are honest, and when all informants are liars, considering $Z = 0$ and $Z = 1$. Since when $Z = 0$ reputation information is not taken into account, the performance in this case does not depend on the quality of the reputation information.

The graphic shows that when $Z = 1$, in the case of a scenario with honest informants, the level of satisfaction obtained by the agent considerable increases with respect to the case in which $Z = 0$. Assuming normality in the data, from the turn 10, the difference is already statistically significant with a 95% of confidence (p_value $\leq$ 0.05), and from the turn 20 on, the difference becomes significant with a 99% of confidence (p_value $\leq$ 0.01).

Also, when $Z = 1$ and in the scenario all informants spread false reputation, the performance of the buyer decreases considerably with respect to the case in which $Z = 0$. In fact, from the very first turns, the difference becomes already significant with a confidence of 99%.

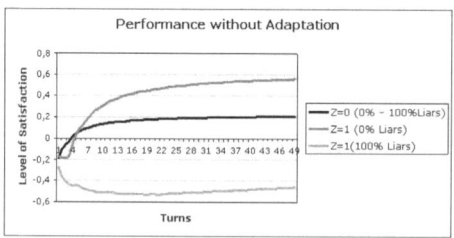

Fig. 5. Level of satisfaction obtained after

---

[4] We use the JASON platform [17], which offers to logic-based agents (prolog-like) a multia-gent communication layer. The source code, together with the exact parameters and the set of desires used to run the experiments can be found at
http://www.iiia.csic.es/~ipinyol/sourceABSS09.zip

These results are quite obvious. Since image information is only created from direct experiences (1 at each turn) and reputation information through communicated reputation (5 at each turn) if the communicated information corresponds to the reality and the agent believes what circulates in the society ($Z = 1$) the buyer should discover faster which are the sellers that accomplish her objectives. As well, if reputation information if mostly false, and the agent believes it, for a long time the buyer would not be able to fulfill her objectives.

5.3 Dynamic Adaptation Experiments

The main idea behind the updating of Z is that in scenarios where mostly false reputation information circulates Z should be tend to 0. On the contrary, scenarios where reputation information is mostly accurate, Z should tend to 1. In this very preliminary paper, we study the effects of an adaptation strategy in the same situations tested in the previous extreme experiments.

The strategy is very simple, but effective. As described in the theoretical part of the paper, metarules can be updated from the beliefs that the agent hold. In our case, we theorize that a good metaprocess for updating Z is aggregate the differences for each agent and role of the image and reputation information hold in the Repage system. So, if most of the image information coincide with reputation information (about the same agent/role), the Z value should increase from the current value (in certain proportion). On the contrary, it should decrease. This algorithm contains the parameter *Increment*, which could be also considered as another degree of freedom.for the sake of simplicity we consider it as a constant value.

Figure 6 shows the performance obtained in both scenarios. It can be observed how the final performance tends to the theoretical optimum in each situation. In both scenarios there is no statistical significant difference between the performance and the theoretical optimum, with p_values higher than 0.2 with most of the points of the graph.

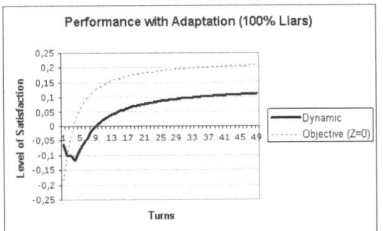 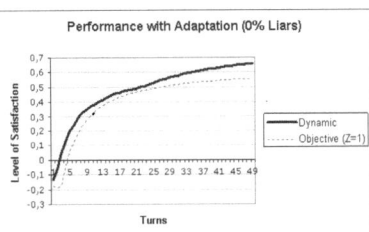

Fig. 6. Level of satisfaction obtained with agents using adaptation in a scenario with 100% of liars (left) and 0% of liars (right). Dot line represents the theoretical best possible performance.

The adaptation process can be clearly observed with the performance of a single execution. Figure 7 shows a typical pattern (usually the period where the level of satisfaction is so low is much shorted. For this reason the final average of 10 executions does not show it) in which after a while, the agent is able increase her level of goal achievement.

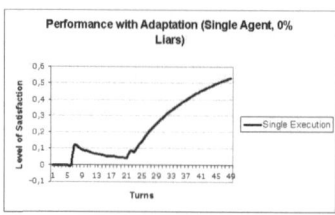

Fig. 7. Performance of a single agent with adaptation in a scenario without liars

6   Conclusions and Future Work

After reading the paper it should be clear the importance of allowing degrees of freedom in the reasoning processes of autonomous agents. As mentioned in the cognitive theory presented by Castelfranchi and Paglieri in [4], cognitive agents are aware of the *reasons* from which certain information if believed, and because of that, they are able to reason about how they reason, and change it if necessary. Thus, we strongly believe that real autonomous agents should be designed taking into account certain degree of granularity. Cognitive designs should be aware that the path that an agent follows to arrive at certain conclusion is as important as the conclusion itself. Therefore, ways to reason about such paths and the capability to modify them should be taken into account, not only for the agent itself, but also for possible explanations to other agents, like in argumentation.

We also encourage the use of logical approaches in the design of cognitive systems. The advantage of such systems is that with a finite set of rules a whole deduction tree can be created, implicitly providing supporting sets. This big advantage has an important counter effect: the static nature of the axiomatization. At a metareasoning level though, similar to belief revision process, certain set of axioms (those which define typology of agents, not that structurally guarantee certain logical properties) can me updated, changing then the whole reasoning tree. We show a possible method to do it in this paper. However, this needs a deeper study in the future.

Getting into the concrete scenario that we faced, it should also be clear that reputation and image information can totally participate in metareasoning processes. We proposed a method in which bridge rules and axioms can be specified as metarules following the idea of degrees of freedom introduced in [15]. We let for future work the proper formalities of the proposed design method. Also, regarding the actual design of BDI+Repage model we plan to provide alternative metaprocesses to update the trust predicate and the relation with the other metarules $M(A_I)$ and $M(A_R)$.

Acknowledgments

This work was supported by the projects AEI (TIN2006-15662-C02-01), AT (CONSOLIDER CSD2007-0022, INGENIO 2010), LiquidPub (STREP FP7-213360), RepBDI (Intramural 2008501136) and by the Generalitat de Catalunya under the grant 2005-SGR-00093.

References

1. Luck, M., McBurney, P., Shehory, O., Willmott, S.: Agent Technology: Computing as Interaction (A Roadmap for Agent Based Computing). AgentLink (2005)
2. Sabater-Mir, J., Paolucci, M., Conte, R.: Repage: Reputation and image among limited autonomous partners. JASSS 9(2) (2006)
3. Pinyol, I., Sabater-Mir, J.: Pragmatic-strategic reputation-based decisions in BDI agents. In: Proc. of the AAMAS 2009, Budapest, Hungary, pp. 1001–1008 (2009)
4. Castelfranchi, C., Paglieri, F.: The role of beliefs in goal dynamics: Prolegomena to a constructive theory of intentions. Synthese 155, 237–263 (2007)
5. Paglieri, F.: Belief dynamics: From formal models to cognitive architectures, and back again. PhD dissertation (2006)
6. Prakken, H.: Formal systems for persuasion dialogue. Knowl. Eng. Rev. 21(2), 163–188 (2006)
7. Conte, R., Paolucci, M.: Reputation in artificial societies: Social beliefs for social order. Kluwer Academic Publishers, Dordrecht (2002)
8. Giunchiglia, F., Serafini, L.: Multilanguage hierarchical logic (or: How we can do without modal logics). Journal of AI 65, 29–70 (1994)
9. Parsons, S., Sierra, C., Jennings, N.: Agents that reason and negotiate by arguing. Journal of Logic and Computation 8(3), 261–292 (1998)
10. Casali, A., Godo, L., Sierra, C.: Graded BDI models for agent architectures. In: Leite, J., Torroni, P. (eds.) CLIMA 2004. LNCS (LNAI), vol. 3487, pp. 126–143. Springer, Heidelberg (2005)
11. Pinyol, I., Sabater-Mir, J., Dellunde, P.: Probabilistic dynamic belief logic for image and reputation. In: Proc. of the CCIA 2008, Empuries, Spain (2008)
12. Fagin, R., Halpern, J.: Reasoning about knowledge and probability. J. ACM 41(2), 340–367 (1994)
13. Demolombe, R.: To trust information sources: a proposal for a modal logical framework. Trust and deception in virtual societies, 111–124 (2001)
14. Rao, A.S., Georgeff, M.P.: Modeling rational agents within a BDI-architecture. In: Allen, J., Fikes, R., Sandewall, E. (eds.) Proc. of KR 1991, pp. 473–484. Morgan Kaufmann Publishers Inc., San Francisco (1991)
15. Artikis, A.: Dynamic protocols for open agent systems. In: Proc. of the AAMAS 2009, Budapest, Hungary, pp. 97–104 (2009)
16. Pinyol, I., Paolucci, M., Sabater-Mir, J., Conte, R.: Beyond accuracy. Reputation for partner selection with lies and retaliation. In: Antunes, L., Paolucci, M., Norling, E. (eds.) MABS 2007. LNCS (LNAI), vol. 5003, pp. 128–140. Springer, Heidelberg (2008)
17. Bordini, R.H., Hübner, J.F., Wooldridge, M.: Programming Multi-Agent Systems in AgentSpeak Using Jason. John Wiley and Sons, Ltd., Chichester (2007)

Experiments on the Acquisition of the Semantics and Grammatical Constructions Required for Communicating Propositional Logic Sentences

Josefina Sierra[1,*] and Josefina Santibáñez[2]

[1] Software Department, Technical University of Catalonia, Spain
jsierra@lsi.upc.edu
[2] Education Department, University of La Rioja, Spain
josefina.santibanez@unirioja.es

Abstract. We describe some experiments which simulate a grounded approach to language acquisition in which a population of autonomous agents without prior linguistic knowledge tries to construct at the same time a conceptualisation of its environment and a shared language. The conceptualisation and language acquisition processes in each individual agent are based on general purpose cognitive capacities, such as categorisation, discrimination, evaluation and induction. The emergence of a shared language in the population results from a process of self-organisation of a particular type of linguistic interaction which takes place among the agents in the population.

The experiments, which extend previous work by addressing the problem of the acquisition of *both the semantics and the syntax of propositional logic,* show that at the end of the simulation runs the agents build different conceptualisations and different grammars. However, these conceptualisations and grammars are compatible enough to guarantee the unambiguous communication of propositional logic sentences.

Furthermore the categorisers of the perceptually grounded and logical categories built during the conceptualisation and language acquisition processes can be used for some forms of common sense reasoning, such as determining whether a sentence is a tautology, a contradiction, a common sense axiom or a merely satisfiable formula.

Keywords: Language acquisition, logical categories, induction, self-organisation.

1   Introduction

This paper addresses the problem of the acquisition of both *the semantics and the syntax* (i.e., lexicon and grammatical constructions) required for constructing and communicating concepts of the same complexity as propositional logic formulas. It describes some experiments in which a population of autonomous

---

* Partially supported by the MICINN SESAAME-BAR (TIN2008-06582-C03-01) and DGICYT MOISES-BAR (TIN2005-08832-C03-03) projects.

A.V. Vasilakos et al. (Eds.): AUTONOMICS 2009, LNICST 23, pp. 236–251, 2010.

agents without prior linguistic knowledge constructs at the same time a conceptualisation of its environment and a shared language. The experiments show that at the end of the simulation runs the agents build different conceptualisations and different grammars. However these conceptualisations and grammars are compatible enough to guarantee the unambiguous communication of meanings of the same complexity as propositional logic formulas.

The research presented in this paper builds up on previous work on the acquisition of *the semantics of logical connectives* [1] by addressing the problem of the acquisition of both the semantics and the syntax of propositional logic. In [1] a grounded approach to the acquisition of logical categories (i.e., the semantics of logical connectives) based on the discrimination of a "subset of objects" from the rest of the objects in a given context is described. Logical categories are constructed by the agents identifying subsets of the range of the truth evaluation process (i.e., sets of Boolean pairs or Boolean values) which result from evaluating a pair of perceptually grounded categories or a single category on a subset of objects. Discrimination is performed characterising a "subset of objects" by a logical formula constructed from perceptually grounded categories which is satisfied by the objects in the subset and not satisfied by the rest of the objects in the context.

The complementary problem of the acquisition of *the syntax of propositional logic* by a population of autonomous agents without prior linguistic knowledge has been addressed independently as well. In [2] an approach to the acquisition of the syntax of propositional logic based on *general purpose cognitive capacities,* such as invention, adoption and induction, and on *self-organisation principles* is proposed. The experiments described in [2] show that a shared language (i.e., a lexicon and a grammar) expressive enough to allow the communication of meanings of the same complexity as propositional logic formulas can emerge in a population of autonomous agents without prior linguistic knowledge. This shared language, although simple, has some interesting properties found in natural languages such as recursion, syntactic categories for propositional sentences and connectives, and partial word order for marking the scope of each connective.

The acquisition of the syntax of subsets of logic has been addressed as well by other authors. In particular [3,4,5] study the emergence of case-based and recursive communication systems in populations of agents without prior linguistic knowledge. However none of these works deals with the problem of the acquisition of both the semantics and the syntax of logic.

The experiments described in this paper extend therefore previous work by using a linguistic interaction (*the evaluation game*) in which the agents must first conceptualise the *topic* (a subset of objects) using the mechanisms proposed in [1] for the acquisition of logical categories, and then construct a *shared language* (a lexicon and a grammar) using the invention, adoption, induction and self-organisation mechanisms proposed in [2].

The rest of the paper is organised as follows. Firstly we describe the mechanisms the agents use in order to conceptualise sensory information. Secondly we consider the process of truth evaluation and explain how logical categories can be discovered by identifying sets of outcomes of the truth evaluation process. Then we focus on the construction and emergence of a shared communication language describing the main steps of *the evaluation game:* conceptualisation, verbalising, interpretation, induction and coordination. Next we present the results of some experiments in which three agents without prior linguistic knowledge build a conceptualisation and a shared language that allows them to construct and communicate meanings of the same complexity as propositional logic formulas. Finally we consider the issue of common sense reasoning and summarise the main ideas we tried to put forward in this paper.

2   Conceptualisation: Basic Definitions

We assume an experimental setting similar to that proposed in *The Talking Heads Experiment* [6]: A set of robotic agents playing language games with each other about scenes perceived through their cameras on a white board in front of them. Figure 1 shows a typical configuration of the white board with several geometric figures pasted on it.

Firstly we describe how the agents conceptualise the sensory information they obtain by looking at the white board and trying to characterise subsets of objects pasted on it.

Sensory Channels. The agents look at one area of the white board by capturing an image of that area with their cameras. They segment the image into coherent units in order to identify the objects which constitute the context of a language game, and use some *sensory channels* to gather information about each segment, such as its horizontal and vertical position, or its light intensity. In the experiments described in this paper we only use three sensory channels: (1) H(o), which computes the horizontal position of an object o; (2) V(o), which computes its vertical position; and (3) L(o), which computes its light intensity. The values returned by the sensory channels are scaled with respect to the area of the white board captured by the agents cameras so that its range is the interval (0.0 1.0).

Perceptually Grounded Categories. The data returned by the sensory channels are values from a continuous domain. To be the basis of a natural language conceptualisation, these values must be transformed into a discrete domain. One form of categorisation consists in dividing up each domain of output values of a particular sensory channel into regions and assigning a *category* to each region [6]. For example, the range of the H channel can be cut into two halves leading to the categories [left] $(0.0 < H(x) < 0.5)$ and [right] $(0.5 < H(x) < 1.0)$. Object 3 in figure 1 has the value H(O3)=0.8 and would therefore be categorised as [right].

Perceptually Grounded Categorisers. At the same time the agents build categories in order to conceptualise sensory information, they construct as well cognitive procedures (called *categorisers*) which allow them to check whether these categories hold or not for a given object.

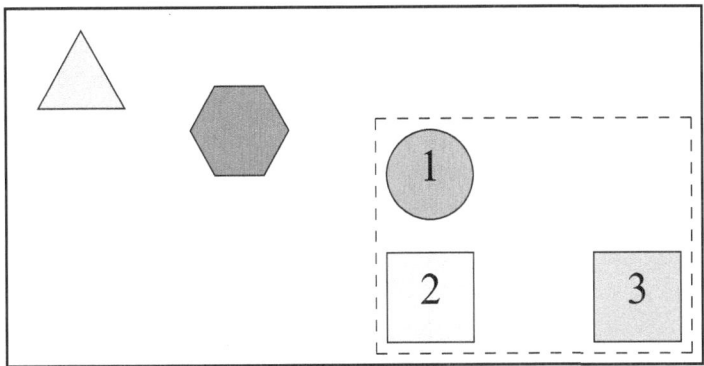

Fig. 1. The area of the white board captured by the agents cameras (i.e., the context of the language game) is the lower right rectangle

Categorisers give grounded meanings [7] to categories (i.e., symbolic representations) by establishing explicit connections between them and reality (external input processed by sensory channels). These connections are learned playing language games [8,6] and allow the agents to check whether a category holds or not for a given object. Most importantly they provide information on the sensory and cognitive processes an agent must go through in order to evaluate a given category.

The behaviour of the categorisers associated with the perceptually grounded categories used in this paper can be described by linear constraints[1]. For example, the behaviour of the categoriser associated with the category [left] can be described as follows: $[left]^C(x) \equiv 0.0 < H(x) < 0.5$.

2.1 Logical Categories

We consider now the process of truth evaluation and describe how logical categories can be constructed by identifying sets of outcomes of the truth evaluation process. Logical categories are important because they allow the generation of structured units of meaning, which correspond to logical formulas, and they set the basis for deductive reasoning.

Evaluation Channel. The *evaluation channel* (denoted by E) is a cognitive process capable of finding the categorisers of a tuple of categories, applying them to an object, and observing their output. If $\boldsymbol{c} = (c_1, \ldots, c_n)$ is a category tuple and o is an object, $E(\boldsymbol{c}, o)$ is a tuple of Boolean values $(v_1, \ldots, v_n)$, where each v_i is the result of applying c_i^C (the categoriser of c_i) to object o. For example, $E(([down], [right]), O1) = (0, 0)$, because $O1$ (object 1 in figure 1) is neither on the lower part nor on the right part of the white board area captured by the agents' cameras.

---

[1] We use the notation $[cat]^C$ to refer to the categoriser that is capable of determining whether category [cat] holds or not for a given object.

Logical Categories and Formulas. Although the evaluation channel can be applied to category tuples of any arity, we consider only unary and binary category tuples. The range of the evaluation channel for single categories is the set of Boolean values $\{0, 1\}$, and its range for category pairs is the set of Boolean pairs $\{(0,0), (0,1), (1,0), (1,1)\}$. By considering all the subsets of these ranges the agents can represent all the Boolean functions of one and two arguments, which correspond to the meanings of all the connectives of propositional logic (i.e., $\neg, \wedge, \vee, \rightarrow$ and $\leftrightarrow$), plus the meanings of other connectives (such as *neither* or *exclusive disjunction*) found in natural languages. For example, the propositional formula $c_1 \vee c_2$ is true for an object o if the result of evaluating the pair of categories (c_1, c_2) on object o is a Boolean pair which belongs to the subset of Boolean pairs $\{(1, 1), (1, 0), (0, 1)\}$.

The sixteen Boolean functions of two arguments which can be constructed using this method are summarised in the following ten connectives in the internal representation of logical categories used by the agents: *and, nand, or, nor, if, nif, oif, noif, iff* and *xor*. Where *and, or, if* and *iff* have the standard interpretation ($\wedge, \vee, \rightarrow$ and $\leftrightarrow$), and the formulas *(A nand B), (A nor B), (A nif B),* (A oif B), (A noif B) and *(A xor B)* are equivalent to $\neg(A \wedge B)$, $\neg(A \vee B)$, $\neg(A \rightarrow B)$, $(B \rightarrow A)$, $\neg(B \rightarrow A)$ and $\neg(A \leftrightarrow B)$, respectively.

The agents construct *logical categories* by identifying subsets of the range of the evaluation channel. The *evaluation game* creates situations in which the agents discover subsets of the range of the evaluation channel, and use them to distinguish a subset of objects from the rest of the objects in a given context. The representation of logical categories as subsets of Boolean tuples is equivalent to the *truth tables* used for defining the semantics of logical connectives.

Logical categories describe properties of propositions, therefore it is natural to apply them to perceptually grounded categories in order to construct structured units of meaning. For example, the formula [not, down] can be constructed by applying the logical category [not] (i.e., $\neg$) to the category [down]. The formula [or, up, right] can be constructed similarly by applying the logical category [or] to the categories [up] and [right][2].

If we consider perceptually grounded categories as propositions, we can observe that the set of concepts that can be constructed by the agents corresponds to the set of formulas of propositional logic, because: (1) a perceptually grounded category is a formula; and (2) if l is an n-ary logical category and F is a list (tuple) of n formulas, then $[l|F]$ is a formula[3].

Logical Categorisers. The categorisers of logical categories are cognitive processes that allow determining whether a logical category holds or not for a tuple of categories and an object. As we have explained above, logical categories can be associated with subsets of the range of the evaluation channel. The behaviour of their categorisers can be described therefore by constraints of the form

---

[2] Notice that we use prefix, Lisp like notation for representing propositional formulas. Thus the list [or, up, right] corresponds to the formula $up \vee right$.

[3] Where l is a logical category, F is a list of formulas and | is the standard list construction operator.

$E(\mathbf{c}, o) \in S_l$, where l is a logical category, S_l is the subset of the range of the evaluation channel for which l holds, E is the evaluation channel, $\mathbf{c}$ is a tuple of categories, and o is an object. For example, the constraint $E((c1, c2), o) \in \{(1, 1)\}$ describes the behaviour of the categoriser of the logical category [and] (i.e., $c1 \wedge c2$).

The evaluation channel can be naturally extended to evaluate arbitrary propositional logic formulas using the categorisers of logical and perceptually grounded categories. The following is an inductive definition of the evaluation channel $E(A, o)$ for an arbitrary formula A of propositional logic:

1. If A is a perceptually grounded category [cat], then $E(A, o) = [cat]^C(o)$.
2. If A is a propositional formula of the form $[l|F]$, where l is a logical category, F is a list of formulas and S_l is the subset of the range of the evaluation channel for which l holds, then $E(A, o) = 1$ if $E(F, o) \in S_l$, and 0 otherwise.

3   Language Acquisition

Language acquisition is seen as a collective process by which a population of autonomous agents without prior linguistic knowledge constructs a *shared language* which allows them to communicate some set of meanings. In order to reach such an agreement the agents interact with each other playing language games. In a typical experiment thousands of language games are played by pairs of agents randomly chosen from a population.

In this paper we use a particular type of language game called the **evaluation game** [2]. The goal of the experiments is to observe the evolution of: (1) the communicative success[4]; (2) the internal grammars constructed by the individual agents; and (3) the external language used by the population. The main steps of *the evaluation game*, which is played by two agents (a *speaker* and a *hearer*), can be summarised as follows.

1. Conceptualisation. Firstly the speaker looks at one area of the white board and directs the attention of the hearer to the same area. The objects in that area constitute *the context* of the language game. Both speaker and hearer use their sensory channels to gather information about each object in the context and store that information so that they can use it in subsequent stages of the game. Then the speaker picks up a subset of the objects in the context which we will call *the topic* of the language game. The rest of the objects in the context constitute *the background*.

The speaker tries to find a unary or binary tuple of categories which distinguishes the topic from the background, i.e., a tuple of categories such that its evaluation on the topic is different from its evaluation on any object in the background. If the speaker cannot find a discriminating tuple of categories, the game fails. Otherwise it tries to find a logical category that is associated with the subset

---

[4] The *communicative success* is the average of successful language games in the last ten language games played by the agents.

of Boolean values or Boolean pairs resulting from evaluating the topic on that category tuple. If it does not have any logical category associated with this subset, it creates a new one. The formula constructed by applying this logical category to the discriminating category tuple constitutes a *conceptualisation* of the topic, because it *characterises the topic as the set of objects in the context which satisfy that formula.*

In general an agent can build several conceptualisations for the same topic. For example, if the context contains objects 1, 2 and 3 in figure 1, and the topic is the subset consisting of objects 1 and 2, the formulas [iff, up, left] and [xor, up, right] could be used as conceptualisations of the topic in an evaluation game.

2. Verbalising. The speaker chooses a conceptualisation (i.e., a discriminating formula) for the topic, generates a sentence that expresses this formula and communicates that sentence to the hearer. If the speaker can generate sentences for several conceptualisations of the topic, it tries to maximise the probability of being understood by other agents selecting the conceptualisation whose associated sentence has the highest score. The algorithm for computing the score of a sentence from the scores of the grammar rules used in its generation is explained in detail in [2].

The agents in the population start with an empty lexicon and grammar. Therefore they cannot generate sentences for most formulas at the early stages of a simulation run. In order to allow language to get off the ground, the agents are allowed to invent new sentences for those meanings they cannot express using their lexicon and grammar. As the agents play language games they learn associations between expressions and meanings, and induce linguistic knowledge from such associations in the form of grammar rules and lexical entries. Once the agents can generate sentences for expressing a particular formula, they select the sentence with the highest score that verbalises a conceptualisation of the topic, and communicate that sentence to the hearer.

3. Interpretation. The hearer tries to interpret the sentence communicated by the speaker. If it can parse the sentence using its lexicon and grammar, it extracts a formula (a meaning) and uses that formula to identify the topic.

At the early stages of a simulation run the hearers cannot usually parse the sentences communicated by the speakers, since they have no prior linguistic knowledge. In this case the speaker points to the topic, the hearer conceptualises the topic using a logical formula, and adopts an association between that formula and the sentence used by the speaker. Notice that the conceptualisations of speaker and hearer may be different, because different formulas can be used to conceptualise the same topic.

At later stages of a simulation run it usually happens that the grammars and lexicons of speaker and hearer are not consistent, because each agent constructs its own grammar from the linguistic interactions it participates in, and it is very unlikely that speaker and hearer share the same history of linguistic interactions unless the population consists only of these two agents. In this case the hearer may be able to parse the sentence communicated by the speaker, but its interpretation of that sentence might be different from the meaning the speaker had in mind. The strategy used to coordinate the grammars of speaker and hearer when this happens is to decrease the score of the rules used by the speaker and the hearer in the processes of generation and parsing, respectively, and allow the hearer to adopt an association between its conceptualisation of the topic and the sentence used by the speaker.

Induction. Besides inventing expressions and adopting associations between sentences and meanings, the agents can use some *induction mechanisms* to extract generalisations from the grammar rules they have learnt so far. The induction mechanisms used in this paper are based on the rules of *simplification and chunk* in [5], although we have extended them so that they can be applied to grammar rules which have scores attached to them following the ideas of [9]. The induction rules are applied whenever the agents invent or adopt a new association to avoid redundancy and increase generality in their grammars.

Instead of giving a formal definition of the induction rules used in the experiments, which can be found in [2], we give an example of their application. We use Definite Clause Grammar for representing the internal grammars constructed by the individual agents. Non-terminals have two arguments attached to them. The first argument conveys semantic information and the second is a *score* in the interval $[0, 1]$ which estimates the usefulness of the grammar rule in previous communication. Suppose an agent's grammar contains the following rules.

$$s(light, S) \rightarrow clair, \{S \ is \ 0.70\} \tag{1}$$

$$s(right, S) \rightarrow droit, \{S \ is \ 0.25\} \tag{2}$$

$$s([and, light, right], S) \rightarrow etclairdroit, \{S \ is \ 0.01\} \tag{3}$$

$$s([or, light, right], S) \rightarrow ouclairdroit, \{S \ is \ 0.01\} \tag{4}$$

The induction rule of **simplification**, applied to 3 and 2, allows generalising grammar rule 3 replacing it with 5. In this case *simplification* assumes that the second argument of the logical category *and* can be any meaning which can be expressed by a 'sentence', because according to rule 2 the syntactic category of the expression 'droit' is s (i.e., sentence).

$$s([and,light,B], S) \rightarrow etclair, s(B,R), \{S \ is \ R{\cdot}0.01\} \tag{5}$$

Simplification, applied to rules 5 and 1, can be used to generalise rule 5 again replacing it with 6. Rule 4 can be generalised as well replacing it with rule 7.

$$s([and,A,B], S) \rightarrow et, s(A,Q), s(B,R), \{S \ is \ Q{\cdot}R{\cdot}0.01\} \tag{6}$$

$$s([or,A,B], S) \rightarrow ou, s(A,Q), s(B,R), \{S \ is \ Q{\cdot}R{\cdot}0.01\} \tag{7}$$

The induction rule of **chunk** replaces a pair of grammar rules such as 6 and 7 by a single rule 8 which is more general, because it makes abstraction of their common structure introducing a syntactic category $c2$ for binary connectives. Rules 9 and 10 state that the expressions *et* and *ou* belong to the syntactic category $c2$.

$$s([C,A,B], S) \rightarrow c2(C,P), s(A,Q), s(B,R), \{S \ is \ P{\cdot}Q{\cdot}R{\cdot}0.01\} \tag{8}$$

$$c2(and, S) \rightarrow et, \{S \ is \ 0.01\} \tag{9}$$

$$c2(or, S) \rightarrow ou, \{S \ is \ 0.01\} \tag{10}$$

4. Coordination. The speaker points to the topic so that the hearer can identify the subset of objects it had in mind, and the hearer communicates the outcome of the evaluation game to the speaker. The game is successful if the hearer can parse the sentence communicated by the speaker, and its interpretation of that sentence identifies the topic (the subset of objects the speaker had in mind) correctly. Otherwise the evaluation game fails. Depending on the outcome of the evaluation game, speaker and hearer take different actions. We have explained some of them already (*invention* and *adoption*), but they can *adapt their grammars* as well adjusting the scores of their grammar rules in order to communicate more successfully in future language games.

Coordination of the agents' grammars is necessary, because different agents can invent different expressions to refer to the same perceptually grounded or logical categories, and because the invention process uses random order to concatenate the expressions associated with the components of a given formula. In order to understand each other, the agents must use a common vocabulary and must order the constituents of compound sentences in sufficiently similar ways as to avoid ambiguous interpretations.

The following **self-organisation mechanisms** help to coordinate the agents' grammars in such a way that they prefer using the grammar rules which are used more often by other agents [6,4].

We consider the case in which the speaker can generate a sentence to express the formula it has chosen as its conceptualisation of the topic using the rules in its grammar. If the speaker can generate several sentences to express that formula, it chooses the sentence with the highest score. The rest of the sentences are called *competing sentences*.

Suppose the hearer can interpret the sentence communicated by the speaker. If the hearer can obtain several formulas (meanings) for that sentence, the meaning with the highest score is selected. The rest of the meanings are called *competing meanings*.

If the topic identified by the hearer is the subset of objects the speaker had in mind, the evaluation game succeeds and both agents adjust the scores of the rules in their grammars. The speaker increases the scores of the grammar rules it used for generating the sentence communicated to the hearer and decreases the scores of the grammar rules it used for generating competing sentences. The hearer increases the scores of the grammar rules it used for obtaining the meaning which identified the topic the speaker had in mind and decreases the scores of the rules it used for obtaining competing meanings. This way the grammar rules that have been used successfully get reinforced, and the rules that have been used for generating competing sentences or competing meanings are inhibited.

If the topic identified by the hearer is different from the subset of objects the speaker had in mind, the evaluation game fails and both agents decrease the scores of the grammar rules they used for generating and interpreting the sentence used by the speaker, respectively. This way the grammar rules that have been used without success are inhibited.

The scores of grammar rules are *updated* as follows. The rule's original score S is replaced with the result of evaluating expression 11 if the score is *increased*, and expression 12 if the score is *decreased*.

$$minimum(1, \ S + 0.1) \tag{11}$$

$$maximum(0, \ S - 0.1) \tag{12}$$

4   Experiments

We describe the results of some experiments in which three agents try to construct at the same time a conceptualisation and a shared language which allow them to discriminate and communicate about subsets of objects pasted on a white board in front of them. In particular, the agents characterise such subsets of objects constructing logical formulas which are true for the objects in the subset and false for the rest of the objects in the context. Such formulas, which are communicated using a shared language, express facts about the relative spatial location and brightness of the objects in the subset with respect to the rest of the objects in the context. These experiments have been implemented using the Ciao Prolog system [10].

Figure 2 shows the evolution of the communicative success for a population of three agents. The *communicative success* is the average of successful language games in the last ten language games played by the agents. Firstly the agents play 700 evaluation games about subsets of objects which can be discriminated using only a single category or the negation of a perceptually grounded category. In this part of the simulation the population reaches a communicative success of 94% after playing 100 games and keeps it over that figure till the end of this part of the simulation. Next the agents play 6000 evaluation games about subsets of objects which require the use of perceptually grounded categories as well as unary and binary logical categories for their discrimination. In this part of the simulation the population reaches a communicative success of 100% after playing 3600 evaluation games and keeps it till the end of the second part of the simulation. The data shown in the figure correspond to the average of ten independent simulation runs with different random seeds.

We analyse now the conceptualisations and grammars built by the agents at the end of a particular simulation run. As we shall see the conceptualisations and grammars constructed by the individual agents are different, however they are compatible enough to guarantee the unambiguous communication of meanings of the same complexity as propositional logic formulas.

Table 1 shows the lexicon constructed by each agent in order to refer to perceptually grounded categories. We can observe that all the agents constructed the perceptually grounded categories (*up, down, right, left, light* and *dark*) and that all of them prefer the same expressions for referring to such categories.

We can observe in table 2 that all the agents constructed the logical category *not*. They all have a recursive grammar rule for expressing formulas constructed using negation and they use the same expression (ci) for referring to the logical

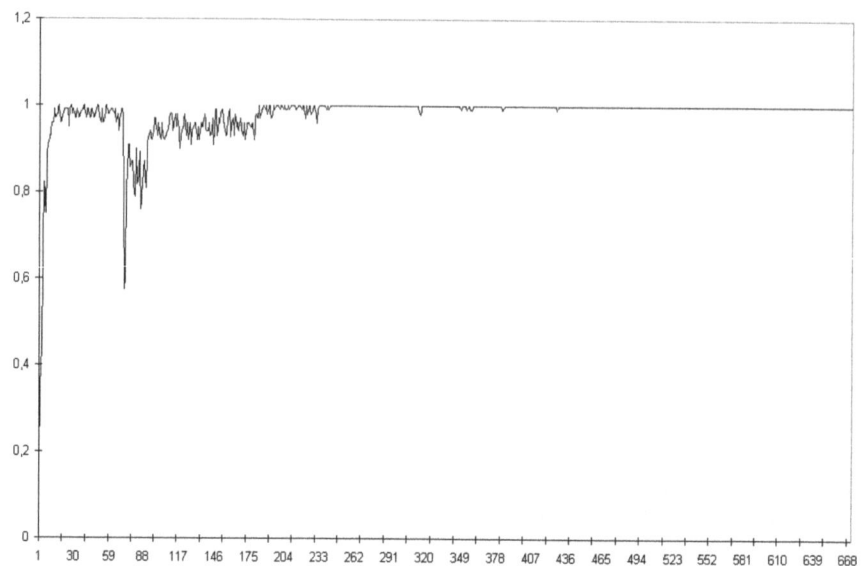

Fig. 2. Evolution of the communicative success for a population of three agents. Firstly the agents play 700 evaluation games which only require the use of perceptually grounded categories and negation for discrimination. Then they play 6000 evaluation games which require the use of all the perceptually grounded and logical categories for discrimination.

Table 1. Lexicon built by each agent to refer to perceptually grounded categories

Lexicon a1	Lexicon a2	Lexicon a3
s(up,1) → n	s(up,1) → n	s(up,1) → n
s(down,1) → b	s(down,1) → b	s(down,1) → b
s(right,1) → w	s(right,1) → w	s(right,1) → w
s(left,1) → dgq	s(left,1) → dgq	s(left,1) → dgq
s(light,1) → fdy	s(light,1) → fdy	s(light,1) → fdy
s(dark,1) → qyp	s(dark,1) → qyp	s(dark,1) → qyp

category *not*. There is a difference however: Agents a2 and a3 use a generic grammar rule based on a syntactic category for unary connectives, whereas agent a1 uses a specific grammar rule for expressing formulas constructed using negation.

We can also see in table 2 that all the agents constructed logical categories for all the **commutative connectives** (*and, nand, or, nor, xor* and *iff*), and that they use the same expressions (*ybd, d, j, sbr, wg* and *q*, respectively) for referring to such connectives.

Although in this particular simulation run all the agents use the same type of grammatical constructions to express formulas constructed using commutative connectives, this is not always the case. In a different simulation run agent a1

Table 2. Grammars built by the individual agents, including grammatical constructions, syntactic categories and lexicons for logical categories

Grammar a1
s([not,X],Q) $\rightarrow$ ci, s(X,P), {Q is P · 1}
s([X,Y,Z],T) $\rightarrow$ c1(X,P), s(Y,Q), s(Z,R), {T is P · Q · R · 1}
c1(and,R) $\rightarrow$ ybd, {R is 1}
c1(nor,R) $\rightarrow$ sbr, {R is 1}
c1(xor,R) $\rightarrow$ wg, {R is 1}
c1(iff,R) $\rightarrow$ q, {R is 1}
c1(if,R) $\rightarrow$ jdgq, {R is 1}
c1(or,R) $\rightarrow$ j, {R is 1}
s([X,Y,Z],T) $\rightarrow$ c2(X,P), s(Z,Q), s(Y,R), {T is P · Q · R · 1}
c2(noif,R) $\rightarrow$ oi, {R is 1}
c2(nand,R) $\rightarrow$ d, {R is 1}

Grammar a2
s([X,Y],R) $\rightarrow$ c1(X,P), s(Y,Q), {R is P · Q · 1}
c1(not,R) $\rightarrow$ ci, {R is 1}
s([X,Y,Z],T) $\rightarrow$ c2(X,P), s(Y,Q), s(Z,R), {T is P · Q · R · 1}
c2(nif,R) $\rightarrow$ oi, {R is 1}
c2(and,R) $\rightarrow$ ybd, {R is 1}
c2(nor,R) $\rightarrow$ sbr, {R is 1}
c2(xor,R) $\rightarrow$ wg, {R is 1}
c2(iff,R) $\rightarrow$ q, {R is 1}
c2(if,R) $\rightarrow$ jdgq, {R is 1}
c2(or,R) $\rightarrow$ j, {R is 1}
s([X,Y,Z],T) $\rightarrow$ c3(X,P), s(Z,Q), s(Y,R), {T is P · Q · R · 1}
c3(nand,R) $\rightarrow$ d, {R is 1}

Grammar a3
s([X,Y],R) $\rightarrow$ c1(X,P), s(Y,Q), {R is P · Q · 1}
c1(not,R) $\rightarrow$ ci, {R is 1}
s([X,Y,Z],T) $\rightarrow$ c2(X,P), s(Y,Q), s(Z,R), {T is P · Q · R · 1}
c2(nif,R) $\rightarrow$ oi, {R is 1}
c2(and,R) $\rightarrow$ ybd, {R is 1}
c2(nor,R) $\rightarrow$ sbr, {R is 1}
c2(xor,R) $\rightarrow$ wg, {R is 1}
c2(iff,R) $\rightarrow$ q, {R is 1}
c2(if,R) $\rightarrow$ jdgq, {R is 1}
c2(or,R) $\rightarrow$ j, {R is 1}
s([X,Y,Z],T) $\rightarrow$ c3(X,P), s(Z,Q), s(Y,R), {T is P · Q · R · 1}
c3(nand,R) $\rightarrow$ d, {R is 1}

used a grammar rule for expressing formulas constructed using *nor* (the negation of a disjunction) which placed the expression associated with the first argument of *nor* in the third position of the sentence, whereas agents a2 and a3 used a grammar rule which placed the same expression in the second position of the sentence. However, given that the expression associated with the connective of a logical formula is always placed in the first position of a sentence by the induction algorithm, the agents have no difficulty in understanding each other. Because the difference in the positions in the sentence of the expressions associated with the arguments of the connective can only generate an interpretation which corresponds to a formula which uses the same connective and which inverts the order of the arguments of such a connective with respect to the formula intended by the speaker. But such a formula is logically equivalent to the one intended by the speaker, because we are assuming that it is constructed using a commutative connective.

The results for **non-commutative connectives** are different however. All the agents constructed the logical category *if*, which corresponds to *implication*, and all of them use the same expression (*jdgq*) for referring to such a logical category. They also use the same grammatical construction for expressing implications, i.e., they all place the expression associated with the antecedent of an implication in the second position of the sentence, and the expression associated with the consequent in the third position.

Agents a2 and a3 constructed the logical category *nif*, whereas agent a1 does not have a grammar rule for expressing such a logical category. Instead of that, agent a1 constructed the logical category *noif* and a grammar rule that allows it to understand correctly the sentences generated by a2 and a3 in order to communicate formulas of the form [*nif*, A, B]. That is, whenever a2 and a3 try to communicate a formula of the form [*nif*, A, B], i.e., $\neg(A \rightarrow B)$, they use the grammar rules

$$s([X,Y,Z],T) \rightarrow c2(X,P), s(Y,Q), s(Z,R), \{T \text{ is } P \cdot Q \cdot R \cdot 1\}$$
$$c2(nif,R) \rightarrow oi, \{R \text{ is } 1\}$$

to generate a sentence. This sentence is parsed by a1 using the grammar rules

$$s([X,Y,Z],T) \rightarrow c2(X,P), s(Z,Q), s(Y,R), \{T \text{ is } P \cdot Q \cdot R \cdot 1\}$$
$$c2(noif,R) \rightarrow oi, \{R \text{ is } 1\}$$

interpreting the formula [*noif*, B, A], i.e., $\neg(B \leftarrow A)$, which is logically equivalent to the formula intended by the speaker. This is so because the grammar rules used by a1 not only use the same expression for referring to the logical connective *noif* than a2 and a3 for referring to *nif*, but they also reverse the order of the expressions associated with the arguments of the connective in the sentence.

On the other hand, given that the formulas [*nif*, A, B] and [*noif*, B, A] are logically equivalent, agent a1 will not be prevented from characterising any subset of objects because of the lack of the logical category *nif*. Since it will always prefer to conceptualise the topic using the second formula. The same holds for agents a2 and a3 with respect to the logical category *noif*.

Finally none of the agents constructed the logical category *oif* nor grammar rules for expressing formulas constructed using such a logical category. But this does not prevent them from characterising any subset of objects, because [*oif*, A, B] is logically equivalent to [*if*, B, A] and all the agents have grammar rules for expressing implications.

5   Intuitive Reasoning

During the processes of conceptualisation and grounded language acquisition the agents build categorisers for perceptually grounded categories (such as *up, down, right, left, light* and *dark*) and for logical categories (*and, nand, or, nor, if, nif, oif, noif, iff* or *xor*). These categorisers allow them to evaluate logical formulas constructed from perceptually grounded categories.

Intuitive reasoning is a process by which the agents discover relationships that hold among the categorisers of perceptually grounded categories and logical categories. For example, an agent may discover that the formula $up \rightarrow \neg down$ is always true, because the categoriser of *down* returns false for a given object whenever the categoriser of *up* returns true for the same object.

It may work as a process of constraint satisfaction in natural agents, by which they try to discover whether there is any combination of values of their sensory channels that satisfies a given formula. It is not clear to us how this process of constraint satisfaction can be implemented in natural agents. It may be the result of a simulation process by which the agents generate possible combinations of values of their sensory channels and check whether they satisfy a given formula. Or it may be grounded on the impossibility of firing simultaneously some categorisers due to the way they are implemented by physically connected neural networks.

In particular intuitive reasoning can be used to perform the following inference tasks which constitute the basis of the logical approach to formalising common sense knowledge and reasoning [11].

1. Using the categorisers of logical categories an agent can determine whether a given formula is a *tautology* (it is always true because of the meaning of its logical symbols) or an *inconsistency* (it is always false for the same reason).
2. Using the categorisers of logical and perceptually grounded categories an agent can discover that a given formula is a *common sense axiom*, i.e., it is always true because of the meaning of the perceptually grounded categories it involves. The formula $up \rightarrow \neg down$, discussed above, is a good example of a common sense axiom. Similarly it can discover that a particular formula (such as $up \wedge down$) is always false, because of the meaning of categories it involves. It can determine as well that certain formulas (such as $up \leftrightarrow left$) are merely *satisfiable,* but that they are not true under all circumstances.
3. Finally the categorisers of logical and perceptually grounded categories can be used as well to discover *domain dependent axioms*. These are logical formulas that are not necessarily true, but which always hold in the particular domain of knowledge or environment the agent interacts with during its development history. This is the case of formula $up \wedge light \rightarrow left$, which is not necessarily true, but it is always true for every subset of objects of the white board shown in figure 1.

The process of determining whether a formula is a tautology, an inconsistency or a common sense axiom by intuitive reasoning can be implemented using constraint satisfaction algorithms, if the behaviour of the categorisers of perceptually grounded and logical categories can be described by constraints. It can also be proved that intuitive reasoning is closed under the operator of *logical consequence* if the behaviour of the categorisers of perceptually grounded categories can be described by linear constraints. That is, if a formula is a logical consequence of a number of common sense axioms which can be discovered using intuitive reasoning, it must also be possible to prove that such a formula is always true using intuitive reasoning. This is a consequence of the fact that the linear constraints describing the behaviour of the categorisers of perceptually grounded categories constitute a logical model, in the sense of model theory semantics [11], of the set of common sense axioms that can be discovered using intuitive reasoning.

6 Conclusions

We have described some experiments which simulate a grounded approach to language acquisition, in which a population of autonomous agents without prior linguistic knowledge tries to construct at the same time a conceptualisation of its environment and a shared language.

These experiments extend previous work by using a linguistic interaction (*the evaluation game*) in which the agents must first conceptualise the *topic* (a subset of objects) using the mechanisms proposed in [1] for the acquisition of logical categories, and then construct a *shared language* (a lexicon and a grammar) using the invention, adoption, induction and self-organisation mechanisms proposed in [2] for the acquisition of the syntax of propositional logic.

The results of the experiments show that at the end of the simulation runs the agents build different conceptualisations and different grammars. However these conceptualisations and grammars are compatible enough to guarantee the unambiguous communication of meanings of the same complexity as propositional logic formulas.

We have also seen that the categorisers of the perceptually grounded and logical categories built during the conceptualisation and language acquisition processes can be used for some forms of common sense reasoning, such as determining whether a sentence is a tautology, a contradiction, a common sense axiom or a merely satisfiable formula – all this in a very restricted domain. However this form of intuitive reasoning requires the agents to be conscious of the fact that they use certain categorisers and of the behaviour of such categorisers.

References

1. Sierra, J.: Grounded models as a basis for intuitive and deductive reasoning: The acquisition of logical categories. In: Proceedings of the European Conference on Artificial Intelligence, pp. 93–97. IOS Press, Amsterdam (2002)
2. Sierra, J., Santibáñez, J.: The acquisition of linguistic competence for communicating propositional logic sentences. In: Artikis, A., O'Hare, G.M.P., Stathis, K., Vouros, G.A. (eds.) ESAW 2007. LNCS (LNAI), vol. 4995, pp. 175–192. Springer, Heidelberg (2008)

3. Steels, L.: The origins of syntax in visually grounded robotic agents. Artificial Intelligence 103(1-2), 133–156 (1998)
4. Batali, J.: The negotiation and acquisition of recursive grammars as a result of competition among exemplars. In: Linguistic Evolution through Language Acquisition, pp. 111–172. Cambridge University Press, Cambridge (2002)
5. Kirby, S.: Learning, bottlenecks and the evolution of recursive syntax. In: Linguistic Evolution through Language Acquisition: Formal and Computational Models, pp. 96–109. Cambridge University Press, Cambridge (2002)
6. Steels, L.: The Talking Heads Experiment. Special Pre-edition for LABORATORIUM. Antwerpen (1999)
7. Harnad, S.: The symbol grounding problem. Physica D (42), 335–346 (1990)
8. Wittgenstein, L.: Philosophical Investigations. Macmillan, New York (1953)
9. Vogt, P.: The emergence of compositional structures in perceptually grounded language games. Artificial Intelligence 167(1-2), 206–242 (2005)
10. Bueno, F., Cabeza, D., Carro, M., Hermenegildo, M., López-García, P., Puebla, G.: The Ciao Prolog system. reference manual. Technical Report CLIP3/97.1, School of Computer Science, Technical University of Madrid, UPM (1997), http://www.clip.dia.fi.upm.es/
11. McCarthy, J.: Formalizing Common Sense. Papers by John McCarthy. Edited by Vladimir Lifschitz. Ablex, Greenwich (1990)

An Autonomic Computing Architecture
for Self-* Web Services

Walid Chainbi[1], Haithem Mezni[2], and Khaled Ghedira[3]

[1] Sousse National School of Engineers/LI3 Sousse, Tunisia
Walid.Chainbi@gmail.com
[2] Jendouba University Campus/LI3 Jendouba, Tunisia
haithem.mezni@fsjegj.rnu.tn
[3] Institut Supérieur de Gestion de Tunis/LI3 Tunis, Tunisia
Khaled.Ghedira@isg.rnu.tn

Abstract. Adaptation in Web services has gained a significant attention and becomes a key feature of Web services. Indeed, in a dynamic environment such as the Web, it's imperative to design an effective system which can continuously adapt itself to the changes (service failure, changing of QoS offering, etc.). However, current Web service standards and technologies don't provide a suitable architecture in which all aspects of self-adaptability can be designed. Moreover, Web Services lack ability to adapt to the changing environment without human intervention. In this paper, we propose an autonomic computing approach for Web services' self-adaptation. More precisely, Web services are considered as autonomic systems, that is, systems that have self-* properties. An agent-based approach is also proposed to deal with the achievement of Web services self-adaptation.

Keywords: Web service, autonomic computing systems, self*-properties.

1 Introduction

With the rapid growth of communication and information technologies, adaptation has gained a significant attention as it becomes a key feature of Web services allowing them to operate and evolve in highly dynamic environments. A flexible and adaptive Web service should be able to adequately react to various changes in these environments to satisfy the new requirements and demands.

When executing Web services, network configurations and QoS offerings may change, new service providers and business relationships may emerge and existing ones may be modified or terminated. The challenge, therefore, is to design robust and responsive systems that address these changes effectively while continually trying to optimize the operations of a service provider.

So, while we obviously need effective methodologies to adapt web services to a dynamically changing environment, we also need elegant principles that would give web services the ability to continue seeking opportunities to improve their behavior and to meet user needs. To meet these goals, we propose an autonomic computing architecture for self-adaptive web services. More precisely, we consider Web services as autonomic computing systems.

A.V. Vasilakos et al. (Eds.): AUTONOMICS 2009, LNICST 23, pp. 252–267, 2010.
© Institute for Computer Sciences, Social-Informatics and Telecommunications Engineering 2010

Autonomic computing is to design and build computing systems that can manage themselves [1]. These systems are sets of called autonomic elements whose interaction produces the self-management capabilities. Such capabilities include: self-configuration, self-healing, self-optimization, and self-protection [2].

- Self-configuration by adapting automatically to the dynamically changing environment.
- Self-optimization by continually seeking opportunities to improve performance and efficiency.
- Self-healing by discovering, diagnosing and reacting to disruptions such as repairing service failure.
- Self-protection by defending against malicious attacks or cascading failures.

Other instantiations of self-managing mechanisms have been also adopted namely autonomy of maintenance by Chainbi [3], and system adaptation and complexity hiding by Tianfield and Unland [4].

The rest of this paper is organized as follows: section 2 gives a background material on autonomic computing and shows how autonomic computing capabilities may be applied in Web services. In section 3, we describe our solution for self-* Web services. Section 4 deals with a case study. In section 5, we compare the proposed study to related work. Section 6 presents some hints justifying the possible implementation of the presented autonomic architecture via an agent-based approach. Section 6 compares the proposed study to related work. The last section presents the conclusion and the future work.

2 Autonomic Computing and Web Services

2.1 Autonomic Computing

Autonomic Computing is started by IBM in 2001 and is inspired by the human body's autonomic nervous system [2]. It is a solution which proposes to reallocate many of the management responsibilities from administrators to the system itself.

Autonomic computing deals with the design and the construction of computing systems that posses inherent self-managing capabilities. Such systems are then considered as autonomic computing systems. Each autonomic system is a collection of autonomic elements – individual systems constituents that contain resources and deliver services to humans and other autonomic elements. It involves two parts: a managed system, and an autonomic manager. A managed system is what the autonomic manager is controlling. An autonomic manager is a component that endows the basic system with self-managing mechanisms such as self-configuration, self-optimization, self-healing, and self-protection.

The autonomic computing architecture starts from the premise that implementing self-managing attributes involves an intelligent control loop [2] (see figure 1). This loop enables the system to be self-*, and is dissected into three parts that share knowledge:

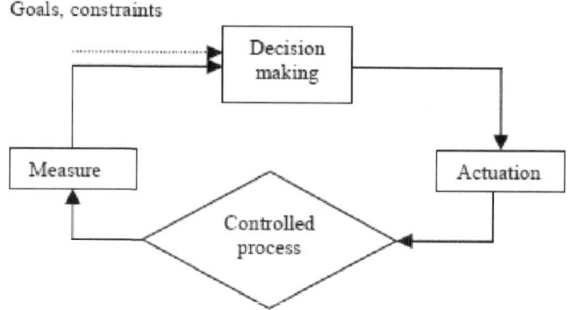

Fig. 1. A control loop (adapted from [4])

- The measure part provides the mechanisms that collect, aggregate, filter, and report details (e.g., metrics) collected from the controlled process.
- The decision part provides the mechanisms to produce the action needed to achieve goals and at the same time respect the constraints.
- The actuation part provides the mechanisms that control the execution of actions.

With this characterization in place, the integration of autonomic computing may be envisioned by two ways, namely (a) embedding the autonomic manager into the Web service, and (b) using autonomic managers as an external layer which provides the autonomic behavior for the Web service. The latter approach treats the Web service as a "black box" surrounded by autonomic managers controlling the state of the Web services and performing actions each of which can configure, optimize, heal, or correct the Web service. This approach requires the development of specific interfaces facilitating the interaction between the autonomic manager part and the managed part of the Web service. In this paper, we adopt the latter approach because it seems to be more appropriate since it separates the monitoring problem from the application specification. Accordingly, it ensures separation of concerns in the development process. Hence, the proposed architecture clearly separates the business logic of a Web service from its adaptation functionality.

2.2 Self-* Properties in Web Services

We define self-adaptive Web services as Web services supporting the autonomic computing properties which are often referred to as self-* properties, and are the followings.

- *Self-configuration*: a Web service may be a set of interacting Web services which in turn may interact with external applications with different interfaces and protocols. A new component Web service has to incorporate itself seamlessly and the rest of the system will adapt itself to its presence. For example, when a new Web service is introduced into a composite Web service, it will automatically learn about and take into account the composition and the configuration of the Web service, so that it can process services mismatches between interfaces and protocols by taking mediation actions.

- *Self-healing*: In case of problems such as service failure or QoS violation, a Web service has to perform recovery actions including retry execution, substitute candidate service, etc.
- *Self-optimization*: Web services have to continually seek opportunities to improve their own performance and efficiency. For example, a component Web service may be substituted with another one guaranteeing a better QoS and taking into account runtime execution context and other constraints.
- *Self-protection*: Web services can interact with other services or Web applications. They have to identify and detect intrusions and defend themselves against attacks (e.g., message alteration or disclosure, availability attacks, etc.) by defining and integrating some security policies.

It is important to note that these capabilities may be heavily interrelated with one another in the Web service adaptation. For example, consider a system that fails to invoke a component Web service. The adaptation process starts by detecting and diagnosing the failure. Then the recovery action is to substitute this component Web service (i.e., self-healing action). The system selects the suitable service based on current state of the environment (i.e., self-optimization action). If the substituted and the substituting services interfaces don't match, some mediation actions have to be taken (i.e., self-configuration action).

3   Autonomic Web Service Architecture

Dealing with self-* in a Web service means simply that the Web service act without the direct intervention of any other external agent (including but not limited to, a human) in order to meet self-* properties. Accordingly, autonomy is required for Web services to self-* themselves to the different events occurring in their environment. Autonomy is also necessary for an autonomic computing system. Indeed, system self-* has to be carried out without requiring the user's consciousness [4].

With this characterization in place, the match process in favor of an autonomic computing system solution is straightforward. A Web service is the system to be managed, and an autonomic manager is required to endow the Web-service with self-* capabilities such as self-configuration, self-healing, self-optimization, and self-protection. Figure 2 represents an autonomic Web service architecture.

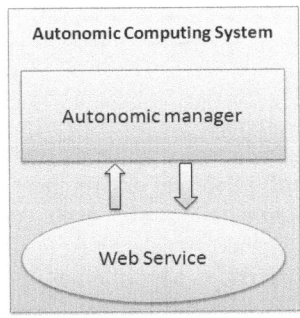

Fig. 2. An autonomic Web service architecture

An autonomic Web services environment may combine a variety of managed resources including autonomic Web services, processes, etc. These resources have different requirements and architectures which need autonomic managers with different capabilities allowing Web services to be self-*.

In case of executing a composite Web service, the management tasks of the component services are shared between a set of autonomic managers. The topology of the autonomic system is specified as a correlation between autonomic managers which perform many tasks such as managing the executing Web service, interacting with registries to select suitable web services or coordinate the work of other autonomic managers. The autonomic managers' work is orchestrated by a special autonomic manager which is responsible for the coordination of the basic Web services autonomic managers. Note that the autonomic manager coordinating the set of services' autonomic managers can be considered as a managed resource in case the associated composite Web service is used as a component service in another process. Figure 3 shows the structure of such a system.

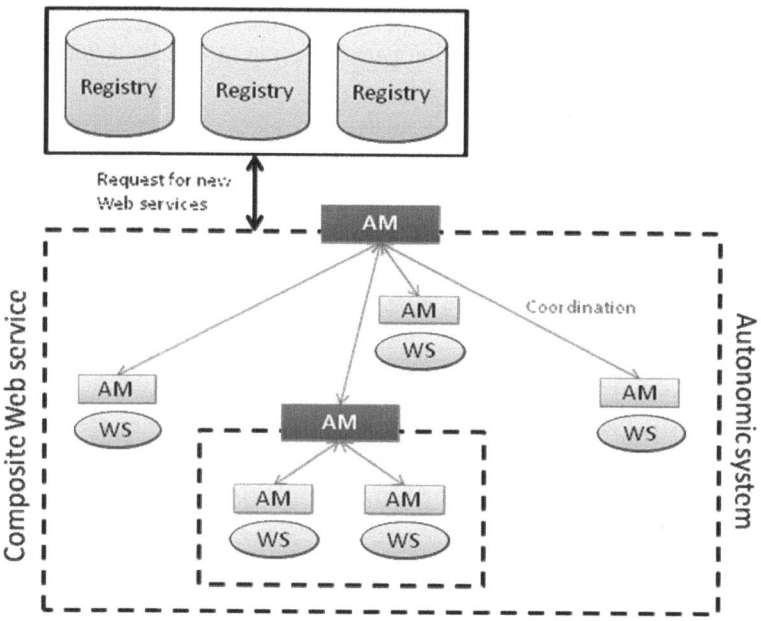

Fig. 3. An autonomic Web service system architecture

An autonomic manager has to interact with its external environment to be able to manage Web services efficiently. External environment includes registries and other autonomic managers. Interaction with registries enable autonomic managers to send a substitution request of a Web service, to select best available Web services for a new composition, to get new Web services opportunities, etc. Autonomic managers may also interact with registries (independently of the managed Web services) when detecting any change in the state of the environment, such as emergence of new nodes, termination of existing ones, etc. Such interaction allows autonomic managers to be aware of the available resources for the adaptation process of the executing Web service or for a future management task.

Autonomic managers may also interact between each others to send information or to perform adaptation. For example, if a Web service's autonomic manager fails to adapt its managed Web service, it may send a request to its coordinating autonomic manager to execute adaptation actions at the composite service level (i.e., adaptation of the whole Web service). The same information's flow may occur between two coordinating autonomic managers in case the managed composite service is composed of some complex Web services.

4   Example of an Adaptation Scenario

The structure of the autonomic Web service system may change in run time to satisfy a self*-property. Figure 4 shows an example of adaptation scenario where a basic Web service is substituted by a composite Web service.

Using the search for music scenario, we show how an executing Web service may be adapted to meet the user needs and we show how the autonomic system is able to adjust its specification according to the changing conditions. In this scenario, the client wishes to listen to music and to download songs in the *rm* format while reading lyrics and information about the artist. Figure 5 shows the sequence diagram related to the adaptation scenario.

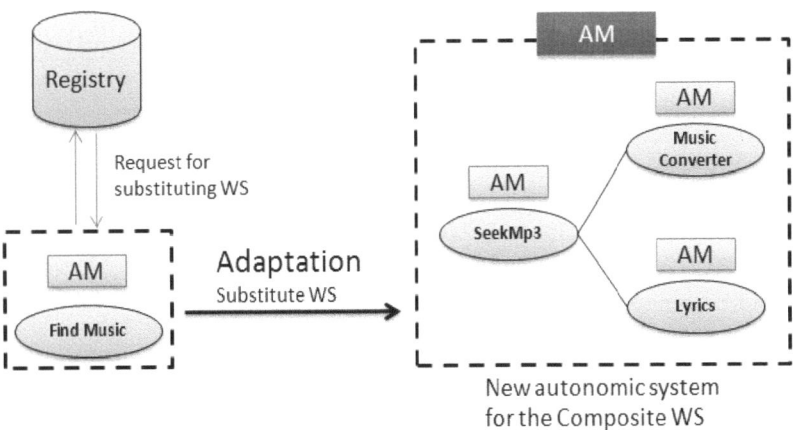

Fig. 4. Example of adaptation scenario

Consider the *FindMusic* Web service invoked by a user to look for a song. The service takes as inputs the song's title or the performer's name. Then, it shows the results according to the user's request. Since the user searches songs with a particular format, he may specify this format in the song's parameters. After selecting the desired song, the Web service proposes to play or to download the song. While listening, users have the possibility to read lyrics and artist information.

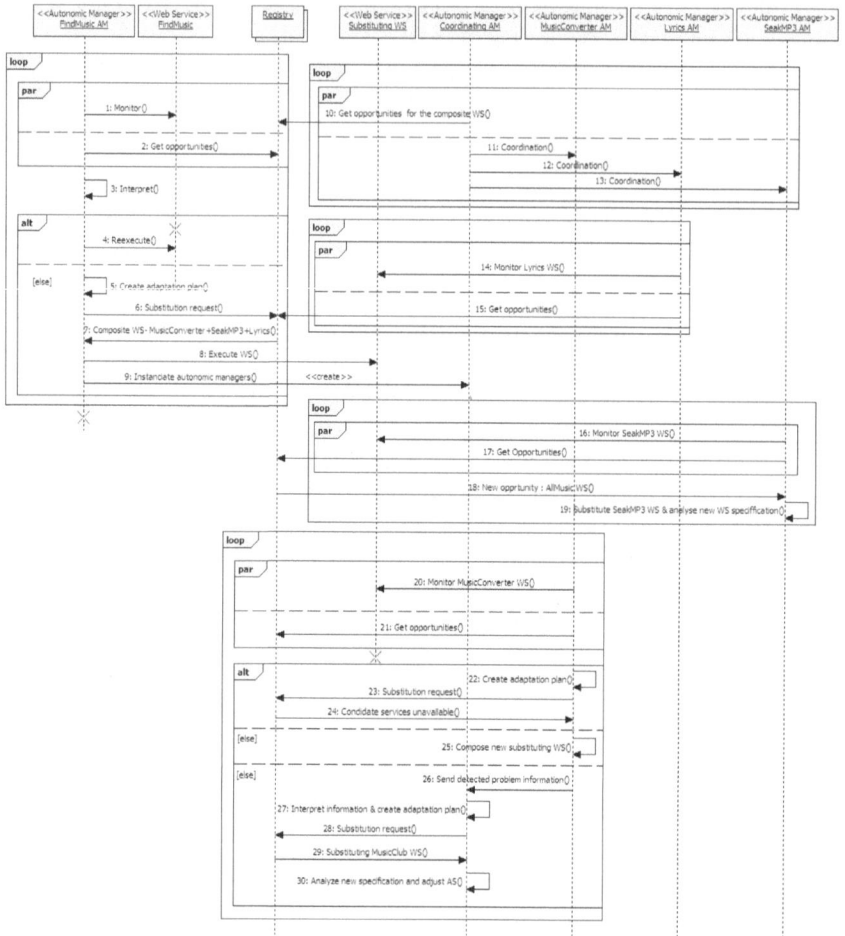

Fig. 5. Sequence diagram of the adaptation scenario

When starting the execution, the *FindMusic* Web service is associated to an autonomic manager, which performs the monitoring task and interaction with the external environment such as requesting registries to look for new opportunities or to execute adaptation actions.

The adaptation process starts when the autonomic system detects an event triggering a self-* action. Such an event may be an invocation failure of the *FindMusic* Web service. Therefore, the autonomic Web service system takes some recovery actions such as retrying execution. If the execution fails again then the autonomic system triggers a recovery action in order to replace this component with another one such as a composite Web service. The autonomic system interacts with Web services registries to perform the selection of the suitable service according to objectives of the failed *FindMusic* service.

Let the selected service be a composite web service, where the components are the *SeekMp3* Web service, the *MusicConverter* Web service and the *Lyrics* Web service. These Web services interact to satisfy user requests. The *SeekMp3* service receives the song title or artist name and returns a result which is a set of songs with different formats. Then *MusicConverter* service, based on the preferred format, is invoked to convert input files (results returned by the *SeekMp3* service) to the desired format. The *Lyrics* service uses the song's title to return lyrics of the song and information about the corresponding artist. Finally, outputs of the *MusicConverter* Web service and the *Lyrics* Web service (songs in *rm* format, lyrics and artist's information) are returned to the user.

The selected composite Web service is made up with three autonomic managers for managing the basic services and an autonomic manager for coordinating services' autonomic managers. The autonomic manager, responsible for monitoring the failed *FindMusic* service, is replaced by these autonomic managers that will manage the substituting composite service. Analysis and design of autonomic Web services is out of the scope of this paper which main content deal with an autonomic architecture to self-* Web services. Some hints are given in section 6.

Note that the desired service (that will substitute the failed *FindMusic* service) may be unavailable or may not exist. The autonomic system, then, has to interact with the registries to look for possible actions such as composing the substituting service. In case of replacing a composite web service by a basic one, the set of autonomic managers associated to the composite service are replaced by a single autonomic manager to manage the substituting basic service.

From a functional perspective, the substituting service meets the user needs and offers the same functionality of the failed *FindMusic* service. Rarely does service' WSDL interfaces match exactly. In our scenario, *SeakMp3* and *MusicConverter* services interfaces don't match. This requires taking some mediation actions to translate between the two service-interface signatures, so that interoperability can be made effective. For this, associated autonomic managers, based on their self-configuration capabilities, should interact to generate an adapter (e.g. a service) that mediates the interactions among the two *SeakMp3* and *MusicConverter* services.

Once the autonomic system is established, it continuously monitors the executing Web service to detect problems while trying to improve its performance. In our work, Web services self-optimization behavior is a combination of monitoring, selection and substitution capabilities. Self-optimization may occur in case of emergence of a new Web service with the same functionality and with a better quality. Regarding our executing Web service, possible optimization actions are: substituting one of the component services (*SeakMp3*, *MusicConverter* or *Lyrics*) or substituting the whole executing composite Web service.

Each autonomic manager must continuously try to improve the whole executing Web service performance by interacting with registries to get services opportunities. Indeed, each Web service's autonomic manager receives opportunities from registries and decides about substituting its associated Web service. In the same way, the coordinating autonomic manager should also interact with registries to look for a better Web service that may replace the whole executing web service.

Suppose that a change in the execution environment occurs (emergence of new Web services similar to the *SeakMp3* service). The autonomic manager associated to the *SeakMp3* service receives the ranked candidate services list from registries and decide to replace the *SeakMp3* service with the basic *AllMusic* service. Then, it analyses the new specification of the executing composite Web service to look for any change. Since the *SeakMp3* Web service is replaced with a basic one, the autonomic manager decides that no changes have occurred in the service specification and keeps the current autonomic system specification.

Consider now, that the *MusicConverter* service is no longer available. The associated autonomic manager handles adaptation accordingly. For this, it implements a set of recovery actions that allows substituting the *MusicConverter* service. If no candidate service is available for substituting, the autonomic manager tries to apply another appropriate recovery action to let the execution successfully terminate. To that end, the autonomic manager features several recovery actions. Possible solutions are (a) composing a new Web service with the same functionality of the failed *MusicConverter* service or (b) replacing the whole executing composite service. In case of adopting the first solution, the autonomic manager uses its automatic service discovery and composition capabilities to interact with registries and perform the necessary repair actions. It may also choose the second solution and looks for assistance from its coordinating autonomic manager. This is by informing the coordinating autonomic manager about the detected problem (*MusicConverter* unavailability) and about the new information collected after trying to execute repair actions (unavailability of substituting services). So, based on information sent by the *MusicConverter* autonomic manager, and after interacting with registries, the self-healing behavior of the coordinating autonomic manager is to substitute the whole executing composite Web service with another one having the same goals. For this, the coordinating autonomic manager contacts the registries to get a set of candidate services similar to the desired one (the whole Web service) and selects the best available service: the *MusicClub* Web service. Once the *MusicClub* service starts to execute, the autonomic system adjust itself according to the new specification of the executing Web service by instantiating new autonomic managers or deactivating existing ones. In case of replacing the composite Web service by a basic one, the set of autonomic managers associated to the composite service are replaced by a single autonomic manager to manage the substituting basic service.

5   Related Work

The main purposes of service adaptation vary from ensuring interoperability to service recovery and optimization and context management. Some approaches address the problem of interoperability due to interfaces and protocols heterogeneity. Recovery deals with techniques for detecting problems in services interaction and searching alternative solutions. Optimization is about discovering and selecting the suitable Web service with respect to QoS offerings and user needs. Finally, solutions for context change aim to optimize the service function of their execution context. Here are some works on service adaptation:

The WS-Diamond Project [5] aims at the development of a framework for self-healing Web services, that is, services able to self monitor, to self-diagnose the causes of a failure, and to self-recover from functional and non-functional failures. In [6], the solution for self-healing of BPEL processes is based on Dynamo, a monitoring framework, together with an AOP extension to ActiveBPEL, and a monitoring and recovery subsystem that uses Drools ECA rules.

In [7], a methodology and a tool for learning the repair strategies of WS to automatically select repair actions are proposed. The methodology is able to incrementally learn its knowledge of repairs, as faults are repaired. Thus, it is at runtime possible to achieve adaptability according to the current fault features and to the history of the previously performed repair actions. In [8], the authors propose a methodology for the automated generation of adaptors capable of solving behavioral mismatches between BPEL processes. [9] introduces PAWS, a framework for flexible and adaptive execution of managed WS-based business processes. In the framework, several modules for service adaptation (mediation engine, optimization and self-healing) are integrated in a coherent way.

In [10], the authors developed a staged approach for adaptive Web service composition and execution (A-WSCE) that cleanly separates the functional and non-functional requirements of a new service, and enables different environmental changes to be absorbed at different stages of composition and execution.

In [11], the authors are focusing on run-time adaptation of non-functional features of a composite Web service by modifying the non-functional features of its component. The aspect oriented programming technology is used for specifying and relating non-functional properties of the Web services as aspects at both levels of component and composite services.

While current approaches address significant subsets of adaptation requirements, they have some drawbacks including the degree of automation, few techniques for capturing non-functional properties, etc. The autonomic approach presented in this paper deals with the different facets of adaptation since its purpose is the design and the construction of Web services that posses inherent self-* capabilities.

Furthermore, there is no existing approach addressing the adaptation cross all the functional layers of the service based systems (i.e., the business process layer, the service composition layer, and the service infrastructure layer) since all the approaches address only a particular functional layer. For example, [10] and [11] deal with the infrastructural layer whether the composition layer was dealt with in [8] and [9]. In addition, existing approaches try to integrate a maximum of requirements in order to have a complete framework. For this purpose, our main concern is to propose a general autonomic architecture that provides self-* capabilities and meets the most important adaptation requirements without affecting services consistency and by preserving the robustness of the applications. Moreover, none of the existing approaches have studied the complexity in the implementation, that is, hiding the complexity from user and how much the adaptation is complex at any time of the application lifetime. Autonomic computing provides self-adaptation while keeping its complexity hidden from the user [4].

In our work, considering Web services as autonomic systems, offers many advantages. First, unlike existing approaches, the management task is shared between a set of autonomic managers, each of them is associated to a Web service. This leads to an effective monitoring and consequently to a high degree of adaptation.

6 Implementation Issues

In this section, we deal with the technical machinery to achieve the self-adaptation. We adopt an agent-based solution for the autonomic Web-service system. The integration of agent-based computing into the framework of autonomic computing may be envisioned by two ways, namely (a) integrating the autonomic cycle into the system, thus in a certain sense embedding the autonomic manager into the managed system and adopting an agent solution for the whole, and (b) using agents as an external layer which provides the autonomic behavior. We propose to adopt an agent solution for the whole namely the managed part of the Web service and the manager part. Consequently, the interaction between the managed and the managing parts of the system become easier. This is mainly due to the homogeneity of the adopted solution (an agent is the unit of design).

In any design process, finding the right models for viewing the problem is a main concern. In general, there will be multiple candidates and the difficult task is picking the most appropriate one. Next, we analyze the high degree of match between the characteristics of agent systems and those of autonomic systems [3].

6.1 Behavioral Match

The match process argument in favor of an agent based solution can be expressed by the fact that an agent is able to deal with the aforementioned actions related to the control loop (see figure 1 §2.1). The term *agent* in computing covers a wide range of behavior and functionality. In general, an agent is an active computational entity that can perceive (through sensors), reason about, and initiate activities (through effectors) in his environment [12]. Normally, an agent has a repertoire of actions available to him. This set of possible actions represents his ability to modify his environment. The types of actions an agent can perform at a point of time include:

- Physical actions are interactions between agents and the spatial environment.
- Communicative actions are interactions between agents. They can be emission or reception actions.
- Private actions are internal functions of an agent. They correspond to an agent exploiting his internal computational resources.
- Decision action can generate communicative, physical and private actions. A decision action can also update the agent's beliefs.

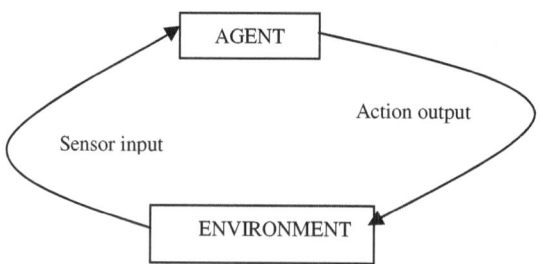

Fig. 6. An ongoing interaction between an agent and his environment

An action may be classified as either configuration, optimization, healing, or protection action depending on the reason of its execution. For example, substituting a candidate service is a physical action which can be optimization action if it is intended to guarantee a better QoS. It can be as well considered as a healing action in case of service failure.

6.2 Complexity Management

Computing systems consisting of software, hardware and communication infrastructure have become ever increasingly complex. If autonomic computing paradigm is to be engineered for such complex systems, hierarchical control architectures are considered as a key technology to rely upon [4, 13, 14]. In such case, a hierarchy of control loops is required to endow the whole system of self-managing mechanisms. Each level of control loop, achieving correspondingly, one of the different control goals which collectively constitute the overall control objectives of the system. Consequently, a multi-agent system solution is envisioned to deal with the self-* capabilities within an autonomic computing system.

For example, numerous autonomic managers in a composite Web service system must work together to deliver autonomic computing to achieve common goals. This is the case of a composite Web service which needs to work with the autonomic managers of the elementary Web services, registries in order for the Web service infrastructure as a whole to become a self-* system.

The argument in favor of a multi-agent system solution can also be described in terms of the ability of such systems to deal with complexity management. Previously, Booch identified three techniques for tackling complexity in software: *decomposition*, *abstraction* and *organization* [15].

- *Decomposition*: the process of dividing large problems into smaller, more manageable chunks each of which can then be dealt with in relative isolation.
- *Abstraction*: the process of defining a simplified model of the system that emphasizes some of the details or properties, while suppressing others.
- *Organization*: the process of identifying and managing the interrelationships between the various problem solving components. This helps designers tackle complexity in two ways. Firstly, by enabling a number of basic components to be grouped together and treated as a higher-level unit of analysis (e.g., the individual components of a subsystem can be treated as a single coherent unit by the parent system). Secondly, by providing a means of describing the high-level relationships between various units (e.g., a number of components may cooperate to provide a particular functionality).

Next, we deal with each technique in turn.

- *Agent-oriented decomposition is an effective way of partitioning the problem space of a complex system*: the agent-oriented approach advocates decomposing problems in terms of autonomous agents that can engage in flexible, high-level interactions. Decomposing a problem in such a way helps the process of engineering complex systems in two main ways. Firstly, it is simply a natural representation for complex systems that are invariably distributed and that invariably have multiple loci of control. This decentralization, in turn, reduces the system's control complexity and results in a lower degree of coupling between components. The fact that agents are active entities means they know for themselves when they should be acting and when they should update their state. Such self-awareness reduces control complexity since the system's control know-how is taken from a centralized repository and localized inside each individual problem solving component [12]. Secondly, since decisions about what actions should be performed are devolved to autonomous entities, selection can be based on the local situation of the problem solver. This means that the agent can attempt to achieve its individual objectives without being forced to perform potentially distracting actions simply because they are requested by some external entity. The fact that agents make decision about the nature and scope of interactions at run-time makes the engineering of complex systems easier. Indeed, the system's inherent complexity means it is impossible to know a priori about all potential links: interactions will occur at unpredictable times, for unpredictable reasons, between unpredictable components. For this reason, it is futile to try and predict or analyze all the possibilities at design time. Rather, it is more realistic to endow the components with the ability to make decisions about the nature and scope of their interactions at run-time. Thus agents are specifically designed to deal with unanticipated requests and they can spontaneously generate requests for assistance whenever appropriate.

- The key abstractions of agent-oriented mindset are a natural means of modeling complex systems: In the case of a complex system, the problem to be characterized consists of subsystems, subsystems components, interactions and organizational relationships. Taking each in turn: firstly, there is a strong degree of correspondence between the notions of subsystems and agent organizations. They both involve a number of constituent components that act and interact according to their role within the larger enterprise. Secondly, the interplay between the subsystems and between their constituent components is most naturally viewed in terms of high level social interactions (e.g., agent systems are described in terms of "cooperating to achieve common objectives" or "negotiating to resolve conflicts"). Thirdly, complex systems involve changing webs of relationships between their various components. They also require collections of components to be treated as a single conceptual unit when viewed from a different level of abstraction. On both levels, the agent-oriented mindset again provides suitable abstractions. A rich set of structures is typically available for explicitly representing and managing organizational relationships such as roles (see [16, 17] for example). Interaction protocols exist for forming new groupings and disbanding unwanted ones (e.g., Sandholm's work [18]). Finally, structures are available for modeling collectives (e.g., teams [19]).

- *The agent-oriented philosophy for dealing with organizational relationships is appropriate for complex systems*: organizational constructs are first-class entities in agent systems. Thus explicit representations are made of organizational relationships and structures. Moreover, agent-based systems have the concomitant computational mechanisms for flexibly forming, maintaining and disbanding organizations. This representational power enables agent-oriented systems to exploit two facets of the nature of complex systems. Firstly, the notion of primitive component can be varied according to the needs of the observer. Thus, at one level, entire subsystems can be viewed as singletons, alternatively, teams or collections of agents can be viewed as primitive components, and so on until the system eventually bottoms out. Secondly, such structures provide a variety of stable intermediate forms that, as already indicated, are essential for the rapid development of complex systems. Their availability means individual agents or organizational groupings can be developed in relative isolation and then added into the system in an incremental manner. This, in turn, ensures there is a smooth growth in functionality.

6.3 Pragmatic Reasons

Autonomic computing denotes a move from the pursuit of high speed, powerful computing capacity to the pursuit of self-managing mechanisms of computing systems. Indeed, today's computing and information infrastructure have reached a level of complexity that is far beyond the capacity of human system administration. For instance, follow the evolution of computers from single machines to modular systems to personal computers networked with larger machines. Along with that growth has came increasingly sophisticated architectures governed by software whose complexity now routinely demands tens of millions of lines of codes. The internet adds yet another layer of complexity by allowing us to connect this world of computers and computing systems with telecommunications networks. In the process, the systems have become increasingly difficult to manage, and ultimately, to use. Inspired by the functioning of the human nervous system which frees our conscious brain from the burden of dealing with some vital functions (such as governing our heart rate and body temperature), autonomic computing is considered as a promising solution to such problems.

As yet, however there is not a successful solution to autonomic computing which can be applied on a significant scale. So far a mature solution has not yet appeared. In part, this is due mainly to the youth of this paradigm and the absence of adequate tools, but our experience suggests that the absence of tools that allow system complexity to be effectively managed is a greater obstacle.

Agent technology is one of the most dynamic and exciting areas in computer science today. Many observers believe that agents represent the most important new paradigm for software development since object-orientation. Agent technology has found currency in diverse applications domains including ambient intelligence; grid computing where multi-agent system approaches enable efficient use of the resources of high-performance computing infrastructure in science, engineering, medical and commercial applications; electronic business, where agent-based approaches support the automation of information-gathering activities and purchase transactions over the internet; the semantic web, where agents are needed both to provide services, and to make best use of the resources available, often in cooperation with others ; and others including resource management, military and manufacturing applications [20].

Agent paradigm has achieved a respectable degree of maturity and there is a widespread acceptance of its advantages: a relatively large community of computerscientists which is familiar with its use now exists. A substantial progress has been made in recent years in providing a theoretical and practical understanding of many aspects of agents and multi-agent systems [21].

7 Conclusion and Future Work

In this paper we adopt autonomic computing paradigm to propose an approach for self-adaptive web services. The basic idea is to consider web services as autonomic systems, that is, systems able to manage and adapt themselves to the changing environment according to a set of goals and policies. As a result, autonomic web services can recover from failure, optimize their performance, configure themselves, etc. without any human intervention. An autonomic distributed architecture is proposed where each component service is associated with one or a set of specific autonomic manager(s). We have also presented in this paper the main reasons to consider agent technology as a suitable candidate to deal with the technical machinery achieving self-adaptation within a Web services system.

Motivated by the fact that self-adaptation systems are recently considered as the trend of the new systems, and by the justified claim that agent-based computing has the potential to be integrated into the framework of autonomic computing we will show, in our future work, how software agents may be used to deal with autonomic Web services systems. More precisely, we envision to develop an agent-based architecture for an autonomic Web services system. An adaptive version of the *Search for Music* scenario, presented in this paper, is currently being implemented by using an agent based approach.

References

1. Ganek, A.G., Corbi, T.A.: The Dawning of the Autonomic Computing Era. J. IBM Systems 42(1), 5–18 (2003)
2. IBM Group.: An Architectural Blueprint for Autonomic Computing, `http://www-03.ibm.com/autonomic/pdfs/AC`
3. Chainbi, W.: Agent Technology for Autonomic Computing. J. Transactions on Systems Science and Applications 1(3), 238–249 (2006)
4. Tianfield, H., Unland, R.: Towards Autonomic Computing Systems. J. Engineering Applications of Artificial Intelligence 17(7), 689–699 (2004)
5. Console, L., Fugini, M.: The WS-Diamond Team: WS-DIAMOND: an Approach to Web Services - DIAgnosability, MONitoring and Diagnosis. In: e-Challenges Conference, The Hague (2007)
6. Baresi, L., Guinea, S., Pasquale, L.: Self-healing BPEL Processes with Dynamo and the JBoss Rule Engine. In: International Workshop on Engineering of Software Services for Pervasive Environments (ESSPE 2007), pp. 11–20 (2007)
7. Pernici, B., Rosati, A.M.: Automatic Learning of Repair Strategies for Web Services. In: 5th European Conference on Web Services (ECOWS 2007), pp. 119–128 (2007)
8. Brogi, A., Popescu, R.: Automated Generation of BPEL Adapters. In: International Conference on Service Oriented Computing (2006)
9. Ardagna, D., Comuzzi, M., Mussi, E., Pernici, B., Plebani, P.: PAWS: A Framework for Executing Adaptive Web-Service Processes. J. IEEE Software 24(6), 39–46 (2007)
10. Chafle, G., Dasgupta, K., Kumar, A., Mittal, S., Srivastava, B.: Adaptation in Web Service Composition and Execution. In: International Conference on Web Services, pp. 549–557 (2006)

11. Narendra, N.C., Ponnalagu, K., Krishnamurthy, J., Ramkumar, R.: Run-Time Adaptation of Non-functional Properties of Composite Web Services Using Aspect-Oriented Programming. In: Krämer, B.J., Lin, K.-J., Narasimhan, P. (eds.) ICSOC 2007. LNCS, vol. 4749, pp. 546–557. Springer, Heidelberg (2007)
12. Jennings, N.: On Agent-based Software Engineering. J. Artificial Intelligence 117(2), 277–296 (2000)
13. Albus, J.S., Meystel, A.M.: Engineering of Mind: an Introduction to the Science of Intelligent Systems. Wiley, New York (2001)
14. Tianfield, H.: Formalized Analysis of Structural Characteristics of Large Complex Systems. J. IEEE Transactions on Systems, Man and Cybernetics. Part A: Systems and Humans 31(6), 59–572 (2001)
15. Booch, G.: Object-Oriented Analysis and Design with Applications. Addison-Wesley, Reading (1994)
16. Baejis, C., Demazeau, Y.: Organizations in Multi-Agent Systems. Journées DAI, Toulouse (1996)
17. Fox, M.S.: An Organizational View of Distributed Systems. J. IEEE Transactions on Systems, Man and Cybernetics 11(1), 70–80 (1981)
18. Sandholm, T.: Distributed Rational Decision Making. Multi-Agent Systems. MIT Press, Cambridge (1985)
19. Tambe, M.: Toward Flexible Teamwork. J. Artificial Intelligence Research 7, 83–124 (1997)
20. Luck, M., McBurney, P., Preist, C.: Agent Technology: Enabling Next Generation Computing. In: AgentLink II (2003)
21. D'inverno, M., Luck, M.: Understanding Agent Systems, 2nd edn. Springer, Heidelberg (2004)

Author Index